T0320608

Physical Optics
of Ocean Water

Physical Optics
of Ocean Water

K. S. Shifrin

Translated by
David Oliver

To the memory of my father, Solomon Nissanovich Shifrin,
who perished in the Siege of Leningrad

Originally published as *Vvedenie v optiku morya*

Library of Congress Cataloging-in-Publication Data

Shifrin, K. S.
    Physical optics of ocean water.

    Bibliography: p.
    Includes index.
    1. Optical oceanography.   I. Title.
II. Series.
GC178.2.S5413   1988   551.46'01   87-18667
ISBN 978-0-88318-529-2

Contents

Preface to the Russian Edition

Ocean optics research now covers a wide range of subjects. It encompasses both traditional problems as well as new areas of research. The traditional problems include the analysis of the optical properties of oceanic waters, light fields in the ocean, light transformation processes at the surface; the newer problems include the development of optical methods for investigating the ocean including satellite remote sensing and the study of radiative air-sea interaction.

The core problem of both old and new research topics is the investigation of the optical properties of oceanic waters in conjunction with their constituents—suspended and dissolved substances. This is the principal task of ocean optics. Its systematic exposition forms the main part of this book.

The large variety of subjects—from the subtle problems of physical optics to biological phenomena—makes ocean optics a complex branch of oceanology. Its study is made particularly difficult in that important findings in the field of ocean optics often occur in publications which are far removed from oceanology, in works on mathematical physics and computing methods, radio physics and astrophysics, physical and colloidal chemistry, biophysics, biology, etc. The disparity in the concepts used in such a wide range of scientific disciplines and the confusion in the terminology employed by different authors further complicate matters. In addition it should be mentioned that a significant number of useful publications are dispersed among collections of works, conference transactions and other sources which are difficult to access. In view of all these factors, it has seemed worthwhile to prepare a monograph which, in a condensed form, gives a systematic account of those branches of science and the facts and theories which may be considered reliably established.

This book is based on a course of lectures which I delivered at the Leningrad Institute of Hydrometeorology in Spring 1976. It consists of three parts. In the first part (Chapters 1 and 2), the basic quantitative characteristics used to describe the optical properties of oceanic waters and the light field in the ocean are introduced and typical values are given. This part also gives a brief summary of the information on dissolved and suspended substances in the ocean—the factors which are responsible for the variability of optical properties. It has to be said from the start that our knowledge in this field is still incomplete and it is to be hoped that optics research will help to develop effective methods for studying the composition of oceanic waters. It is probably in the development of these methods that the oceanographer can expect most from ocean optics.

In the second part (Chapters 3 to 5), the optical properties of various models of the ocean water are examined. Chapter 3 deals with the optical properties of pure water, pure seawater, dissolved organic substances and suspended particles. The characteristics concerned are the optical constants $n(\lambda)$ and $k(\lambda)$ for these substances and the molecular scattering of light in water. Chapter 4 is concerned with the optics of suspensions; to be more exact it is concerned with the simplest model for suspensions—homogeneous spheres. Exact formulae are given, cases of approximation important for ocean optics are examined and the results of calculations analysed. In Chapter 5, data on actual ocean water are examined—absorption and dispersion of light as well as the distinguishing features of the spectral, vertical and geographical variation of these characteristics. It is interesting that although suspended particles in the ocean are quite unlike homogeneous spheres, the homogeneous sphere model satisfactorily describes many observations. Of course, this does not apply to polarization properties where this model is unable to explain even the simplest polarization characteristic, the degree of polarization $p(\gamma)$.

In the third part (Chapter 6), the inverse problems are investigated—the determination of the composition of suspended matter from the characteristics of light scattering. General questions which arise in solving this problem are examined and actual procedures for studying oceanic suspensions which are currently used or hold promise for the future are considered. Again, it is also found that the homogeneous sphere model manages to give a satisfactory account.

This does not complete the list of subjects treated in this book, much less, of course, does it exhaust the range of subjects which belongs to this theme and could be included. The condensed form forced us to select what we thought was most important, and it was our aim to strike a sensible compromise between the interests of our different readers—oceanographers, physicists and engineers.

I wish to thank academician I. V. Obreimov and the corresponding member of the USSR Academy of Sciences A. S. Monin for their constant support of this work. I also wish to thank the Associate Member of the USSR Academy of Sciences V. V. Bogorodsky, Doctor of physico-mathematical sciences O. V. Kopelevich, Doctor of physico-mathematical sciences G. G. Neuimin, Doctor of geographical sciences Yu. E. Ochakovsky, and Doctor of physico-mathematical sciences A. Ya. Perel'man for discussions on individual questions.

K. S. Shifrin
Leningrad

Preface to the English Edition

This book is dedicated to the physical optics of ocean water. Two different types of problem are treated: the effects of dissolved and suspended substances in seawater on its optical properties, which we call the direct problem, and how it is possible to derive the composition of seawater from the optical characteristics, which we call the inverse problem. Apart from a treatise on the physics, the book systematically examines the numerical values of optical properties, the average dispersions and variation with depth, etc. These data are necessary in order to obtain a correct idea of the scale of these phenomena as well as for different practical problems.

The importance of the direct problems can be understood by everybody. The potential possibilities of the inverse problems may be illustrated by the study of ocean suspensions. For example, it is shown in Chapter 6 that the spectrum of suspended particles obtained by inverting the volume scattering function agrees well with the data obtained by the Coulter counter whereas standard microscopic analysis gives significantly lower values. We also remember that yellow substance, an important component of ocean water, was discovered by K. Kalle in the ultraviolet absorption spectrum although prior to this chemists had spent many years carefully analysing the composition of ocean water.

The methods of hydrooptical measurements and the regional characteristics of the phenomena are dealt with in the book only to the extent that they are necessary for the understanding of the questions under study and for an estimate of the range of variation of the characteristics in the world's oceans.

The attentive reader will notice that the extent of our knowledge of the different phenomena governing ocean optics varies considerably. In some extremely straightforward matters involving quite simple properties, the data of different researchers differ appreciably. In these cases, I have attempted to present the situation as it is at the moment.

The author would like to thank Derek Pilgrim for his help in preparing the English edition.

I hope that this book will be useful to scientific and research workers and practitioners in the field of oceanography and allied disciplines, engineers engaged in designing optical apparatus and undergraduates and research students.

K. S. Shifrin
Leningrad

Chapter 1

Hydrooptical characteristics

1.1. Optical properties of ocean water

1.1.1. Definitions

Traditionally, a system of primary and secondary characteristics has been devised for the quantitative description of optical phenomena in oceanology. The primary hydrooptical characteristics, or inherent optical properties, are a set of quantities characterizing the optical properties of ocean water itself. They define the composition and condition of the water in the same way as other physical properties such as temperature, salinity, etc. The secondary hydro-optical characteristics, or apparent optical properties, are quantities describing the light field in the ocean and in the atmosphere above the ocean. In contrast to the primary characteristics, they depend not only on the properties of the water but also on the nature of the illumination on to the surface of the sea. Examples of these are spectral and integral irradiance at different depths in the sea, the distribution factor for radiance, the degree of polarization of the light, the colour index of the water, etc. A knowledge of these quantities is necessary, for example, in the investigation of photosynthesis and a number of other problems.

The inherent properties define the conditions for the propagation of light in the sea. They provide information on the particles suspended in the water and organic substances dissolved in it, and hydrooptical measurements can be used to determine the concentration of dissolved substances and particles in the water from the optical properties. These may be called inverse problems in hydrooptics.

In theory, the apparent properties, such as the properties of the field in a medium, may be calculated for any given illumination characteristics and inherent properties. However, in practice, such calculations meet considerable difficulties because the inherent properties are usually not known in sufficient detail and also because of the difficulties involved in solving the equation of transfer for real conditions in the sea. For this reason, light fields in the ocean are usually measured directly but the measurements yield only the average values of the inherent properties.

The relationships between the different optical properties are widely used in ocean optics. They make it possible to restrict measurements to a small number of quantities, the remaining quantities being determined from the measured quantities. In recent years, interest has grown in the apparent properties in conjunction with the development of remote sensing methods for studying the ocean. In these methods, apparent properties, such as the water-leaving irradiance, are measured directly. By double inversion, information can be deduced about the suspended matter in the ocean, the concentration of dissolved substances, etc.

We examine both groups of characteristics below, the relationship between them, typical values and their range of variation. We pay special attention to the depth of visibility of a standard white disk z_b and the colour grade N on the chromaticity scale for seawater. These two apparent properties were the first quantities to be systematically studied in ocean optics and considerable experimental data have been accumulated on them.

The propagation of light in ocean waters is accompanied by its absorption and scattering. These phenomena may be described with the help of two coefficients, the absorption coefficient a and scattering coefficient b, and a single function, the volume scattering function $\beta(\gamma)$. The physical meaning of these coefficients as well as some of their derivatives can be seen from the formulae given in Table 1.1. Strictly speaking, this system is incomplete. It does not take into account the changes in polarization during scattering. A complete description requires the introduction of the scattering matrix. We define this in Chapter 4.

In Table 1.1, the following notation is used: Φ is the flux of parallel monochromatic radiation illuminating an elementary volume dv, whose thickness in the direction of propagation of the initial beam of radiation is dl; $d\Phi_a$, $d\Phi_b$ and $d\Phi_c$ are the elementary radiation fluxes, absorbed, scattered and attenuated on passing through the volume dv, respectively; γ is the scattering angle (the angle between the direction of the incident and scattered light); E_n is the normal irradiance produced by the flux Φ at the surface of the volume dv; $dI(\theta)$ is the radiant intensity scattered by the volume dv in the direction θ; $\Phi(r)$ is the radiation flux passing through a layer of finite thickness r of the medium; $\Phi(0)$ is the flux on entering the medium.

The formula for the coefficient a in Table 1.1 is simply defined. Let a parallel beam of light pass through the water of the elementary layer dr. Thus the amount of energy of the beam absorbed by this layer $d\Phi_a$ will be proportional to the intensity of the beam Φ and path length dr:

$$d\Phi_a = - a \cdot \Phi \cdot dr \tag{1.1}$$

The proportionality coefficient a is called the absorption coefficient. It characterizes the properties of the given layer of water. The coefficients b and c have a similar meaning. Also three coefficients have the dimension L^{-1} and are measured in m^{-1}. The total attenuation of the beam $d\Phi_c$ is equal to the sum of $d\Phi_a$ and $d\Phi_b$:

$$d\Phi_c = d\Phi_a + d\Phi_b = -(a + b) \cdot \Phi \cdot dr = - c \cdot \Phi \cdot dr; \quad c = a + b \tag{1.2}$$

During the interaction of the beam of photons with the volume element of the substance, some of the photons a are converted from light into heat. Another part b is scattered, ie remains as light but changes direction. Therefore, the ratio $\Lambda = b/c$ is called the photon survival probability during the elementary interaction or the single scattering albedo. For a purely absorbing medium, $\Lambda = 0$, and for a purely scattering medium, $\Lambda = 1$. The infrared region and the band of minimum absorption by water in the region of $\lambda \simeq 500$ nm are approximations of these theoretical cases.

Strictly, all characteristics given in Table 1.1 must be applied to monochromatic radiation, ie they should be suffixed (λ); we have omitted this for the sake of simplicity. However, these characteristics are often also used for wide spectral bandwidths which are related to a mean effective wavelength λ_{eff}. Its exact value is determined by the spectral sensitivity of the apparatus used.

The first 11 coefficients in Table 1.1 describe the properties of a substance at a given point in the medium. It follows from Equation (1.2) that for a radiation flux $\Phi(r)$ passing through water in a layer of thickness r:

$$\Phi(r) = \Phi(0) \cdot T = \Phi(0) \cdot e^{-\tau} = \Phi(0) \cdot e^{-\bar{c}r} \tag{1.3}$$

where $\bar{c}$ is the average value of $c(r)$ in the layer r. Equation (1.3) defines the last two characteristics in Table 1.1. They characterize integral properties of the layer of thickness r.

In deriving Equation (1.3), we used a non-rigorous representation of a parallel beam of radiation.* Strictly speaking, Equation (1.3) must be applied to the angular density of the radiation fluxes, ie the radiance:

$$L(r) = L(0) \cdot T = L(0) \cdot e^{-\tau} = L(0) \cdot e^{-\bar{c}r} \tag{1.4}$$

Equation (1.4) describes the attenuation of the radiance of the beam as it is propagated in a turbid medium. In it, $L(0)$ is the radiance of the incident beam, $L(r)$ is the radiance of the same beam after traversing a path r in the medium. Equation (1.4) is called Bouguer's Law. This is one of the most important equations in the optics of turbid media.**

In light technology and also in hydrooptics, Equation (1.4) is sometimes written in the form:

$$L(r) = L(0) \cdot 10^{-c_{10}r}$$

$$c_{10} = M \cdot c, \text{ where } M = \log_{10}e = 0.4343 \ldots \tag{1.5}$$

At that time, hydrooptical investigations were being carried out in connection with underwater illumination calculations. The log to base ten coefficients a_{10} and b_{10} are defined in a similar manner. In what follows, we use only the natural log values of the coefficients but in earlier studies, on the optics of the sea, decimal

---

* In photometry, it is stated that the energy flux $dF = L \cdot d\omega$ where L is the luminance (or brightness) of the beam and $d\omega$ is the solid angle in which the beam is propagated. Consequently, a parallel beam does not carry energy since when $d\omega = 0$, $dF = 0$.

** Bouguer (1698–1768) was one of the most remarkable scientists of the 18th century. Today, his name is mainly linked with the early work in scientific photometry. However, he was responsible for key results in geodesy (measurement of the shape of the earth), hydrography, naval architecture and many other sciences. His manual on navigation was reprinted many times in Russian in the 18th century. Bouguer's corrections and reductions feature in all courses on marine gravimetry to this day.

Table 1.1. OPTICAL PROPERTIES OF OCEAN WATER

No.	Name	Symbol	Formula
	Basic properties		
1	Absorption coefficient of radiation	a	$a = -\dfrac{1}{\Phi} \cdot \dfrac{d\Phi a}{dr}$
2	Scattering coefficient of radiation	b	$b = -\dfrac{1}{\Phi} \cdot \dfrac{d\Phi b}{dr}$
3	Volume scattering function	$\beta(\gamma)$	$\beta(\gamma) = \dfrac{4\pi \cdot b(\gamma)}{b}$
	Derived properties		
4	Attenuation coefficient of radiation	c	$c = -\dfrac{1}{\Phi} \cdot \dfrac{d\Phi_c}{dr}$
5	Transmittance	θ	$\theta = e^{-c}$
6	Photon survival probability	Λ	$\Lambda = \dfrac{b}{c} = \dfrac{b}{a+b}$
7	Scattering coefficient in a given direction	$b(\gamma)$	$b(\gamma) = \dfrac{1}{E_n} \cdot \dfrac{dI(\gamma)}{dv}$
8	Forward scattering coefficient	b_f	$b_f = 2\pi \displaystyle\int_0^{\pi/2} b(\gamma) \cdot \sin\gamma \cdot d\gamma$
9	Backward scattering coefficient	b_b	$b_b = 2\pi \displaystyle\int_{\pi/2}^{\pi} b(\gamma) \cdot \sin\gamma \cdot d\gamma$
10	Scattering asymmetry coefficient	G	$G = \dfrac{b_f}{b_b}$
11	Scattering asymmetry factor	g	$g = \dfrac{1}{2} \displaystyle\int_0^{\pi} \beta(\gamma) \cdot \cos\gamma \cdot \sin\gamma \cdot d\gamma$
12	Optical thickness of a layer of aqueous medium	τ	$\tau = \displaystyle\int_0^r c(r) \cdot dr$
13	Transmittance of a layer of aqueous medium	T	$T = \dfrac{\Phi(r)}{\Phi(0)} = e^{-\tau}$

Dimension	Measurement unit	Remarks
L^{-1}	m^{-1}	
L^{-1}	m^{-1}	$b = 2\pi \int_0^\pi b(\gamma) \cdot \sin \gamma \cdot \mathrm{d}\gamma$
Dimensionless		$\dfrac{1}{2} \int_0^\pi \beta(\gamma) \cdot \sin \gamma \cdot \mathrm{d}\gamma = 1$
L^{-1}	m^{-1}	$c = a + b$
Dimensionless		
Dimensionless		
$L^{-1} \cdot \text{sr}^{-1}$	$\text{m}^{-1} \cdot \text{sr}^{-1}$	
L^{-1}	m^{-1}	
L^{-1}	m^{-1}	
Dimensionless		
Dimensionless		$g = \overline{\cos \gamma}$
Dimensionless		$\gamma = \bar{c} \cdot r$
Dimensionless		$T = \mathrm{e}^{-\bar{c} \cdot r}$

log coefficients were used and even today some authors use them. This must be borne in mind when comparing the data of different authors.

Ocean water scatters light extremely non-uniformly in different directions. The function $\beta(\gamma)$, showing the distribution of radiance of scattered light over the scattering angles γ, is called the volume scattering function. Its derivation is shown in Fig. 1.1.

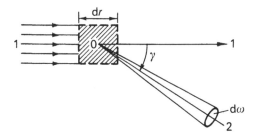

Fig. 1.1. Derivation of the equation for the volume scattering function. The hatched area is the scattering volume: 1–0–1 is the direction of the incident beam; 0–2 is the scattered beam; γ is the scattering angle.

The incident flux Φ produces a normal illuminance $E_n = \Phi/S$ at the volume element $dv = S \cdot dr$ (S is the cross-section, dr is the length of the element along the direction of propagation). The luminous intensity dI scattered by the element dv in the scattering direction γ will be:

$$dI = b(\gamma) \cdot E_n \cdot dv \qquad (1.5')$$

The proportionality factor $b(\gamma)$ is called the scattering coefficient in the given direction. It is apparent that the amount of light scattered by the element dv will be equal to the integral of dI over all directions: $d\Phi = \int_{4\pi} dI \cdot d\omega$. On the other hand, $d\Phi = d\Phi_b = b\Phi \, dr$. Equating these expressions for $d\Phi$, we find that:

$$b = \int_{4\pi} b(\gamma) \cdot d\omega = \int_0^{2\pi} d\phi \cdot \int_0^\pi b(\gamma) \cdot \sin \gamma \cdot d\gamma = 2\pi \int_0^\pi b(\gamma) \cdot \sin \gamma \cdot d\gamma \quad (1.6)$$

Equation (1.6) establishes a relationship between the total scattering coefficient b and the scattering coefficient in the given direction $b(\gamma)$. We divide both sides of Equation (1.6) by b. Thus we have:

$$\int_{4\pi} [b(\gamma)/b] d\omega = 1 \qquad (1.7)$$

In this equation, the expression $[b(\gamma)/b]d\omega$ may be treated as the probability of the scattering of a photon in the element of solid angle $d\omega$. It is customary to write Equation (1.7) in another form:

$$\int \beta(\gamma) \cdot d\omega/4\pi = 1; \quad \beta(\gamma) = 4\pi \cdot b(\gamma)/b \qquad (1.8)$$

Here, $d\omega/4\pi$ is a dimensionless quantity (the solid angle $d\omega$ expressed as a fraction of the total solid angle 4π); $\beta(\gamma)$, the volume scattering function, is the scattering density probability at an angle γ. The volume scattering function $\beta(\gamma)$ satisfies the equation:

$$\frac{1}{2}\int_0^\pi \beta(\gamma)\cdot\sin\gamma\cdot d\gamma = 1 \tag{1.9}$$

This equation is called the normalization condition of the volume scattering function. The representation of the function in the form given in Equation (1.8) is very convenient. For example, for an isotropically scattering medium, assuming $\beta(\gamma) = \text{constant} = k$, then:

$$\frac{1}{2}k\int_0^\pi \sin\gamma\cdot d\gamma = k = 1 \tag{1.9'}$$

ie for such a medium, the scattering density probability $\beta(\gamma) = 1$ for all angles.

We may add that strictly speaking, all derived characteristics in Table 1.1 are redundant since they derive from the three main characteristics. Nevertheless, we have given them because they are widely used in the optical literature. Furthermore, it is often these derived characteristics which are measured directly in the sea and the principal characteristics are calculated from them.

1.1.2. Typical values

In order to make estimates, it is useful to have an idea of the typical values and range of variation of the studied characteristics right from the start. In Table 1.2, we give typical values for $\lambda = 0.550\,\mu m$ for the principal coefficients under average ocean conditions and we compare them with similar values for a cloudless atmosphere above the sea. The values for the ocean coefficients were taken from [67, 68]. Although the measurements in [67, 68] were made in the Mediterranean Sea, they agree well with average ocean conditions. The data for the atmosphere above the sea were determined on the basis of [101]. They were obtained by analysing observations of the spectral transmittance of the

Table 1.2 COMPARISON OF THE OPTICAL PROPERTIES OF THE OCEAN AND THE ATMOSPHERE

Medium	Coefficients (m^{-1})			Λ	Reduced volume scattering function $\beta(\gamma)/\beta(90°)$					
	a	b	c		0	5	10	45	90	$\overline{\cos\gamma}$
Ocean	0.07	0.16	0.23	0.7	10^4	10^3	250	8	1	0.95
Atmosphere	0	2.10^{-4}	2.10^{-4}	1.0	600	200	100	10	1	0.51

atmosphere $T(\lambda)$ and the brightness of the sky $L(\phi)$ on the shores of the Black Sea and the Baltic Sea. If we take into account that the density of water is approximately 1000 times higher than the density of air, then the average attenuation of the light beam relative to the mass of the traversed substance is approximately the same ($\sim 2\,\mathrm{kg}^{-1}$) for the ocean and the atmosphere.

We now consider spectral variation. We limit ourselves to the attenuation coefficient c. First of all we note that seawater contains three optically active components: pure water, dissolved substances (inorganic and organic), suspended matter (mineral and organic). The effect of these components on the optical properties of seawater is different; it changes as a function of the concentration of each component and differs for different wavelengths. A typical picture showing the dependence of the attenuation coefficient c on the wavelength λ for different components of seawater is shown in Fig. 1.2.

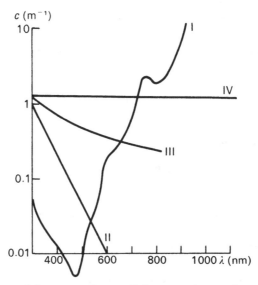

Fig. 1.2. Dependence of the attenuation coefficient c on the wavelength λ_{nm} for different components of seawater.

Curve I shows the spectral distributions of the attenuation coefficient for pure water. The sharp minimum lies in the blue region. Curve II is the absorption spectrum for dissolved organic substances. The absorption sharply increases in the blue and ultraviolet regions. Curve III is the attenuation spectrum for small, mainly mineral particles. Here, in the blue region of the spectrum, the attenuation of light is greatest. Curve IV is the attenuation spectrum for large particles of biological origin (diatom algae, foraminifera, organic detritus, etc). The attenuation of light by them is spectrally non-selective as it is in cloud droplets. The curve of c_1 is given in absolute units; the units for the other curves depend on the concentration of the corresponding component.

In Fig. 1.3, $c(\lambda)$ is shown for two characteristic examples: transparent ocean waters and the turbid waters of the Baltic Sea (based on [110]). For each of these systems, measurements at $\lambda = 550$ nm give: $c_I = 0.036$ m^{-1} (the attenuation of pure water is of course the same in both cases); $c_{II} = 0.010$, 0.030 m^{-1} (in the Baltic Sea, dissolved organic matter, yellow substance, is significantly higher and correspondingly c_{II} is three times larger); $c_{III} = 0.01$, 0.11 m^{-1}, $c_{IV} = 0.050$, 0.200 m^{-1} (the scattering particles in the Baltic Sea are also significantly larger than in the ocean). The total attenuation coefficients $c = \sum_{i=1}^{4} c_i$ will be 0.11 and 0.38 m^{-1}, respectively, ie in these cases, c of pure ocean water was almost four times smaller than the c of the waters of the Baltic Sea.

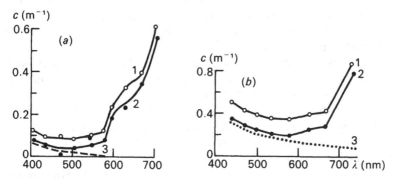

Fig. 1.3. Attenuation of light in transparent ocean water (continental slope) (a) and in the Baltic Sea (b). 1, unfiltered sample; 2, filtered through a fine filter; 3, yellow substance and fine suspended matter.

It can be seen from Table 1.2 and Fig. 1.2 that in certain areas the optics of the ocean and of the atmosphere are considerably different: (1) in contrast to air, which has a weak effect on the colour of objects when they are seen through the atmosphere, pure ocean water has a very strong effect. Any object dropped overboard acquires a blue colour—it is seen through a layer of water which is effectively, a blue filter. The author himself observed this striking phenomenon in the transparent waters to the North-West of the island of Rarotonga on the fifth voyage of the scientific research ship 'Dmitry Mendelyev'. On being lowered overboard, the transmittance meter gradually changed its colour from a bright brownish red colour in air via grey to bright blue; (2) in the atmosphere the number of scattering events may be extremely large; in the ocean, because the photon survival probability Λ is appreciably less than unity, the light beam cannot undergo very much scattering. After being scattered six times, its intensity will be only 10% of the incident light; (3) just as important, is the large difference in the shape of the volume scattering function—in the ocean it is 20 times more elongated than in the atmosphere; this considerably complicates measurements of the transmittance, calculations of the light field, etc.

1.2. Properties of the light field in the ocean

1.2.1. Definitions

Sunlight penetrates the sea to great depths. With the help of sensitive photoelectric instruments, daylight may be observed at depths of the order of 1000 m. The same characteristics are used to describe the light field in the sea as are used in the photometry of any turbid medium. For a complete description of the light field taking into account the state of polarization, it is necessary to define four parameters (I, Q, U, W) which form the so-called Stokes vector parameters of a light beam. We first limit ourselves to a simpler system which is based on the radiance of the radiation L, to be more exact, the spectral density of the radiance L_λ. This quantity is a function of the point of observation r, the direction of propagation of the radiation ε [ε is a unit direction vector and we sometimes write it as (θ, ϕ)] and the moment in time $t : L = L_\lambda(r, \varepsilon, t)$. In the following, for the sake of simplicity we omit the symbol λ in the same way as we omitted it in the optical characteristics.

Two systems of units are used to described the light field: energy and light units. The energy units are based on a black-body detector and the light units on the average human eye. They are related by the maximum luminous efficiency coefficient V_{max} which occurs in a waveband centred at about 554 nm where the eye possesses maximum sensitivity. According to data from direct measurements, $V_{max} = 683$ lm/W. The reciprocal quantity $M = 1/V_{max} = 1.466 \times 10^{-3}$ W/lm is called the mechanical equivalent of light.

Table 1.3 gives the properties of the light field in the sea. Each of the variables is given in energy and in light units. We remember that the luminous intensity I is one of the basic SI units. Its unit of measurement is the candela (cd). The lumen and lux are also SI units, although they are derived from the candela (1m $=$ cd · sr, lx $=$ cd · sr · m^{-2}).

The radiance is a basic property of the light field studied in hydrooptics. We remember that in photometry, the radiance of an elementary area dA is defined by the radiation energy flux $d^2\Phi$ emitted by the area in the direction (γ, ϕ) in the elementary solid angle $d\omega$:

$$d^2\Phi = L(\theta, \phi) \cdot \cos \theta \cdot dA \cdot d\omega \qquad (1.10)$$

If we remember that in the electromagnetic field, the energy flux density is determined by the Poynting vector, then it follows from Equation (1.10) that in electrodynamics, the radiance is the average value of the modulus of the Poynting vector of a beam of plane waves propagated within a solid angle $d\omega$. In photometry, another definition of radiance (luminance) is given. We imagine that we have placed a radiation receiver normal to the axis of the incident beam. Then, the (normal) irradiance produced by the radiation on the receiver, will be:

$$dE_n = d^2\Phi/dA = L \cdot d\omega \qquad (1.11)$$

In this definition, which of course is equivalent to the two previous definitions, the radiance is a property of the radiation beam itself and is not related to the irradiating surface [21].

For a full description of the light field at a given point in the sea, it is necessary to define the radiance distribution, the scalar function $L(\varepsilon) = L(\theta, \phi)$ in the total solid angle 4π. We note that this function may have any degree of complexity in its structure. Knowing its value for certain directions, we cannot say anything about its value in any other direction, ie it does not have tensor properties. Knowing the radiance distribution, it is possible to determine any property of the light field at a given point. Of course, in going from the radiance distribution to the various integral averaged properties, our understanding of the light field is reduced. We begin with complete knowledge and finish with incomplete knowledge but we are forced to do this because to give a complete definition of the light field is extremely difficult. At the same time, in many cases the use of averaged properties turns out to be adequate.

After the radiance distribution, the next property (in terms of the detail of the description of the field) is the irradiance distribution. If we place a small area oriented in any direction at the point in question, then we obtain for its irradiance:

$$E(\theta, \phi) = \int_{2\pi} \cos \alpha \cdot dE_n = \int_{2\pi} L(\theta', \phi') \cdot \cos \alpha \cdot d\omega' \qquad (1.12)$$

The angles (θ, ϕ) denote the direction of the normal to the area, and the angles (θ', ϕ') the direction of the ray incident on the area and α the angle between these directions. If we plot the value of the irradiance $E(\theta, \phi)$ along the direction (θ, ϕ) and rotate the elementary area in any way so that the normal to it describes the angle 4π, then we obtain the irradiance distribution.

The radiance and irradiance fields have significantly different structures. The radiance field contains full information on the light field. The irradiance field is smoothed. The functions $E(\theta, \phi)$ and $L(\theta, \phi)$ are related by Equation (1.12) which incorporates integral smoothing with a kernel $\cos \alpha$. This smoothing is very strong, so it is impossible to invert equation (1.12), ie to determine the radiance field from the irradiance field at a given point; the problem is ambiguous [21, p. 42].

An important quantity characterizing the light field is associated with the irradiance field, the radiation transfer vector **H**. The quantity **H** characterizes the radiation flux across the area. It is defined by the equation:

$$H = \int_{4\pi} L(\theta', \phi') \cdot \cos \alpha \cdot d\omega' = E_1 - E_2 \qquad (1.13)$$

where E_1 and E_2 are the values of the irradiance of the area on different sides. Thus, the radiation flux across the area is equal to the difference in the irradiances of this area. According to Equation (1.12), the irradiance differs only from the light pressure by a numerical factor. Thus, the radiant flux is proportional to the resultant pressure produced by the light on the area. We place a system of

Table 1.3. PROPERTIES OF THE LIGHT FIELD IN THE SEA

No.	Name	Symbol	Formula	Dimension	Measurement unit	Remarks
			Basic properties			
1	Radiance (luminance)	L	$L = \dfrac{d^2\Phi}{d\omega \cdot dA}$	$MT^{-3} \cdot sr^{-1}$ $L^{-2} J \cdot sr^{-1}$	$W/(sr \cdot m^2)$; cd/m^2	$d^2\Phi$ is the energy flux transported by the radiation propagating in a small solid angle $d\omega$ whose axis is perpendicular to the irradiated area dA
			Derived properties			
2	Degree of polarization	p	$p = \dfrac{L_{max} - L_{min}}{L_{max} + L_{min}}$	Dimensionless		L_{max} and L_{min} are the maximum and minimum radiance of two mutually perpendicular polarized components
3	Irradiance of a horizontal area from above (from below)	$E_d, (E_u)$	$E_d, (E_u) = + (-) \displaystyle\int L(\theta, \phi) \cdot \cos \theta \cdot d\omega$ With respect to the upper (lower) hemisphere	MT^{-3} $L^{-2} J$	W/m^2 lx	Flux in the sea across a unit horizontal area downwards (upwards); θ and ϕ are the vertical and azimuth angles
4	Scalar irradiance (illuminace)	E_0	$E_0 = \displaystyle\int_{4\pi} L \cdot d\omega$	MT^{-3} $L^{-2} J$	W/m^2 lx	Radiation energy density ρ (amount of radiant energy in unit volume) related to E_0 by the equation

No.	Quantity	Symbol	Defining equation	Dimensions	Units	Remarks
5	Downward (upward) scalar irradiance (illuminance) from above (below)	E_{od}, (E_{ou})	$E_{od}, (E_{ou}) = +(-)\int L(\theta,\phi)\cdot d\omega$ with respect to the upper (lower) hemisphere	MT^{-3} $L^{-2}J$	W/m² lx	
6	Spherical irradiance (illuminance)	E_s	$\pi r^2 \int$ $E_s = \dfrac{\int_{4\pi} L\cdot d\omega}{4\pi r^2} = \dfrac{E_0}{4}$	MT^{-3} $L^{-2}J$	W/m² lx	Average illumination of an infinitesimal sphere of radius r. Obviously $E_S = c\cdot\rho/4$
7	Radiation transfer vector (light vector)	**H**	$H_x = E_{x,1} - E_{x,2}$ $H_y = E_{y,1} - E_{y,2}$ $H_z = E_{z,1} - E_{z,2}$	MT^{-3} L^{-3}	W/m² lx	Vector whose projections on the axes are equal to energy fluxes passing through the areas normal to the axes
8	Vertical attenuation coefficient for irradiance	$K(z)$	$K(z) = -\dfrac{1}{E}\cdot\dfrac{dE}{dz}$	L^{-1} $\dfrac{dE}{dz}$	m⁻¹	Defined for all six types of illumination indicated in 3–6. Denoted by K_d, K_u, K_0, K_{od}, K_{ou} and K_S, respectively
9	Transmission of irradiance (illuminance)	$\eta(z)$	$\eta(z) = \dfrac{E_d(z)}{E_d(0)}$	Dimensionless		Ratio of the downward irradiance at a depth z and directly below the surface

Table 1.3 Cont.

No.	Name	Symbol	Formula	Dimension	Measurement unit	Remarks
			Derived properties			
10	Average cosine of radiance propagating in the sea	$\bar{\mu}(z)$	$\bar{\mu}(z) = \dfrac{\displaystyle\int_{4\pi} L(z, \theta, \phi) \cdot \cos\theta \cdot d\omega}{\cos\theta \displaystyle\int_{4\pi} L(z, \theta, \phi) \cdot d\omega}$	Dimensionless		Together with $\bar{\mu}$, the quantity $\bar{U}(z) = 1/\bar{\mu}(z)$ is often defined. It is called the angular distribution factor of radiance
11	Irradiance reflectance (albedo) of sea layers	$R(z)$	$R(z) = \dfrac{E_u(z)}{E_d(z)}$	Dimensionless		Ratio of upwelling to downwelling irradiance
12	Radiance factor of sea layers	$\rho(\lambda, \theta, \phi)$	$\rho(\lambda, \theta, \phi) = \dfrac{L(\lambda, \theta, \phi)}{L_0(\lambda)}$	Dimensionless		Ratio of radiance emanating from the sea $L(\lambda, \theta, \phi)$ to the radiance of an ideal scatterer L_0 illuminated in the same way
13	Colour index of seawater	I	$I = \dfrac{L(180°, 550\text{ nm})}{L(180°, 450\text{ nm})}$	Dimensionless		Ratio of the nadir radiance of the sea at $\lambda = 550$ nm and $\lambda = 450$ nm
14	Colour grade of the sea	N		Dimensionless	Grades on colour scale	
15	Depth of visibility of a standard white disc (Secchi depth)	Zs		L	m	

14

rectangular coordinates at the point in question and introduce radiation fluxes in the direction of the coordinate axes:

$$H_x = \int L(\theta', \phi') \cdot \cos \theta_x \cdot d\omega'$$

$$H_y = \int L(\theta', \phi') \cdot \cos \theta_y \cdot d\omega' \qquad (1.14)$$

$$H_z = \int L(\theta', \phi') \cdot \cos \theta_z \cdot d\omega'$$

where θ_x, θ_y and θ_z denote the angles formed by the ray with the x, y and z axes. It is not difficult to show that the flux across any area with normal ε can be expressed in terms of the orthogonal components H_x, H_y and H_z in the following manner:

$$H_x \cdot \varepsilon_x + H_y \cdot \varepsilon_y + H_z \cdot \varepsilon_z = (\mathbf{H}, \varepsilon) = H_\varepsilon \qquad (1.15)$$

Thus, the radiation flux in a certain direction is the projection of the vector $\mathbf{H}$ with the components given in Equation (1.14) in this direction.

Apart from the irradiance, other integral characteristics of the field are also used: the volume density of radiation energy ρ and the spherical irradiance at the given point of the field E_S. In order to define ρ, we first imagine that we are dealing with a narrow beam which arrives from the angle $d\omega$ at the area dA. The flux transported by it will be: $d^2\Phi = dA \cdot L \cdot d\omega$.

It occupies a volume $dv = x \cdot dA$; thus, $\rho = d^2\Phi/dv = (1/c) \cdot L \cdot d\omega$. It is apparent that in any light field:

$$\rho = \frac{1}{c} \int_{4\pi} L \cdot d\omega = \frac{1}{c} \cdot E_0 \qquad (1.16)$$

As far as E_S is concerned, in order to define it, we point out that the elementary flux arriving at a small sphere of radius r is equal to $\pi r^2 L(\theta, \phi) \, d\omega$. The total flux at the sphere $\Phi = \int_{4\pi} \pi r^2 L(\theta, \phi) \, d\omega$, and the average irradiance at the sphere is:

$$E_S = \frac{\Phi}{4\pi r^2} = \frac{1}{4} \int_{4\pi} L \cdot d\omega = \frac{1}{4} E_0 = \frac{c}{4} \rho \qquad (1.17)$$

The characteristics of the light field given above were derived by A.A. Gershun [21], N. Jerlov [25] and R. Preisendorfer [170]. It is not difficult to measure them in the sea, it is only necessary to have an underwater photometer with the appropriate attachment. These measurements are shown in Fig. 1.4. The relative simplicity of these measurements is why these quantities are systematically studied in ocean optics. There are empirical dependences between the various characteristics so than knowing one of them, it is possible to determine the others approximately. These empirical relationships are extremely useful. They are examined in [25, 39, 164, chapter 4]. We point out some of them later on.

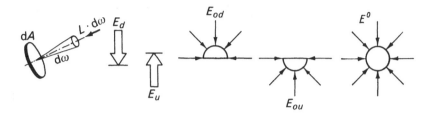

Fig. 1.4. Measurements of the characteristics of the light field: radiance and irradiance (vector downward (upward), 2π-scalar, 4π-scalar).

1.2.2. Typical values

The radiation below the surface of the sea is attenuated by absorption (infrared and the red region of the spectrum) and scattering. The coefficient K_d depends on λ and weakly on z. This means, that approximately:

$$E_d(z, \lambda) = E_d(0, \lambda) \cdot e^{-K_d(\lambda) \cdot z} \qquad (1.18)$$

It is not difficult to show that the coefficient $K_d(z) = U_d(z) \cdot (a + b_b)$, where $U_d(z)$ describes the angular distribution of the downwelling irradiance. In homogeneous waters, this factor is constant and Equation (1.18) will be strictly true if the irradiance distribution does not vary with depth. On passing through the sea, collimated light rapidly disappears according to Equation (1.4) and remains only as diffused light, the self-illumination of the medium. In addition, the radiance distribution in water gradually changes; its maximum shifts from the surface direction of the refracted sunlight towards the vertical. The factor U_d also slowly changes with depth. In fact, experiments show that starting at a certain level, the irradiance decreases with depth approximately exponentially. The coefficient b_b is usually appreciably smaller than a (see p. 26) and the quantity U_d at high solar altitude is close to unity so that the coefficient $\mathbf{K}_d(\lambda)$ is close to the absorption coefficient $a(\lambda)$.

The spectral transmittance of the surface waters of the oceans was studied by N. Jerlov. Generalization of the observational data allowed him to distinguish three basic optical types of water; I, II and III, to which two intermediate types were added later, IA and IB. The values of $K_d(\lambda)$ for different types of waters are given in Table 1.4 [25]. These values (Jerlov water types) are used for the optical classification of the waters of the world ocean (for more details see Chapter 5). We have given only data for K_d. There is little data on the other K coefficients but as A.A. Ivanoff pointed out [28] they are all very close to each other. This means that in estimates, the data in Table 1.4 may be used for any $K(K_u, K_o, K_{od}, K_{ou}, K_s)$.

The most transparent section is situated at $\lambda = 460$ nm in the blue region. With increasing depth, the spectrum of transmitted light narrows and the maximum of the spectral distribution curve in pure ocean waters is shifted to the region $\lambda_{max} = 460$ nm. A typical variation in the spectral composition of the irradiance with depth is shown in Fig. 1.5 (100% is taken as $E_d(\lambda_{max})$ immediately below the

Table 1.4. VERTICAL (DIFFUSE) ATTENUATION COEFFICIENT $K_d(\lambda)$ (m^{-1})

	λ(nm)								
Water type	310	350	400	450	500	550	600	650	700
I	0.15	0.062	0.028	0.019	0.028	0.063	0.24	0.36	0.56
IA	0.19	0.078	0.038	0.026	0.031	0.067	0.24	0.37	0.57
IB	0.22	0.10	0.051	0.036	0.042	0.073	0.25	0.38	0.58
II	0.37	0.17	0.094	0.067	0.070	0.089	0.26	0.40	0.62
III	0.65	0.31	0.19	0.13	0.12	0.12	0.30	0.45	0.65

surface). In sunny weather in homogeneous waters, K_d somewhat decreases with depth and reaches a limiting value at asymptotic state. There is practically only blue light remaining at depths exceeding 100 m. Therefore, a single coefficient K_d (based on $\lambda = 460$ nm) can be used for these depths. Typical depth values of K_d for pure waters in open ocean regions are given in Table 1.5 [25].

The absolute values of the irradiance at different depths depends on the irradiance at the surface and if $K_d(\lambda)$ is known then they can be easily calculated from Equation (1.18). The coefficient K_d is always smaller than c. It has been found from simultaneous measurements of K_d and c for $\lambda = 520$–540 nm [38] that approximately:

$$K_d = 0.20c \pm 0.02 \tag{1.19}$$

In fact, for the open ocean, we find $c = 0.23$ m^{-1} and $K_d = 0.05$ m^{-1} in Tables 1.2 and 1.4. The coefficient K_d, as other apparent optical properties, depends on the structure of the light field. As the altitude of the sun changes, it varies according to $K = K^0 \cdot \sec \theta$, where K^0 is the coefficient when the sun is at the zenith and θ is the angle of refraction of sunlight. The variation in irradiance with depth for five types of water classified by Jerlov (as a percentage of the surface irradiance) at $\lambda = 465$ nm is shown in Fig. 1.6.

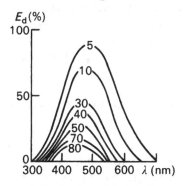

Fig. 1.5. Variation in the spectral distribution of the irradiance with depth (100% is taken as $E_d(\lambda_{max})$ immediately below the surface). Depths in metres are indicated on the curves.

Table 1.5. COEFFICIENT OF VERTICAL ATTENUATION K_d FOR GREAT DEPTHS

Region	Depth range (m)	K_d (m^{-1})
Sargasso Sea	100–400	0·040
	400–500	0·038
Northern part of the Atlantic Ocean	100–350	0·031
Western part of the Indian Ocean	200–800	0·022–0·033
Near Tahiti	100–400	0·034
World's oceans as a whole		0·03–0·04

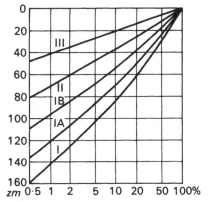

Fig. 1.6. Variation in irradiance with depth for five types of water classified by Jerlov (as a percentage of the surface irradiance) at $\lambda = 465$ nm.

The Jerlov standards are determined from observations of the irradiance with broad-band glass filters. Using the data of their own observations of $E_d(\lambda)$ and also the data of A. Morel and L. Prieur obtained by instruments with high spectral resolution, V.N. Pelevin and V.A. Rutkovskaya recently suggested a more detailed optical classification of waters [70]. They construct a nomogram which makes it possible to determine $K_d(\lambda)$ from the value of K_d (500 nm).

1.2.3. Absorption coefficient and divergence of the light vector: Gershun's formula

We have already noted that the relationships between different hydrooptical characteristics are widely used in hydrooptics. The majority of them are empirical relationships such as Equation (1.19) whose area of application is not accurately known. In this sense, exact equations which hold true under any conditions are especially important. One of these is the relationship between the coefficient a

and the divergence of the light vector **H**. It can be very simply derived. On absorption, the energy of the electromagnetic field is converted into heat. Let q be the dissipative function of the system. It is equal to the amount of heat which is produced in unit volume in unit time. According to the definition of the coefficient a, the amount of heat dQ which is produced in the volume $dv = dr \cdot dA$ from the elementary radiation beam $L \cdot d\omega$ will be $dQ = a \cdot dA \cdot L \cdot d\omega \cdot dr$, in other words: $dq = dQ/dv = a \cdot L \cdot d\omega$. If the volume in question is irradiated on all sides with radiance $L(\theta, \phi)$, then the total amount of heat which is generated in unit volume will be equal to:

$$q = a \int_{4\pi} L(\theta, \phi) \cdot d\omega = a \cdot E_0$$

We now define the volume V in the light field enclosed by the surface S. The flux of light energy from this volume to the outside is equal to $\int_S H_n \cdot dS$, where H_n is the projection of the vector **H** on the external normal to dS. According to the theorem of Gauss:

$$\int_S H_n \cdot dS = \int_V \text{div } \mathbf{H} \cdot dv$$

ie div **H** is the density of the energy sources or sinks of the electromagnetic field. Under steady-state conditions, obviously:

$$q = -\text{div } \mathbf{H}$$

Thus, we have:

$$a = -(1/E_0) \cdot \text{div } \mathbf{H} \tag{1.20}$$

Equation (1.20) is called Gershun's equation. It is used in hydrooptics to determine a. In the ocean, the underwater irradiance depends only on the depth z, ie div $\mathbf{H} = d(E_d - E_u)/dz$. Consequently:

$$a = -(1/E_0)[(dE_d/dz) - (dE_u/dz)] \approx -(1/E_0)(dE_d/dz) \tag{1.21}$$

since usually $dE_u/dz \ll dE/dz$*. Thus in order to determine a, it is necessary to measure E_0 and the gradient of the downwelling irradiance E_d.

1.3. Depth of visibility of a white disc and the colour of the sea

1.3.1. White (Secchi) disc

The simplest method of estimating the transmittance of seawater from the depth of visibility of the white sea bottom was described as early as the beginning of the

---

*The coefficients K_d and K_a are approximately the same which means that the derivatives in Equation (1.21) are in the same ratio as the quantity $E_u/E_d = R$. The irradiance reflectance R is usually approximately 2%.

18th century by P. Bouguer. O. Kotzebu in 1817 studied the transmittance of the waters of the Pacific Ocean by the depth of visibility of a white plate and a piece of red material. Secchi and Cialdi in 1865 systematically used white and coloured discs to estimate the transmittance of the waters of the Mediterranean. These observations were gradually standardized [38, 39]. A metal disc of diameter 30 cm, painted white with oil paint with a weight of 50 kg attached to it (to prevent drift) is lowered overboard into the sea. The depths of disappearance Zs_2 and reappearance Zs_1 of the disc are noted $(Zs_2 > Zs_1)$. The Secchi depth Zs is defined as $(Zs_2 + Zs_1)/2$.

Because of their simplicity, observations with the disc have become widespread. They have covered the entire areas of the oceans in the world. By the end of 1976, there were approximately 300,000 observations, about half of them on the oceans and the other half on seas. The accuracy of the determination of Z_s, when the sea is calm is ± 0.5 m and at force 4 it is $\pm (1-2\,m)$; taking measurements in very rough seas is not recommended. For high values of Zs, ie in transparent waters, the conditions of illumination and sea disturbance have a strong effect on the value of Zs. With variable illumination and patches of light the threshold of the eye's contrast sensitivity μ rises appreciably and Zs drops by as much as 20%, according to the data given in [52]. In order to reduce this effect, it is recommended in [5, 52] that the observations of the disc are made through a box with a glass bottom. We examine the values of Zs in different seas and oceans in Chapter 5. Here, we present only the material in Table 1.6. Generally speaking, we note that Zs varies from values of the order of 1 m in rivers and lakes to 67 m. This record value was found by V.M. Pavlov on the fifth voyage of the scientific research ship 'Dmitry Mendeleyev' in April 1971 [69]. The station was situated in the zone of the southern tropic convergence in the Pacific Ocean to the North-West of the island Rarotonga, at a point with coordinates 19°04' latitude south, 162°36' longitude west. Usually the record is given as $Zs = 66.5$ m. It was observed by O. Krümmel in 1987 in the Sargasso Sea. However, these observations were made with a 2 m disc. On changing to a standard disc, Zs should be reduced by 6% which gives $Zs = 62$ m.

We now derive an equation relating Zs to the optical characteristics of the water. The contrast which the disc at a depth z makes with the surrounding background, will be determined as usual by the expression:

$$C(z) = [L_s(z) - L_b(z)]/L_s(z)$$

We assume for the sake of simplicity that the point of observation is directly below the surface of the water. In this case, the radiance of the disc and the

Table 1.6. SECCHI DEPTHS Zs (m)

Gulf of Panama	10	Black Sea	25
Baltic Sea	13	Bay of Bengal	45
Barents Sea	18	Sargasso Sea	62

background arriving at the observer's eye, is:

$$L_s = L_s^0 e^{-cz} + L^*(z)$$

$$L_b = L_b^0 e^{-cz} + L^*(z)$$

Where $L^*(z)$ is the veiling radiance created due to light back scattering by the water column above the Secchi depth. L_s^0 and L_b^0 are the radiance of the disc and background at the depth z—the level of the disc. It is proportional to the irradiance $E_d(z)$ at this level:

$$L_s^0 = \frac{1}{\pi} \rho_s E_d(z); \quad L_b^0 = \frac{1}{\pi} \rho_b E_d(z)$$

where ρ_s and ρ_b are the reflectances of the disc and background.
   Finally, we obtain:

$$C(z) = C(0)/(1 + f)$$

$$C(0) = (\rho_s - \rho_b)/\rho_s; \quad f = (\pi L^*/\rho_s)[e^{cz}E_d(z)]$$

In the theory of oblique visibility, the quantity f is called the haze coefficient and $C(0)$ the initial contrast.
   At depths close to the limiting depth, $f \gg 1$. Consequently,

$$C(z) = C(0) \cdot (\rho_s E_d(0)/\pi L^*)e^{-(c + K_d)z}$$

since $E_d(z) = E_d(0)e^{-K_d z}$. Strictly speaking, this equation is true for large z. Substituting $\tilde{C}$ for $C(0)\rho_s E_d(0)/\pi L^*$, we easily find that:

$$C(z) = \tilde{C}e^{-(c + K_d)z} \quad (1.22)$$

At limiting depths, $L^* =$ constant and $\tilde{C}$ is independent of z.
   The variable z becomes equal to Zs when:

$$C(z) = \mu$$

(μ is the threshold of contrast sensitivity of the eye).
   Thus, we obtain for Zs:

$$Zs = F/(c + K_d) \quad (1.23)$$

The constant F_s incorporates an allowance for the loss in contrast which will occur at the sea surface. If we take $K_d = \eta \cdot c$ we obtain:

$$Zs = k_c/c \quad (1.24)$$

The dependence given in Equation (1.24) has been studied by a number of authors. A.A. Gershun [5] compared the Zs data with the values of c in the middle part of the visible spectrum at more than 30 stations. The observations were made on the inland seas of the USSR. It was found that $k_c = 8$ (Fig. 1.7). According to the same data, $K_d \cdot Zs \simeq 2$. Thus, according to Gershun, $K_d = 0.25c$. In [58], the correlation between the mean value of $c(z)$ for the layer $0 - Zs$ and the quantity Zs was investigated. Here, 111 stations were used covering practically all the regions of the world's oceans. The author found the same value $k_C = 8$ with a correlation coefficient $r = 0.94 \pm 0.01$.

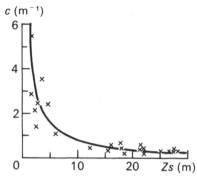

Fig. 1.7. Relationship between Zs and c. The crosses are the experimental data and the solid line is from Equation (1.24) with $k = 8$.

In [38], the correlation coefficient r between Zs and $(c + K_d)$ and between Zs and c (the values of c and K_d were taken for $\lambda = 520$ nm) was also calculated for different regions of the ocean from the material generated by approximately 100 observation stations. For Equation (1.23), r was equal to 0.95 with $F = 8.15$; for Equation (1.24), r was somewhat lower ($= 0.88$) with $k = 7.0$. Taking into account Equation (1.19) the values of k and F are practically equal ($k = F/1.2$). In [38], it is emphasized that the relationship between Zs and c and K_d is more stable if the observations on Zs are made in steady homogeneous conditions with a calm sea and dull weather.

A more complete theory of Zs, taking into account the aureole arising around the disc and the dependence of the threshold k on the width of the smearing zone, is given in [52]. For us it is important that it also confirms the possibility of using Equations (1.23) and (1.24) under real conditions.

1.3.2. Colour of the sea

It is known that different seas do not have the same colour. The saturated blue of tropical and subtropical regions of the ocean turns into bluish-green and green at higher latitudes and to shades of whitish green in turbid coastal waters. For example, there is an appreciable difference between the colour of the waters of the Baltic Sea, Red Sea and White Sea. This gave rise to the idea of using the visible colour of seawater as an oceanological characteristic.

At the end of the last century, Forel and Ule suggested a special chromaticity scale (a set of test tubes with coloured liquids) for defining the colour of the sea. It was previously called the Forel-Ule scale from the names of the scientists who suggested it [F. Forel (1895), W. Ule (1892)]. The colour grades for ocean water are given in Table 1.7.

The determination of the colour of the sea involves visually selecting a test tube, containing a solution whose colour is closer than the others to the colour of the sea. The colour of the sea is then designated by the number of the corresponding test tube on the scale, by the colour grade N. A blue solution was

Table 1.7. COLOUR SCALE FOR OCEAN WATER

Colour grade N	Percentage		Colour grade N	Percentage		
	Blue	Yellow		Blue	Yellow	Brown
I	100	0	XI	35	65	0
II	98	2	XII	35	60	5
III	95	5	XIII	35	55	10
IV	91	9	XIV	35	50	15
V	86	14	XV	35	45	20
VI	80	20	XVI	35	40	25
VII	73	27	XVII	35	35	30
VIII	65	35	XVIII	35	30	35
IX	56	44	XIX	35	25	40
X	46	54	XX	35	20	45
XI	35	65	XXI	35	15	50

prepared by Forel from a mixture of copper sulphate and ammonia (this produces the complex salt $Cu(NH_3)_4SO_4 \cdot H_2O$), and a yellow liquid from a 0.5% solution of potassium dichromate. Ule added a brown colour which was prepared from cobalt sulphate and ammonia.

As the colour grade N increases from I to XI, the percentage of the blue solution decreases to 35%. For $N > XI$, it is constant but the percentage of the yellow solution decreases and the contribution of the brown solution increases.

However, the visible colour of the sea changes not only from one region to another but also when the illumination conditions are changed and even when the angle of observation is changed. This considerably complicated the original simple idea of the colour index for different kinds of waters. However, the simplicity of the observation of the colour of the sea has meant that even to this day, colour is a widely used oceanological characteristic. In order to reduce the distortion due to the angle of observation and the hydrometeorological conditions, the observation procedure has been standardized. On observation, the colour of the test tube is compared with the colour of a column of ocean water examined against the background of a standard white disc lowered to $Zs/2$. We note, for example, that for waters of the Mediterranean Sea and open oceans, N is equal to I–II, for the Caspian Sea it is equal to VII–IX, for the estuaries of rivers in the Baltic Sea it is equal to XII and for branches of the Ob-Yenisey currents in the Kara Sea it is XIX.

We briefly dwell on the theory of colour vision. It is known that any shade of colour may be obtained by mixing three primary colours in specific proportions. This is because the light-sensitive elements in the retina of the eye contain three types of pigments with different absorption spectra. One type absorbs red light well, another green light and the third blue light. The light flux incident in the eye produces a reaction in the light-sensitive elements of the retina which depends on the spectral composition of the incident light.

If all types of anomalous conditions are excluded, for example very small angular dimensions of the object, etc, then the colour perception of the human eye may be described quantitatively with the help of the so-called chromaticity coordinates x, y and z. These coordinates are defined by the formulae:

$$x = X/(X + Y + Z); \quad y = Y/(X + Y + Z); \quad z = Z/(X + Y + Z) \qquad (1.25)$$

The variables X, Y and Z are calculated from the spectral density of the radiation flux $E(\lambda)$ arriving at the observer's eye from the formulae:

$$X = \int_{\lambda_1}^{\lambda_2} E(\lambda) \cdot \bar{x}(\lambda) \cdot d\lambda; \quad Y = \int_{\lambda_1}^{\lambda_2} E(\lambda) \cdot \bar{y}(\lambda) \cdot d\lambda; \quad Z = \int_{\lambda_1}^{\lambda_2} E(\lambda) \cdot \bar{z}(\lambda) \cdot d\lambda$$

$$\qquad (1.26)$$

where $\bar{x}(\lambda)$, $\bar{y}(\lambda)$ and $\bar{z}(\lambda)$ are the so-called specific chromaticity coordinates. These are curves describing the spectral sensitivity of the colour receptors of the eye. The limits of integration λ_1 and λ_2 are determined by the bandwidth of visible light and are equal to 380 and 780 nm. Since $x + y + z = 1$, then in order to specify uniquely the position of any colour in colour space it is sufficient to know only two coordinates, say x and y.

Let us calculate the coordinates x and y of visible monochromatic radiation and plot them successively from 380 to 780 nm on the (x, y) plane. They form a curve, the so-called chromaticity diagram. If we joint the extreme points, the violet and red ends of the spectrum, with a straight line, the closed figure which is obtained describes the set of all real colours (Fig. 1.8).

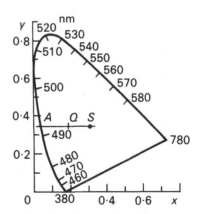

Fig. 1.8. Chromaticity diagram

Apart from the specification of the colour coordinates x and y, there is another method of defining colour. It is based on the concept of the colour white and on the fact that any colour may be obtained by mixing white and monochromatic colours in a specific proportion. This does not contradict the rule of mixing colours: since the colour white is the sum of two so-called complementary colours. The colour white is formed by a homogeneous mixture of three primary colours. The following must therefore apply: $E(\lambda) = $ constant; $X = Y = Z = 1$ and $x = y = z = 1/3$. In the chromaticity diagram (Fig. 1.8), it is represented by the point $S(x = y = 1/3)$. The position of any point Q on the plane is uniquely defined by specifying A, the point of intersection of the straight line SQ with the chromaticity diagram and the ratio $P = QS/AS$. The variable P is called the purity of the colour. It defines the ratio in which the pure colour A is mixed with the colour white S. For monochromatic radiation, the point Q coincides with the point A. In this case, $P = 100\%$. Thus, another method of defining colour involves defining the colour of the radiation (point A) and the purity of the colour P.

For an illustration of the above, we turn to Table 1.8. This gives data on the colour and colour purity of the upwelling radiation in the surface layers of some seas [25].

Table 1.8. COLOUR AND COLOUR PURITY OF SOME SEAS

Region	Coordinates	Solar altitude	Colour A (nm)	Colour purity $P(\%)$
Pacific Ocean	1°20′ lat. south 167°23′ long. east	61	473	85
Indian Ocean	11°25′ lat. south 102°08′ long. east	80	474	84
Mediterranean Sea	33°54′ lat. north 28°17′ long. east	74	473	83
Black Sea	43°17′ lat. north 40°04′ long. east	49	492	51
Baltic Sea	60° lat. north 19° long. east	55	540	24

It can be seen from Table 1.8 that in oceans as well as in the Mediterranean Sea the colour is blue and of high purity; in the Baltic the colour is muddy green (low purity).

We now consider the relationship between the colour of the sea and the optical properties of seawater. On observation from the side of a ship, the total radiance of the sea is given by $L = L_r + L_b$. It consists of the radiance of the reflected light L_r and of the light back scattered by the sea L_b. The ratio between the radiances L_r and L_b depends on the angles of observation (θ, ϕ), on the altitude of the sun and the metereological conditions, on the scattering coefficients and the absorption of light in seawater. The colour of the sea varies as a function of all these parameters. The radiance of the reflected light L_r does not depend on the optical properties of seawater. This is the main obstacle to using colour as an oceanological characteristic. Therefore, it is necessary to observe the colour of the sea in such a way that the fraction L_r is minimal. It is reduced by observing vertically downwards, in clear weather, etc. If the contribution of L_r is small or changes to a small extent, then the variability of the colour of different seas and oceans will be caused by the variability of the spectral composition of the light beam emerging from the sea. The reflectance of the sea is determined from the formula (for references see [12]):

$$R(\lambda) = \frac{b(\lambda, 180°)}{2[a(\lambda) + b_b(\lambda)]} \quad (1.27)$$

Here $b(\lambda, 180°)$ is the scattering coefficient for $\theta = 180°$; $a(\lambda)$ and $b_b(\lambda)$ are the coefficients of absorption and back scattering (Table 1.1).

The actual colour of the sea depends on the spectral variation of $b(\lambda, 180°)$, $a(\lambda)$ and $b_b(\lambda)$. As has been shown in [12], the value of $b_b(\lambda)/[a(\lambda) + b_b(\lambda)]$ is approximately 15–20% in the spectral region 0.39–0.61 μm. Therefore, the quantity $b_b(\lambda)$ is sometimes neglected and Equation (1.27) is written in the form:

$$R(\lambda) = \frac{b(\lambda, 180°)}{2a(\lambda)} \quad (1.28)$$

Furthermore, since the variation of $a(\lambda)$ is significantly higher than $b(\lambda, 180°)$, the crude assumption is often made that the actual colour of the sea is a characteristic of the spectral variation of the absorption coefficient $a(\lambda)$. The purer and more transparent the water, the more the minimum of $a(\lambda)$ is shifted towards the blue region. The saturated blue colours (N = I–II) are the colours of transparent water of the open ocean. This means, that as the Secchi depth increases, namely Zs, the value N must decrease. Such a relationship has already been pointed out in [5]. Of course, the relationship between Zs and N is rather crude because Zs is a function of c, while the colour and colour grade N are functions of $a(\lambda)$.

In [88], the authors established empirical relationships between the colour index I and Zs and I and the colour grade N:

$$I = 123Zs^{-2.02}$$

$$I = 6.54 \times 10^{-3}N^{2.67} \quad (1.29)$$

Naturally, the relationship between the index I and the colour grade N is closer. It can be seen from the graph in [88] that the scattering of points is small, the correlation coefficient $r = 0.97 \pm 0.02$. The relationship between I and Zs is not so good. The authors state that in this case $r = 0.93 \pm 0.02$. From the data of [88], it can be seen that the region where both relationships apply at the same time is II < N < XII, 40 > Zs > 5 m. In this region:

$$Zs = 137N^{-1.32}; \ N = 40Zs^{-0.75} \quad (1.30)$$

From Equation (1.30), we deduce the following (Table 1.9).

Table 1.9. RELATIONSHIP BETWEEN Zs AND THE COLOUR GRADE N

N	II	IV	VI	VIII	X	XII	XIV
Zs in m	55	22	13	8.8	6.6	5.1	4.2

The relationship between Zs and the value N is illustrated by the charts of isolines for these variables in the south western part of the Kara Sea drawn up by K.A. Gomoyunov (see [5, Fig. 38a, 38b]). The charts agree well with each other and with the chart of currents in this region. In these optical charts, three branches of the Ob-Yenisey current running parallel to the west can be clearly seen.

Chapter 2

Factors determining the optical properties of ocean water

2.1. Pure water and dissolved substances

2.1.1 Pure water

Ocean water is a complex physicochemical biological system. It contains dissolved substances, suspensions and a multitude of different living organisms. Due to the suspensions and all kinds of other heterogeneities, ocean water scatters light strongly, ie it is a turbid medium. Its optical properties depend upon its composition and on its physical state (temperature, pressure, etc). Below, we discuss substances dissolved and suspended in ocean water, their distribution over ocean surfaces and their variation with depth. All these areas are studied in the chemistry, biology and geology of the ocean. We briefly summarize the basic information, paying special attention to those characteristics which are essential for optics. For more detailed information, the reader is referred to the relevant volumes of the *Oceanology* series [65, 66].

When using data on the composition of ocean water for the analysis of optical properties, it is necessary to take into account that the methods for studying the composition of water are still very primitive. This applies not only to such delicate components as living organisms and dissolved organic substances but also to particles of mineral suspensions. The fine particles of suspensions have a considerable effect on optics but are poorly assessed by standard methods. Therefore, it is usually impossible to compare directly optical characteristics and composition data obtained by standard methods. Furthermore, it is often only indirect optical methods which give reliable data on the composition of water. Nevertheless, when analysing the optics of ocean water, one cannot ignore the data on the composition of water obtained in oceanology (even if incomplete). Of course these data have to be considered bearing in mind the actual content.

The main factors affecting the optics of ocean water are the optically active components: pure water, dissolved substances (inorganic and organic) and suspended material (mineral and organic). Apart from these factors, air bubbles and heterogeneities arising as a result of turbulence affect the propagation of light in the sea. However, bubbles in appreciable quantities are only observed in the uppermost surface layer of the sea. The contribution of turbulent heterogeneities to the characteristics studied in this work is very small. Therefore, we are interested only in the three main components. We add that the optical properties of water may be considered as the superimposition of the characteristics of its individual components.

When we speak of pure water, we mean a chemically pure substance. It is a mixture of seveal isotopes of water molecules with different molecular weights. At present, 36 different isotopes have been obtained in the laboratory. Some of them have a very short lifetime and in practice are not encountered in natural waters. Apart from the main isotope, $H_2^{16}O$, molecules of heavy oxygen water, $H_2^{18}O$ and $H_2^{17}O$, and heavy water, HDO, are encountered in natural waters. The fraction of different isotopes in natural water depends on its origin. For the above-mentioned three isotopes, it fluctuates about 2, 0.4, 0.3°/oo, respectively. Detailed data on this subject are given in [32]. Although, the optical properties of individual isotopes have been inadequately studied, it is known that they are appreciably different. Thus, according to the data given in [178], the absorption spectrum of normal water has a blurred band with a minimum in the blue region at 475 nm, with a value for the absorption coefficient of $a = 1.8 \times 10^{-6} m^{-1}$, whereas in heavy water ($D_2$0) the minimum is shifted to the orange region at 600 nm with $a = 1.9 \times 10^{-6} m^{-1}$. This shift is easily explained by the difference in the vibrational frequencies of the OH and OD bonds. Below, we give Equation (3.18), from which it can be seen that the fundamental frequencies of H_2O and D_2O molecules are shifted by a factor of $\sqrt{2}$ and all the harmonics are correspondingly shifted; it is the superimposition of these which gives the blurred bands. In fact, the ratio $\lambda_{min}(D_2O)/\lambda_{min}(H_2O) \approx \sqrt{2}$.

It should also be expected that the lighter molecules will polarize more strongly in an electric field. For example, for $\lambda = 589$ nm, $T = 20°C$ and $p = 10^5$ Pa, the refractive index of H_2O is larger than that of D_2O by 49×10^{-4}, ie the polarizability of the H_2O molecules is 15% higher than that of D_2O. Due to the low concentration of isotope solutions in naturally occurring water, the effect of isotopes may normally be neglected in studies of the optics of the ocean. Nevertheless it is useful to have an idea of the magnitude of the effect.

The range of physical conditions under which water is found in the ocean is as follows: pressure $(1-1100) \times 10^5$ Pa, temperature from $-4°C$ to $36°C$. The variation in the optical properties in this range is relatively small; we consider it below. The main variation in the optical properties is caused by the variation of the dissolved substances and particle suspensions.

2.1.2 Inorganic salts and gases

Seawater has inorganic salts, gases and organic compounds dissolved in it. Gases are pesent in water in negligible quantities and although they are important for the biology and geochemistry of the ocean, they have little effect upon ocean optics. Only dissolved oxygen has some importance, and indirectly nitrogen. The most important factors affecting the optics of ocean water are inorganic salts and organic compounds. The history of research into the chemistry of the ocean is given by J. Riley ([136, chapter 1]). He names the founder of the chemistry of the ocean as Robert Boyle (1670). Scientists such as R. Hook and A. Lavoisier also worked in this area. Dissolved substances have an effect on the absorption as well as on the scattering properties of water. They are particularly important in the absorption of visible and ultraviolet radiation where pure water absorbs weakly by itself. In the red part of the visible spectrum and in the infrared region, the effect of dissolved substances on absorption is small; they simply go unnoticed against the background of strong absorption by the water itself. The effect of dissolved substances on scattering is related to thermal fluctuations in the concentration of the solution. These fluctuations increase the molecular scattering by approximately 30%.

We consider now some of the substances dissolved in ocean water. We start with inorganic salts. These are substances such as NaCl, KCl, $MgCl_2$, $MgSO_4$ and $CaSO_4$. They make up the majority of dissolved substances. Due to the high dielectric constant, water has a high dissociation capacity. Therefore, salt molecules dissociate in it and are found in the form of ions. In the 1880s, an expedition on 'Challenger' discovered a remarkable property of ocean water; although the concentration of dissolved salts in the ocean varies appreciably, the relative ratios between the main salts are constant. The constancy of the relative composition of salts in the ocean was noted at the beginning of the 19th century by A. Marse (1819) and was finally established by V. Dittmar after an analysis of samples collected on 'Challenger'. Some authors therefore call it the Marse principle and others the Dittmar law. The constancy of relative concentrations is violated in the vicinity of estuaries since the salt composition of river water markedly differs from the composition of ocean water and it is also violated in seas isolated from the ocean (eg Caspian Sea, Red Sea). We emphasize that this principle only relates to the main salts and does not apply to the small components. Their contribution, it is true, does not exceed 0.01% of the total amount of dissolved substances.

As noted by A.S. Monin [60], recent investigations carried out with the help of highly sensitive apparatus have shown that the constancy of the main ionic components of ocean water is not as rigorous as had been observed hitherto. Although the practical importance of this effect is probably small and a precise value of the variations is still under study, it should nevertheless be noted.

The value of the salinity S completely characterizes the composition of salts dissolved in the ocean. Therefore the value of S (which is usually expressed parts per thousand) may be determined from the concentration of one of the components. For this purpose, the concentration of Cl^- ions is usually measured as their content is approximately 55% of all substances dissolved in water. The salinity S is determined from the chloride content $Cl^o/_{oo}$ using the following empirical formula:

$$S = 1.80655 \ Cl \qquad (2.1)$$

This refined formula has been used since 1967 when it was recommended by the Intergovernmental Oceanographical Commission (IOC).* Before this, Knudsen's formula (1902) had been used for more than 60 years:

$$S = 0.030 + 1.8050 \ Cl \qquad (2.2)$$

Besides Cl^- ions, the most important ions dissolved in ocean water are Na^+, SO_4^{2-} Mg^{2+}, Ca^{2+} and K^+. Apart from the determination of the chloride content of water which can be carried out with high accuracy (of the order of 10^{-2}‰) by the precipitation of silver nitrate, the measurement of electrical conductivity is now used widely for the determination of S. The electrical conductivity is directly related to the salinity; the accuracy of determining S in this way is 5×10^{-3}‰

The concentration of dissolved substances g_i is defined in two ways in the chemistry of the ocean. For the main components, their ratio to the chloride content is specified, the co-called chloride coefficients (g_i/Cl). Like salinity, the chloride content is given in grams per kilogram of ocean water, ie in ppt. For gases, microelements and biogenenic elements, the concentration is expressed in milligrams (or micrograms) per litre of ocean water at a temperature of 20°C. The concentration in kilograms can be obtained from concentration in litres by dividing by the density. Earlier, milligram atoms (mg-At) per litre at $T = 20°C$ were used for the concentration. The expression of the concentration in milligram atoms or microgram atoms is convenient because it does not depend on the chemical compound to which the element concerned belongs, whereas when units of milligram per litre are used it is necessary to specify the formula of the compound which contains the element.

The distribution of salinity over the surface of the world ocean is essentially regulated by two processes: the ratio between evaporation E cm/year and precipitation P cm/year and the mixing of the surface waters with the underlying layers. For coastal regions it is also important to take into consideration river discharge and for polar regions the melting and formation of ice. Figure 2.1 gives a comparison of the latitudinal variation of the difference $(E - P)$ with the salinity of the surface waters S. We can see that over an enormous range of latitudes from 40° latitude north to 60° latitude south, the path of the two curves is very similar. The close relationship between the fields of S and $(E - P)$ can be convincingly shown by comparing the world maps of both fields. This is done in [14]. In subtropical regions, the evaporation E significantly exceeds the precipitation P; and the salinity is at a maximum. In the equatorial region, it is the opposite picture; the precipitation P is large and the salinity of the surface waters falls. In high latitutdes, in particular in the northern hemisphere, S is appreciably reduced. In regions above 70° latitutde north it is equal to 32–30‰

On the whole, the salinity in the open ocean varies from 33 to 37‰, its average value being approximately 34.7‰. The overall picture described by the graphs

---

* The IOC was set up in 1960 within the framework of UNESCO. In January 1979, 103 countries belonged to the IOC, including the USSR and the Ukranian SSR.

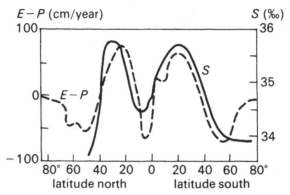

Fig. 2.1. Latitudinal variation of the difference between the average yearly values of the evaporation and precipitation ($E - P$) and the salinity S of surface waters.

in Fig. 2.1 is considerably changed for individual seas where the exchange of water with the ocean is difficult. Thus, the salinity of the Red Sea exceeds 40°/oo, whereas in the northern part of the Baltic Sea it is equal to only 4°/oo.

We examine the variation of salinity with depth. Since denser waters must be found lower down and the density increases with increasing S, one might think that the salinity must also increase with depth. In fact this is not always true. The fact is that positive stratification in the ocean is maintained by a significant reduction of temperature with depth. Therefore, the distribution of salinity often has very complex characteristics and is associated with the overall circulation of water. The distributions of $S(z)$ are shown in Fig. 2.2 (in accordance with [80]). We can see that at high latitudes where the temperature gradient is small, S

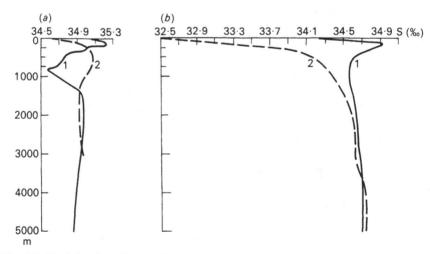

Fig. 2.2. Variation in salinity with depth: 1, low latitudes (average over the zone 0–10° latitude north); 2, high latitudes (zone 50–60° latitude north). (a) Atlantic Ocean; (b) Pacific Ocean.

actually increases with depth. At low and mid latitudes, a layer of low salinity is observed at depths of 500–1500 m. This occurs not only in the Atlantic and Pacific Oceans to which Fig. 2.2 refers, but also in the Indian Ocean. The intermediate layer of low salinity (from 34 to 34.8‰) is called the Antarctic Intermediate Water (AIW). This thick water mass is formed in the Atlantic Ocean in the zone 45–65° latitude south in the Weddell Sea and in the areas of low salinity adjacent to it. At depths of 500–1500 m it is spread far to the north in all three oceans (for more details see [14, 80]). The optical properties of this water were recently subject to special investigation during the tenth voyage of the scientific research ship 'Dmitry Mendeleyev' [16]. New measurements did not confirm the idea expressed previously that AIW has significant turbidity. In the data given in [16], no reduced transmittance was found in the area of the AIW.

Apart from the total salinity S, the concentration of bromides, the anion Br^-, is also of interest in marine optics. In the reference book [74], the average ratio of Br^- to Cl^- over the ocean is given as 3.48×10^{-3} and its range of observed values as $(3.25–4.4) \times 10^{-3}$. In [74], the average values and ranges are also specified for individual seas. Due to the lack of refinement of the analysis methods, the data on the concentration of Br^- obtained by different authors differ appreciably from each other. Therefore, in [136] it is recommended to consider bromide as a conservative substance and to take the chloride coefficient for it as a constant equal to 3.48×10^{-3}.

Since the salinity has a strong effect on the density and thus determines the dynamics of the ocean, the study of the salinity field has for a long time been one of the main objectives of physical oceanography. Detailed data on this subject can be found in [14, 61, 80].

Due to the constancy of the relative composition of dissolved salts in actual ocean water, the model of pure seawater or ocean water has become very popular in oceanography. This is taken to be pure water in which mineral salts are dissolved. It does not contain suspended particles or dissolved organic matter. Its salinity may vary. Pure ocean water is a two-component solution containing only pure water and 'salt'.

We now turn to the dissolved gases. All gases existing in the atmosphere are present in dissolved form in ocean water. The dissolution of gases in water is governed by Henry's law which states that the concentration of the dissolved gas is proportional to its partial pressure in the atmosphere. The coefficient of proportionality, Henry's constant, depends upon the type of gas, the temperature and the salinity. The content of gases q (in ml) on average in one litre of ocean water under normal conditions at $T = 20°C$ and $S = 35‰$ is given in Table 2.1 based on data from different authors.

Only dissolved oxygen is of interest in ocean optics. It absorbs in the far ultraviolet region. Its geographical distribution in the surface waters of the world ocean is mainly determined by the distribution of the temperature of the surface layer; when the water is heated, Henry's constant characterizing the concentration of the dissolved gas is decreased. For O_2, it varies from 8.5‰ (per volume) in the polar regions to 4.4‰ in the equatorial region (see [66]). The

Table 2.1. AVERAGE CONTENT OF GASES IN SEAWATER $q^{\circ}/_{oo}$

Gas	Nitrogen	Oxygen	Argon	Carbon dioxide
q	9.51	5.20	0.25	0.24

solubility of O_2 is also somewhat reduced when the salinity increases. The effect of salinity however is significantly smaller than that of temperature. As far as the variation in concentration of dissolved oxygen with depth is concerned, an interesting fact is observed; there is a marked minimum at depths of 500–1000 m. This can be clearly seen from the data given in Fig. 2.3 where the results are given for three oceans (in accordance with [80]). It is possible that the reason for this minimum is the intense absorption of O_2 associated with the oxidation of organic residues.

Other gases dissolved in water, eg nitrogen, argon, carbon dioxide, do not have a direct effect upon ocean optics. However, dissolved nitrogen is an important link in the nitrogen conversion cycle in the ocean. In this cycle, nitrogen from organic matter is converted to mineral forms. In the decomposition

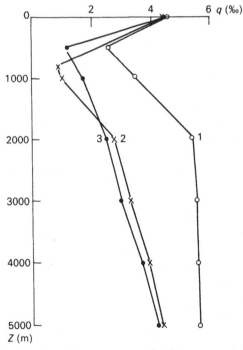

Fig. 2.3. Vertical distribution of the concentration of dissolved oxygen (in $^{\circ}/_{oo}$ by volume) (average over the zone 0–10° latitude north). 1, Atlantic Ocean; 2, Indian Ocean; 3, Pacific Ocean.

of nitrogenous organic substances, ammonia in the form of NH_4^+ is produced and then nitrites NO_2^- with the help of the so-called nitrifying bacteria and finally nitrates NO_3^-. Nitrates are the final product of the biological cycle. The nitrate ion absorbs appreciably in the ultraviolet region. Thus, dissolved nitrogen has mediating effect on ocean optics. The concentration of nitrates increases with depth. It reaches a maximum at depths of the order of 1 km. Examples of the vertical distribution of nitrates in the three main oceans are given in [28].

A typical concentration of nitrates in water (with respect to nitrogen) is approximately 10 μM/l. For more details on the distribution of O_2 and nitrates see [28, 66, 74].

2.1.3. Organic compounds. Yellow substance

Ocean water contains a multitude of different organic substances. These substances are found in two forms. They either belong to the suspended particles, in particular plankton, or are dissolved as molecules in ocean water. Considering the enormous variety of substances in the chemistry of the ocean, it is customary to estimate the concentration of organic substances in the ocean on the basis of the amount of carbon contained in the water. It is clear that the conversion of carbon to the substance depends on the actual chemical formula but for the sake of estimates it is usually assumed that the organic substances contained in seawater consist of approximately 50% organic carbon [77]. The amount of organic carbon in the open ocean fluctuates in the range 0.2–2.7 mg/l and its average concentration is approximately 1.4 mg/l. However, in individual inland seas it is significantly higher.

Most organic chemicals in the ocean are in the form of dissolved substances [77]. Even in water rich in phytoplankton, the amount of dissolved organic forms is seven to eight times higher, and at large depths approximately 1000 times higher than the amount of carbon in particles. Note that N.B. Vassoevich distinguishes not two but three forms of organic substance in the ocean: living, suspended and dissolved matter and gives their average proportions as 1 : 10 : 650 (see [77, p. 4]). Dissolved organic matter plays an important part in the biological cycle of matter in the ocean. This is an intermediate stage through which living organisms pass in their conversion to biogenic inorganic salts with the help of bacteria. The cycle is closed by autotrophic plants in which chlorophyll and other pigments from biogenic inorganic salts resynthesize complex organic compounds.

Strictly speaking, what is called dissolved organic matter is not a molecular solution in the precise sense of the word. As E.A. Romankevich pointed out, a dissolved substance in the chemistry of the ocean covers everything which passes through a filter with a pore size of 0.45–1 μm. This involves true molecular solutions as well as fine suspensions. It is interesting that when filters were used with pores of 5×10^{-10} m (measurements in the equatorial region of the Pacific Ocean) it was found that approximately 40% of what is usually assigned to dissolved substances should be considered as suspended matter (see [77, p. 28, 48]).

The chemical composition of dissolved organic substances is extremely diverse. E. Duursma defines four main groups of chemical compounds which can be formed in the ocean water by decomposition of organisms (see [136, chapter 11]): (1) organic substances not containing nitrogen (apart from lipids); (2) nitrogen-containing compounds; (3) lipids (fatty acids); (4) complex substances including humic acids. He gives a list of about 150 different individual organic substances which are actually identified in the waters of the ocean and indicates the region where they were found. Their concentrations vary from traces to milligrams per litre. The entire list is divided into seven sections: (1) carbohydrates; (2) proteins and their derivatives; (3) aliphatic carboxyls; (4) biologically active substances (vitamins and hormones); (5) humic acids; (6) phenols; (7) hydrocarbons.

In more recent reviews referred to in the reference book [74], other classifications are presented. Thus, for the northern part of the Pacific Ocean, the data in Table 2.2 are given in [74].

Table 2.2. CONCENTRATION OF DIFFERENT CLASSES OF ORGANIC COMPOUNDS IN DISSOLVED ORGANIC MATTER (μg C/l)

No.	Class of compound	0–300 m	300–3000 m
1	Amino acids (free + bound)	25	25
2	Aromatic compounds (substituted phenols)	1	–
3	Vitamins ($B_1 + B_{12} + H$)	10^{-2}	10^{-2}
4	Fatty acids (free + bound)	40	10
5	Urea (free)	20	2
6	Sugars (free)	10	10
	Sum of the 6 classes	100	50
	% of $\Sigma\, C_{org}$	10	10
	$\Sigma\, C_{org}$	1000	500

In the reference book [74], a long list of individual organic substances is also given, not completely coinciding with the data given in [136]. It should be pointed out that on the whole the exact composition of dissolved organic forms is poorly known. The main difficulty is that many substances are present in minute concentrations which are extremely difficult to measure. It is likely that a high percentage of organic substances in ocean water has yet to be identified.

It is an important aspect of the optics of the ocean that most dissolved organic substances do not have any absorption bands in the visible part of the spectrum. The only exception is the group of compounds belonging to the fifth category in Duursma's classification: humic acids, substances with similar properties to the soil humuses. This group of substances was discovered by K. Kalle in 1938. Kalle called it 'Gelbstoff'—yellow substance [152]. This name yellow substance derives from its light-absorbing properties. Absorption by yellow substance sharply increases in the direction of short waves and this is responsible for its yellowish colour. We remind ourselves that the colour of a substance whose absorption band lies in the visible region is complementary to the colour of the absorbed light. If blue light (480–435 nm) or violet light (435–300 nm) is absorbed, then the substance has a yellow colour (595–580 nm) or yellowish green colour (580–560 nm). It is likely that the formation of yellow substance is connected with the destruction of pigments (chlorophylls and carotinoids) in green algae. Chlorophyll has two strong absorption bands, a blue band and a red band. On decomposition, chlorophyll a disappears first and with it the related red absorption band, leaving only the broad blue absorption band which is characteristic of yellow substance. The exact chemical composition of yellow substance is not precisely known. Apparently, it is a complex mixture of humic type compounds. They are formed from hydrocarbons and amino acids by the so-called Maier reaction. According to the data given in [149], yellow substance consists of two main groups of compounds: (1) phenol humic acids with a light to dark brown colour; (2) carbohydrate humic acids or melanoids with a light to yellowish gold colour. The latter are more stable than the phenol humic compounds.

In seawater, yellow substance (as any organic compounds) may be formed by two routes: (1) fresh water runoff from the land (mainly by rivers); (2) direct formation in the sea through the decomposition of plankton organisms. Organic substances produced in the ocean by photosynthesizing plants are called autochthonous and those entering the ocean from the land are called alloch-thonous (Greek autochthonous—animal or plant forms which are indigenous, allochthonous—brought from outside). The active role of humic compounds in river discharges as a source of yellow substance in the sea can clearly be seen from the example of the northern seas. In these seas, the river discharge is considerable and therefore they are rich in yellow substance. Fresh water runoff from the land is especially important in coastal waters: the absorption in the short-wave region characterizing the content of yellow substance rapidly falls on moving away from the shore, ie as the salinity increases [153]. In open regions of the ocean, autochthonous yellow substance is produced. In this case, yellow or brown melanoids are formed. There are different estimates of the ratio between autochthonous and allochthonous proportions of yellow substance. In [149], it is considered that the main component is allochthonous (river). In contradiction to this, N. Jerlov [25] points out that there is an appreciable amount of yellow substance in the zone of upwelling waters to the west of South America where there is practically no influx of waters from the land. We may add that according to the estimate in [77, p. 27] the autochthonous component as a whole makes up 95% of the organic matter of the ocean. It is likely that a similar situation applies to yellow substance but there are no exact data.

In [180], it was asserted that a large part of yellow substance is not dissolved but adsorbed on to small suspension particles. According to other estimates, this proportion reaches 45%.

Although, as already mentioned, the chemical composition of yellow substance is not precisely known, there is some evidence of differences between marine and continental humic compounds. It has been shown by infrared spectroscopy methods and nuclear magnetic resonance that marine humic compounds have more of an aliphatic and less of an aromatic nature than soil humic compounds. This is explained by the fact that in seawater there is practically no lignin which plays an important part in the formation of soil humus. A structural formula has been suggested for marine humic compounds. Its 'building blocks' are important biosynthetic molecules in the sea such as amino acids, sugars, amino sugars and fatty acids. In all probability, the structure also includes carotinoids, chlorine pigments, hydrocarbons and phenols [48, 153].

At present, there are no direct methods in oceanography for isolating yellow substance and determining its concentration. It is therefore customary to characterize the content of yellow substance in terms of its optical manifestation. In [48], the value of the absorption coefficient of the yellow substance at a wavelength of 390 nm (a_y (390)) was used (this value is found from the difference between the excess absorption by seawater and the absorption by phytoplankton pigments). It is of interest to deduce the relationship between yellow substance and the total amount of dissolved organic substances and also the relationship between these substances and another component of dissolved organic matter, fluorescent organic matter [35, 153]. The latter, as the absorbing component, is not isolated by chemical methods but its concentration may be deduced from the intensity of its fluorescence. In [18], values were calculated from observations in the Indian Ocean for the correlation coefficients between a_{net} (390)* and C_{org}^{sol} (based on 41 samples) and also between a_{net} (390) and the concentration of dissolved carbohydrates (based on 59 samples). The coefficients 0.39 and 0.29 are evidence for the presence of a significant relationship between the variables considered although their low value indicates that the proportion of yellow substance to the total mass of dissolved organic substances is extremely variable. It is clear from the various data that it is small. Recently, O.V. Kopelevich, using Nyquist's results (see [149]) and the data given in [18], found that in the layer 0–100 m it fluctuates from 0.1 to 8%, in the layer > 100 m, from 1.5 to 4% and on average it is 2.5%.

K. Kalle also found that some compounds belonging to yellow substance fluoresce with a blue colour when they are excited with ultraviolet radiation. The fluorescent component forms only a part of yellow substance (everything which fluoresces absorbs but not everything which absorbs fluoresces). The intensity of fluorescence usually increases significantly with depth (except in very productive regions) [28]; if a_y (390) increases with depth then this is very slight. Thus, typically in the oceans, there is an increase with depth in the fraction of the fluorescent component of yellow substance, ie an accumulation with depth, of the apparently more stable component of dissolved organic matter. A similar

---

*The variable $a_{net}(\lambda)$ is the absorption coefficient which is in excess of that of distilled water.

situation was observed by Kalle in the Baltic Sea on moving from coastal waters to the open sea: the fluorescence/absorption ratio increased on moving away from the shore with increasing salinity [153].

2.2 Suspensions

2.2.1 Methods of investigation

Marine suspensions are extremely varied. They can be terrigenous particles carried into the sea by rivers and winds, phytoplankton cells, bacteria, detritus (residues of decomposed cells of phytoplankton and skeletons of zooplankton), particles of volcanic or even cosmic origin. Suspensions have an effect on absorption but their important influence is on the scattering of light in the ocean. The effect on absorption is two-fold: (1) the suspension particles themselves absorb light slightly; (2) as a result of scattering they increase the photon path-length which leads to an additional absorption as in the multi-path cell. The main effect, of course, is scattering; suspended material increases the scattering coefficient of pure ocean water to 20 times that of perfectly clear water. A similar situation exists in the atmosphere. Scattering in the actual 'pure' atmosphere is 20 times higher than in an atmosphere free of aerosol. The amount of scattering of light by suspensions is determined by the number of particles, their size, their shape, orientation and refractive indices. From the optical point of view, the main interest is the range of sizes from one hundredth of a micron to tens of microns. Finer material is at a transitional stage with respect to truly soluble matter; these particles are too small to have a significant effect on the optical properties of seawater; there are few large particles and therefore their effect may also be neglected.

The radius r of particles which make the main contribution to the scattering of light in ocean water may be determined by examining the function under the integral in the equations determining the scattering of polydisperse systems. A similar problem is considered in [125] for particles of an atmosphere aerosol. For the scattering coefficient in the case of a polydisperse system with exponential distribution, at $v = 3$ (Chapter 4), the contribution of the various fractions is determined by the monodisperse specific scattering coefficient b_{sp} cm^2/g. This coefficient is based on unit mass and hence the density of the suspension ρ: $b_{sp} = b/\rho$.

If ρ_0 is the density of the suspension particles, then from the equations of Chapter 4, we find:

$$b_{sp}(r) = \frac{3}{2\rho_0} \frac{K(2\pi r/\lambda)}{r} \tag{2.3}$$

The scattering cross-section K for small particles is $\sim r^4$ and for large particles tends towards 2 (Chapter 4). Therefore, $b_{sp} \sim r^3$ for small particles and $b_{sp} \approx 1/r$ for large particles. The function $b_{sp}(r)$ represents the imaginary case in which a piece of matter is broken down into small identical particles and the transmittance of the resulting suspension is studied. In this case, there is an optimum size for the particles r_{opt} for which the system most strongly scatters light (see [90, p. 174]). The graphs of $b_{sp}(r)$ for $m = 1.15$ and $m = 1.02*$ are shown in Fig. 2.4.

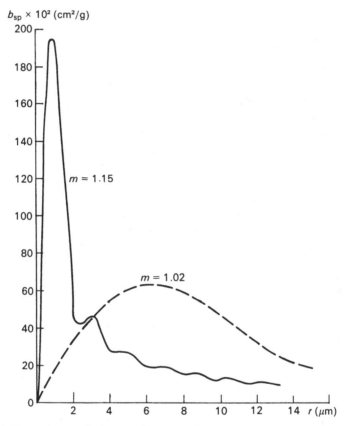

Fig. 2.4. Dependence of the specific scattering coefficient b_{sp} of a monodisperse suspension for terrigenous ($m = 1.15$) and biogenic particles ($m = 1.02$) on the radius of the particles r. The values on the b_{sp} axis relate to $\lambda = 0.55\,\mu m$, the density of terrigenous particles is taken as 2.65 g/cm^3 and that of biogenic particles as 1 g/cm^3.

The curves in Fig. 2.4 characterize the contribution to the scattering of light made by different suspension fractions in terms of size for terrigenous and for biogenic particles assuming that the mass of substance in these fractions is the same. The shape of the curve of $b_{sp}(r)$ depends essentially on m. For terrigenous particles, there is a sharp maximum in the neighbourhood of $r_{opt} \approx 0.8\,\mu m$. As m increases,

_____

* m is the complex refractive index which is properly introduced in Chapter 3 (Equation 3.2).

this maximum is shifted in the direction of smaller r (at $m = 1.33$ it lies at $r = 0.5 \, \mu m$) (see [90, p. 174]). The presence of a sharp maximum of $b_{sp}(r)$ has been confirmed in many experiments, for example in [31] where the transparency of kaolin and clay silt suspensions in distilled water was investigated. For biogenic particles with $r_{opt} \approx 7 \, \mu m$, the maximum is appreciably flatter. When $v \neq 3$, the graph is deformed but still has a sharp maximum. This means that the main contribution to the scattering is made by particles over a defined range of sizes Δr. A similar 'monochromatization' also takes place for the scattering coefficient in a given direction $b(\gamma)$. Exact values of r_{opt} and Δr depend in this case on the angle γ. The vibrational nature of the scattering cross-section $K(\rho)$ leads to the presence of damping vibrations in the graph of $b_{sp}(r)$.

In actual water, the contribution of the different fractions varies and this shifts r_{opt}.

Information about suspensions available to us is not very reliable and the data from different authors are often contradictory. The main reason for this is the absence of any reliable measurement method.

First, we examine the methods for investigating marine suspensions. Starting with the standard methods which are actually used at present in practice in oceanography. We do not describe here methods based on light scattering. They are examined in Chapter 6. Here, we only point out that if we work in homogeneous water, then the relationship between the scattering coefficient $b(\gamma)$ and the suspension concentration C is extremely stable, the correlation coefficient of ~ 0.8 and the accuracy of determination of C using light scattering are no worse than the accuracy of standard methods. The effectiveness of optical methods is significantly higher.

From the start, difficulties arise with the extraction of the suspended particles from the water. Two methods are used for obtaining these samples: filtration and separation [6]. Membrane filters may be used with different pore diameters. The finest filters which are now produced have a minimum pore diameter of approximately $10^{-3} \, \mu m$. Such filters retain even the finest colloid particles. However, filtration through them proceeds very slowly and the investigation of such fine suspensions requires special analytical methods which are not yet in widespread use. In practice, coarser filters are used with pore sizes from 0.35 to 1.2 μm and above. The disadvantage of filters is that they become clogged during operation; it is very difficult to allow for this. Deformation of the soft biological particles may occur on vacuum suction of water as well as when 'stripping' them from the filter during preparation of the samples for microscopy. The second method for extracting a suspension is to pass ocean water through an ultracentrifuge. Separators work continuously and process large volumes of surface water producing representative samples of the suspended material (tens and hundreds of grams). The disadvantage of separation is that the suspension is divided into fractions not on the basis of size but on the basis of the settling diameter. In this way, small particles and large but light particles are removed together with the water. But it is impossible to assess exactly how the collected sample differs from the unperturbed suspension. From the aspect of ocean optics, both these methods are bad; they distort the content of the suspension and it is the finely dispersed fraction which undergoes maximum distortion, ie the very fraction which is most important for us.

The next step is to determine the concentration and granulometric composition of this suspension. For this, the separated samples are dried at $T = 105°C$, weighed and then subjected to hydro-mechanical analysis (separation by the settling velocity in a vertical column). The fractions separated by this method are not directly related to the size as the settling velocity also depends upon the shape and density of the particles. The most important part of the suspension contains particles of light organic detritus which settle so slowly that in practice this analysis is totally unsuitable. The final stage of all standard methods is to count the particle samples under the microscope. This counting takes a long time and there are considerable errors. These errors usually lead to an underestimate of the numerical concentration. The observer simply does not see the majority of the particles whose dimensions lie outside the limit of the resolving capacity of the microscope. Moreover, the aggregation of small particles and the masking of small particles by large ones will also lead to underestimation.

In order to assess the accuracy of the filtration method, a comparison was made by directly counting the number of nanoplankton particles in a sample of seawater (without fixing with formalin and subsequent concentration of the deposit) and by counting the number on a filter [11]. Use was made of a normal optical microscope to compare the fractions of radius 1–2.5 μm. Measurements have been carried out by T.F. Narusevich in the tropical zone of the Indian Ocean.

Table 2.3 gives the results of the comparison. It can be seen from Table 2.3 that the number of particles counted by the filter method is only 25 % of that observed by direct counting. Table 2.3 also gives data obtained by the small angle light scattering method. We discuss this in Chapter 6.

A clear understanding of the drawbacks of standard methods for collecting and examining suspensions led to the search for new methods. Among these, we can mention the application of electron microscopy, optical methods and the automatic Coulter counter. Electron microscopy methods have not yet gained

Table 2.3. NUMBER OF PARTICLES (10^6/l)

Level (m)	Method		
	Direct counting	Filter count	Small angle Light scattering
75	1.0	0.27	1.8
10	1.5	0.47	1.2
0	2.6	–	4.8
0	2.1	0.39	–

wide acceptance because of their complexity. We consider optical methods in Chapter 6.

We look at the Coulter counter which has turned out to be a convenient and simple instrument and is widely used in oceanography. The instrument measures the change in electrical conductivity of a small volume of water when suspended particles are introduced into it. Initially the instrument was developed as a high-speed automatic counter and sizer for blood cells [141]. It is manufactured in the USA by Coulter Electronics and in 1965 began to be systematically used by oceanographers for studying marine suspensions. The operating principle of the instrument can be understood from Fig. 2.5. A potential difference U is applied across two electrodes C and D. Electrode C is situated inside a non-conducting glass tube B and electrode D is situated outside this tube in the vessel E which contains the mixture under study. There is a small opening A of diameter d in the wall of the glass tube B of wall thickness l. The opening A forms a calibrating cylinder; its volume V is equal to $(\pi/4)\,d^2 l$. The current strength I in the circuit is determined from the expression $I = U/(R + r)$ where r and R are the internal and external resistance of the circuit, respectively. Since the electrodes C and D are

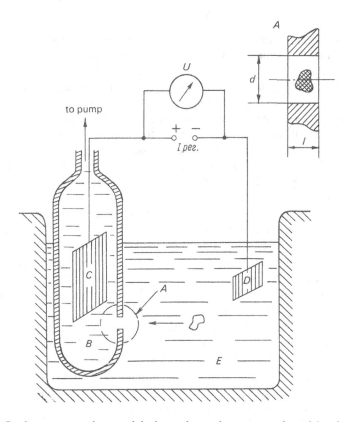

Fig. 2.5. Coulter counter: the top right inset shows the passage of particles through the calibrating cylinder A.

large, the entire potential drop in the external circuit IR is concentrated in the cylinder A. Its resistance is:

$$R = (1/k)(l^2/V) \qquad (2.4)$$

where k is the specific electrical conductivity of ocean water.

To make a measurement, a calibrated volume of water is pumped through the glass tube B, and drawn through the opening A. Particles of the marine suspension also pass through the cylinder A together with the water. The specific electrical conductivity of seawater k at $S = 35^\circ/\text{oo}$ and $T = 20^\circ C$ is approximately $5\,S/m$ [74] but it is significantly smaller for suspended particles, eg $k = 5 \times 10^{-7}$, 5×10^{-5} and $5 \times 10^{-3}\,S/m$ for limestone, dry soil and moist soil, respectively.

Assuming, for example, that k of suspended particles is close to the conductivity of moist soil, we can see that in estimates it may be neglected altogether. This means that when a suspended particle of volume v enters the measuring cylinder A, the volume of liquid in the cylinder is reduced by v. If we assume that in this case the resistance of the liquid in the cylinder A is described by Equation (2.4), then it will increase by the amount:

$$\Delta R/R = v/V \qquad (2.5)$$

In the instrument, described in [141], the cylinder A has $d = 10^{-1}\,\text{mm}$, $1 = 1/15\,\text{mm}$ and a volume of $V = 5.2 \times 10^{-7}\,\text{cm}^3$. The volume of a spherical particle of radius $5\,\mu m$ will be $v = 5.2 \times 10^{-10}\,\text{cm}^3$. Thus, the change in resistance will be $1\,\%$, which can be detected and measured easily. These changes are output on an oscillograph. The instrument enables more than 6000 particles to be counted per second with an interval of 15 seconds. The recording system is designed so that it automatically outputs the number of pulses exceeding a certain reference level. The reference level can be changed; in this way we obtain $F_1(v)$, a cumulative curve of the distribution of particles by volume. The circuit may also include a pulse discriminator by which means together with the function $F_1(v)$ the instrument is able to output a differential distribution function $f_v = dF_1(v)/dv$. The number of particles counted in one measurement is 100 times higher than in normal counting using a microscope which reduces the statistical measurement error by a factor of about 10. Using the Coulter counter, dispersion analysis of suspensions may be carried out immediately after the sample has been taken, on board ship, avoiding errors associated with storage and possible 'ageing' of the samples or the laborious visual counting of particles under the microscope and other errors described above.

Unfortunately, even the Coulter counter is not free from a number of drawbacks. The main one is that the change in resistance $\Delta R/R$ recorded by the instrument for particles which are not very small does not only depend on the volume of the particle but also on its shape and on its orientation in the stream. This can be immediately seen if it is assumed for example that the measured particle is a disc of a diameter d equal in diameter to the opening A. In this case, the particle simply breaks the circuit and $\Delta R/R$ will be equal to ∞ independently of the volume of the particle. In general, if one imagines that the particle is a cylinder whose axis is parallel to the axis of the opening A, then it is not difficult to show that the exact equation for $\Delta R/R$ is:

$$\Delta R/R = \eta v/V, \ \eta = (1 - \varepsilon)/[1 - S_0(1 - \varepsilon)/S] \qquad (2.6)$$

where $\varepsilon = k'/k$ is the ratio of the electrical conductivities of the particle substance k' and seawater k and S_0/S is the ratio of the cross-sectional area of the particle S_0 and the opening S.

When $\varepsilon \ll 1$, Equation (2.6) gives $\eta = S/(S - S_0)$ and we can see that $\eta = 1$ only for very small particles ($S_0 \ll S$). For a large particle, for example of diameter $d = 50\,\mu$m, the value of η for the instrument in [141] is equal to 4/3. Thus, for large particles, the value measured by the Coulter counter is an effective volume. It depends on the shape of the particle and its orientation inside the measuring cylinder. For small particles, the measurements are limited by the sensitivity of the instrument. In practice, the lower limit lies in the neighbourhood of a particle of radius about $0.5\,\mu$m. For such particles, the measured value of $\Delta R/R$ is 10^{-6} which is the noise limit of the actual instrument. Another difficulty arises if two or more particles enter the cylinder A at the same time. In this case, there is the risk of confusing two particles with one large particle. For ocean water with a suspension concentration of 0.5 g/l and an average radius of the particles of 5 μm, density $\sigma = 1$ g/cm^3, the average numerical concentration is 0.35×10^3 l/cm^3. This means that the average number of particles in the measured volume will be $\bar{n} = 5 \times 10^{-4}$, ie the probability of such an error is negligibly small. The risk becomes significant for a suspension of particles with an average radius of about $0.5\,\mu$m when $\bar{n} = 0.5$. However, such distributions lie outside the operating range of the counter. In practice, it is also necessary to take into account that the simultaneous passage of two particles is unlikely. If they are displaced in time, then this will be observed not as a doubling of the height of the pulse but as a broadening which may be eliminated by appropriate processing. The author of [141] recommends that at high cell concentrations the blood sample should be diluted with a standard solution. This procedure may also be applied to ocean water but for bulk measurements this requires considerable volumes of standard pure water.

2.2.2. Investigation results: suspended material in the world's oceans

We examine the basic results on suspended materials in the world's oceans. According to Fig. 2.4, this means that we are interested in small terrigenous particles ($m = 1.15$) of radius $a < 1$–$2\,\mu$m and significantly larger biological particles ($m = 1.02$) with radii of the order of 5–20 μm.

Suspended material in the ocean may be divided into groups based on different characteristics: genetic (in terms of origin), material composition, size, optical constants, etc. In terms of origin, suspended material may be biogenic, terrigenous, hydrogenic, aeolian, cosmogenic, etc. Biogenic suspensions consist of detritus (dead) and living particles: phytoplankton, zooplankton, bacteria and fish. Terrigenous suspensions are mineral particles of the lithosphere arriving in the sea as the result of coastal erosion and river discharge; hydrogenic suspensions arise in the water itself due to the precipitation of dissolved substances; aeolian suspensions are carried by air currents and are deposited in the sea; cosmogenic suspensions are meteorite particles, etc.

The annual influx of suspended material into the ocean q is given in Table 2.4. In Table 2.4, the sources are combined into two large groups: biogenic and terrigenous and in addition suspensions are divided into those arriving in the ocean from outside, allochthonous, and those formed inside, autochthonous. The total content of suspended material in ocean water is $Q \approx 1370 \times 10^9$ tonnes. This is an order of magnitude higher than the amount entering the ocean in a year. Since, in the steady state, $q = \alpha Q$ where αQ is the intensity of discharge into the sea, the constant α characterizing the rate of disappearance of suspended material is ~ 0.1 per year.

Table 2·4 gives only data on suspended material. Apart from suspended material, dissolved organic substances enter the ocean, according to the data given in [77] approximately 3.2×10^9 tonnes per year and these may be sources of hydrogenic suspensions. Suspensions are distributed in the waters of the oceans in the following manner: 51.6% in the Pacific Ocean, 23.6% in the Atlantic Ocean, 21.3% in the Indian Ocean and 0.8% in the Arctic Ocean. It can be seen from Table 2.4 that the main source of suspended material in the ocean is phytoplankton. It contributes approximately 80% of all incoming material (this approximately coincides with the autochthonous percentage). However, if coastal waters are ignored where the composition of suspended material strongly

Table 2.4. ANNUAL SUPPLY OF DEPOSIT MATERIAL TO THE OCEAN

No.	Material	Supply (10^9 tonnes per year)	Fraction of total supply (%)
	Biogenic (dry weight)	115.5	82.4
1	Phytoplankton	110.0	78.5
2	Other sources	5.5	3.9
	Terrigenous	24.6	17.6
1	River discharge	18.5	13.2
2	Vulcanogenic	2.5	1.8
3	Aeolian	1.6	1.14
4	Glacial	1.5	1.06
5	Erosion of shores and sea bottom	0.5	0.4
6	Cosmogenic	0.01–0.08	0.06
7	Anthropogenic	0.01	0.01
	Overall total	140.1	100.0
	of which		
	autochthonous suspensions	113.0	80.7
	allochthonous suspensions	27.1	19.3

depends on river discharge and shore erosion, living plankton only makes up approximately 10% of suspensions in open regions of the ocean. Here, suspended material mainly consists of detritus and mineral particles. In Fig. 2.6, we give an example from [28] illustrating this statement. In this example, the ratio C_{ph}/C_{susp} fluctuates in the range 0.02–0.32 and on average is equal to 0.12.

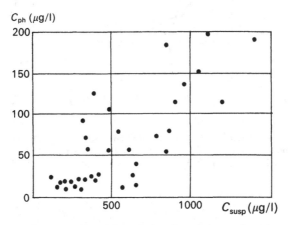

Fig. 2.6. Correlation between the total concentration of suspensions in ocean water C_{susp} and the concentration of phytoplankton C_{ph} (based on 34 samples taken from the Gulf of California at depths down to 51 m).

The main chemical component of suspended material is organic carbon C_{carb}. In [6], it is stated that on average the percentage of C_{carb} in the ocean is approximately 35%. Examples of the vertical distribution of the fraction C_{carb} in the total mass of suspended material are given in Fig. 2.7 (based on [28]). Down to a depth of 75 m, it amounts to 25–40% and at large depths it falls to 15–20%. The main elements making up the terrigenous (mineral) component of suspensions are silicon, aluminium and iron, most often in the form of oxides: SiO_2, Al_2O_3 and Fe_2O_3. According to [6] their percentage of the total mass of suspended material is on average 9.1, 1.4 and 1.2%. An example of the vertical variation of the percentage of these substances is given in Fig. 2.7. In contrast to organic carbon, their contribution increases with depth. Figure 2.7 illustrates the process of mineralization of suspended material the particles descend, and it is apparent that the unstable fractions in the suspension gradually go into solution. As a result, the chemical compositions of ocean deposits and suspensions are appreciably different. According to the data of [28], approximately 70% of suspended material in the 0–75 m layer belongs to the unstable fraction. It is decomposed by bacteria in the layer up to 300 m.

We now look at the form of the particles. Phytoplankton are unicellular and

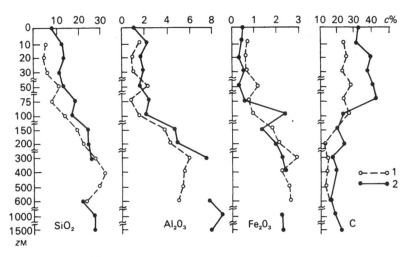

Fig. 2.7. Variation with depth of the fractions: organic carbon; oxides of Si, Al, Fe in the total mass of suspended material. The data refer to the western basin of the Mediterranean Sea (latitude $40°14'N$, longitude $5°34'E$), June 1 and November 2 (1969).

multicellular microscopic plant organisms. They contain a large number of different groups: diatoms, dinoflagellates, coccolithophorids, silicoflagellates, cryptomonadina, Chrysophyceae, green algae, blue-green algae, etc. The groups are divided into classes and the classes into species. Thus, diatoms making up the main group of phytoplankton number approximately 10,000 species which are differentiated into two main classes: Centrales (radically symmetric) and Pennales (bilaterally symmetric). In order to define a species, there are special identification manuals for marine algae in which the shape, size and other characteristics of the cells of the given species are indicated. For the sake of illustration, Fig. 2.8 shows some typical species together with a scale indicating their size. Diatoms have a hard shell made of silica ($SiO_2 \cdot nH_2O$), consisting of two equal halves (flaps), one of which covers the other as a lid covers a box. Dinoflagellates have two flagella: one serves to rotate the cell; it goes round its middle; the other is directed to the rear and helps movement. Coccolithophorids are also biflagellate organisms. They are covered by a large number of limestone ($CaCO_3$) coccoliths.

At mid and high latitutudes, the composition of species and the concentration of phytoplankton undergo significant seasonal fluctuations. Data on these fluctuations are given in Fig. 2.9. In early spring the increasing amount of solar radiation leads to an increase in the number of phytoplankton. The increase in zooplankton grazing following this coupled with calm weather (which reduces turbulence and the resupply of nutrients from deep waters) lead to a minimum in the middle of summer. In autumn, storms develop in mid latitudes. This increases the flow of nutritious substances and brings about the September maximum.

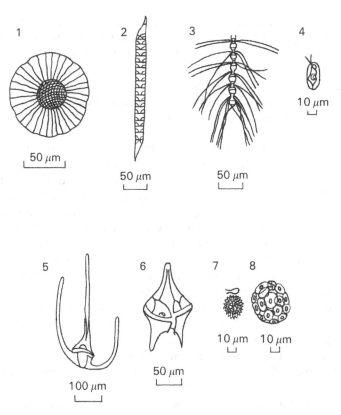

Fig. 2.8. Some species of phytoplankton: diatoms (1–3), dinoflagellates (4–6), coccolitho-phorids (7, 8).

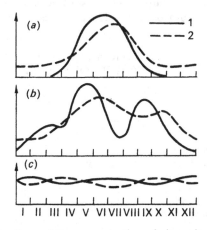

Fig. 2.9. Seasonal fluctuations of the concentration of phytoplankton and zooplankton in three latitudinal zones: (a) high latitudes; (b) mid latitudes; (c) tropical latitudes. 1, phytoplankton; 2, zooplankton.

Once autumn has advanced, the sharp decrease in solar radiation and the growth in zooplankton grazing leads to the winter minimum. In tropical latitudes, the seasonal variation is small. Here, of course, the abundance of phytoplankton and zooplankton are in counter phase.

The shapes of detritus particles themselves are indeterminate. Round particles, elongated rods, with and without spurs and irregular platelets with uneven edges are encountered. Often, conglomerates are observed consisting of two or more particles [77].

The data of different authors on the concentration of suspensions C in the ocean vary considerably. According to the data of [53], the average value of C is 0.82–2.5 mg/l with $C = 1$ mg/l for the ocean as a whole. The authors of [138] give values of 0.05–0.5 mg/l for surface waters of the open ocean and for deep waters only 1–250 μg/l. Nearer the shore, C may increase by one to two orders of magnitude. It is a fact that approximately 90% of terrigenous suspensions arriving in the sea from river discharge and as a result of coastal erosion settle in the coastal regions. The value of the average concentration $C = 1$ mg/l is apparently considerably overestimated. In [7], the authors carefully compared optical methods (Chapter 6) with microscopical analysis data. The measurements were carried out in the eastern part of the tropical zone of the Pacific Ocean. The average value of the concentration of suspensions was approximately 0.14–0.2 mg/l. Taking into consideration river estuaries and littorals, the range of fluctuation of C in the ocean is very large. It may reach four to five orders of magnitude. In the surface waters of the open ocean, the fluctuations are relatively small with a factor of only three to five.

For the optics of the ocean, besides the concentration, the size of the suspended particles is also important. Data on typical particles of biogenic and terrigenous suspensions are given in Table 2.5. The distribution of particles in terms of size $n(r)$ has been studied by a large number of workers. This question is examined in detail in Chapter 6. Here, we only mention some general results. The characteristic feature of the spectrum of suspended particles in the ocean is the sharp

Table 2.5. CHARACTERISTIC SIZES OF PARTICLES OF
BIOGENIC AND TERRIGENOUS SUSPENSIONS

Suspension	Size (μm)
Biogenic	> 1
1. Organic detritus	1–20
2. Siliceous and carbonate residues of	
plankton organisms	1–50
3. Diatom algae	10–50
4. Coccolithophorids	5–15
5. Foraminifers	50
6. Bacteria	2
Terrigenous	< 1

increase in the number of small particles. Many authors suggest that in different seasons and in different regions, the distribution may be modelled by a power law (Chapter 4): $n(r) = C \times r^{-\nu}$ with the exponent ν ranging from 3 to 6. As an example of the distribution of phytoplankton cells in terms of size, we consider Fig. 2.10 based on the data of [37]. In [37], five filters were used with pore size ranging from 3 to 0.4 μm; Fig. 2.10 shows the overall data for all filters. The measurements were carried out in the Pacific Ocean in a region crossing the equator and 160° longitude west. The overwhelming majority of the cells have sizes of 2–5 μm, with another weak maximum at approximately 20 μm and a third at approximately 150 μm. From the optical point of view, all these cells should be regarded as large particles. Although these data have been obtained by microscopic counting of the particles on filters and it is possible that the smallest particles were not retained by the filters, they nevertheless agree well with a known fact in marine biology: the 'richer' the region, ie the higher the production of phytoplankton in it, the smaller are the cells on average. In productive zones, where there is intense primary production, nanoplankton (particles of 2–5 μm) and ultraplankton (particle size $< 2 \mu$m) make up the overwhelming majority of phytocenosis. In relatively poor arid zones, the cells are larger. The data of [37] relate to such a zone.

The data of different authors vary on the distribution of cells in terms of size; it is possible that these differences are linked to the biological cycle. G.I. Semina [65] and a number of other authors believe that the distribution is symmetric, close to the normal distribution. This distribution was assumed in the tables [123] for biological particles. There is information that there is a significant number of ultramicroscopic biological particles with radii in the range 0.01–0.4 μm. The curves in Fig. 2.10 indicate that these distributions are multimodal; each mode corresponds to a specific type of particle. Distributions are shown in Fig. 2.11 based on the data of [7] for the entire suspension. The distribution by volume has two maxima: one for particles with $r = 0.5$–1 μm and the other for particles with $r = 20$–25 μm. The first maximum coresponds to the terrigenous component and the second to the biogenic component. According to measurements with the Coulter counter, the total number of particles whose radius is above 0.5 μm is 10^5–10^8/l. Microscopic counting gives

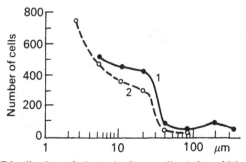

Fig. 2.10. Distribution of phytoplankton cells: 1, by width; 2, by length.

a significantly lower value for the concentration: 10^4–$10^6/1$; the reason for this discrepancy is discussed in Chapter 6. It is shown there that the results obtained from optical data agree with the Coulter counter data.

The data on the distribution of particle size with radii less than 0.5 μm are very sparse. The results obtained by electron microscope investigations show that an exponential distribution with $v = 2.65$ also holds true for these small particles; the total concentration of particles in the radius range 0.01–1 μm is 5×10^9–$10^{10}/1$.

Suspensions in coastal regions appreciably differ from pure ocean suspensions. Data on suspensions in the coastal region of the Black Sea (they were obtained by the inversion of scattering functions of Section 6.7), at depths of 1–4 m are given in Fig. 6.22 and in Table 6.12. The distribution of particles of the terrigenous fraction in terms of size over the radius range 0.1–5 μm is almost uniform, the average modal radius of large organic suspended particles being approximately 15 μm and the average mass concentration approximately 3.5 mg/l. This value is many times greater than the concentration of suspended particles in waters of the open ocean and exceeds the concentration in shelf zones of the ocean by a factor of two to three. Inshore the deposits are stirred up as the result of wind disturbance, coastal currents and for other reasons.

The spatial distribution of suspensions in the surface layer of the ocean is governed by latitudinal and circumcontinental factors. Latitudinal zoning is associated with the distribution of primary production, the rate of photosynthesis. This rate depends on the intensity of incident solar radiation, ie the latitude of the location. Apart from the latitude, the variation of distribution of suspended material is also governed by the fact that the suspension concentration decreases on moving away from coast. This is associated with the reduction in the effect of

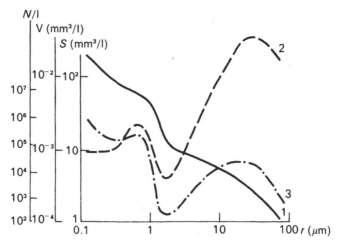

Fig. 2.11. Distribution by size of the number (1), volume (2) and surface area (3) of suspended particles.

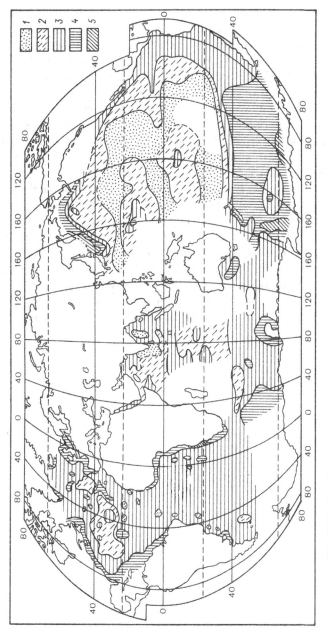

Fig. 2.12. Distribution of the suspended material concentration C in mg/l in the surface waters of the world's oceans (according to [6]). 1, less than 0.25; 2, 0.25–0.50; 3, 0.50–1.0; 4, 1.0–2.0; 5, more than 2.00.

upwelling along the coast on the continental shelf. It has a specific importance in the eastern parts of the Pacific and Atlantic Oceans. The occurrence of upwelling provides nutrients to the photosynthetic layer, ie increases the rate of photosynthesis. Furthermore, on moving offshore, the influence of terrigenous sources on the concentration of suspended materials (shore erosion, river estuaries) is reduced. The three distributions have much in common; the primary production according to Coblents-Mishke [65], the number of phytoplankton cells according to Semina [65] and the distribution of suspensions in the surface waters according to the data of membrane ultrafiltration shown in Fig. 2.12 [6]. In all of them, the influence of the rich Antartic-adjoining and northern zones and the eastern areas of the Atlantic and Pacific Oceans can be seen.

The main feature of the vertical distribution of suspensions in the deep regions of the ocean is the decrease in concentration with depth. If there is a marked discontinuity in density, apart from the surface maximum, an appreciable concentration maximum is usually also observed in the vicinity of the discontinuity. The first maximum in the upper layer is caused by active biogenic production. The second is associated with the reduction in the turbulent exchange at the pycnocline. In some cases, incidentally, it may have a biological nature. As a result of the stirring of deposits, a suspension maximum in the vicinity of the bottom is often observed also. It is especially marked in shallow-water shelf regions (continental and island) where the stirring of sediments is governed by waves. The stirred sediments are often in the form of long tongues and clouds of turbid water extending along the slope into adjacent ravines and troughs.

Chapter 3

Molecular optics of ocean water

3.1. Optical constants

A full theoretical analysis of the optical properties of such a complex system as ocean water is impossible at the present time. The normal approach involves the successive examination of a number of models: pure water, pure seawater, actual ocean water (containing dissolved organic substances and suspended particles). In this chapter, we examine questions which can be related to the molecular optics of ocean water. This includes the optics of pure water, the optics of pure seawater, the optical properties of dissolved organic substances and of the material of suspended particles. An analysis of the scattering and absorption of light by suspensions is examined in Chapter 4. This method of classification is based on the fact that the optics of pure substances and molecular solutions are essentially different from the optics of colloidal systems such as actual ocean water.

We systematically study the processes taking place in a macroscopically homogeneous substance: absorption, refraction and molecular scattering and the effect on these phenomena of the substances dissolved in pure seawater. In the first part of the chapter we described the processes occurring in an optically homogeneous substance: absorption and refraction. We present experimental data of the optical constants $n(\lambda)$ and $k(\lambda)$ and provide a qualitative explanation for them.

Due to molecular chaos, macroscopically homogeneous substances are optically heterogeneous; they scatter light. In the second part of the chapter we examine the molecular scattering of light by seawater.

We start with the simplest case. We take a look at the picture of propagation of electromagnetic waves in a homogeneous body. It is known that in the case of pure periodic functions of time of the form $e^{i\omega t}$, where ω is the frequency of the field, the Maxwell equations for the components of the electric field E and magnetic field B are reduced to the wave equations:

$$\Delta \mathbf{E} + q^2 \mathbf{E} = 0; \quad \Delta \mathbf{B} + q^2 \mathbf{B} = 0 \tag{3.1}$$

Here, Δ is the Laplace operator and the complex wave number q is related to the wave number for a vacuum q_0 and the complex refractive index m by the formulae:

$$q = mq_0; \; q_0 = 2\pi/\lambda_0 = \omega/c; \; m = \sqrt{\varepsilon\mu - i \cdot 4\pi\sigma/\omega} \qquad (3.2)$$

We use the standard notation: λ_0 and c are the wavelength and the velocity of light in a vacuum, respectively; the macroscopic constants ε, μ and σ are physical characteristics of matter: the dielectric constant, the magnetic permeability and electrical conductivity. It follows from Equations (3.1) and (3.2) that all the optics of the medium and all the characteristics of its behaviour in an electromagnetic field are determined by the index m.

For physical reasons, it is convenient to separate out the real and imaginary parts of m. For this purpose, m is usually written in one of the two following ways:

$$m = n(1 - ix); \; m = n - ik; \; k = nx \qquad (3.3)$$

where n is the refractive index, x is the absorption index; and k is the dimensionless electrodynamic absorption coefficient.

The dielectric constant ε is also a complex quantity in the general case. It is usually written as:

$$\varepsilon = \varepsilon' - i\varepsilon'' \qquad (3.4)$$

where ε' and ε'' are the real and imaginary parts of ε, respectively. In optics, $\mu = 1$ and therefore we omit it in the following discussion.

Thus, for m^2, we have:

$$m^2 = \varepsilon' - i(\varepsilon'' + 4\pi\sigma/\omega)$$
$$\varepsilon' = n^2(1 - x^2) = n^2 - k^2 \qquad (3.5)$$
$$\varepsilon'' + 4\pi\sigma/\omega = 2n^2x = 2nk$$

In order to explain the physical meaning of the quantities n and k, we must remember that we are studying waves in a homogeneous body. In this case, the index m will be constant. In electrodynamics, it can be proved that the solutions of Equation (3.1) in an infinite homogeneous medium are plane waves of the form:

$$e^{i(\omega t - qmz)} = e^{-kq_0z} \, e^{i(\omega t - q_0nz)} \qquad (3.6)$$

Here, we have assumed that the waves are propagated in the direction of the z axis and we have omitted the constant amplitudes which are not important in the following discussion.

Equation (3.6) is a damped plane wave with wavelength $\lambda' = \lambda/n$. The amplitude of the wave attenuates to the value $e^{-a'z}(a' = q_0k)$ on travelling over z units of length.

Thus, the two parts of m play essentially different roles in the propagation of waves. The refractive index n determines the phase advance and the electro-dynamic absorption coefficient k the reduction in amplitude of the wave. The intensity of light is expressed in terms of the square of the field. It also decays exponentially with the index:

$$a = 2a' = (4\pi/\lambda_0)k \qquad (3.7)$$

In courses on optics, the quantity a is called the absorption coefficient. It has the dimensionality L^{-1}. This is the standard Bouguer absorption characteristic.

We note that the absorption capacity of a substance is described by three quantities: the absorption index x, the electrodynamic absorption coefficient k and the absorption coefficient a. These quantities are defined by Equations (3.3) and (3.7). Some authors designate the absorption coefficient a by x. Sometimes in the courses on optics [8] the electrodynamic absorption coefficient k is also designated by x. In that case two different physical quantities are called and designated in the same way. However, it is always easy to understand what is meant: the electrodynamic coefficient is dimensionless; the Bouguer's coefficient is expressed in cm^{-1}.

The quantities n and k are essentially dependent upon the frequency. The actual form of the functions $n(\omega)$ and $k(\omega)$ is determined by the characteristics of the spectra of the individual molecules and the transformation of these spectra when the molecules are combined into a macroscopic medium. The theory of the functions $n(\omega)$ and $k(\omega)$ is based on the electronic theory of matter. The calculations here are very complex and as a rule only give a qualitative picture. In the optics of disperse systems where the exact values of $n(\omega)$ and $k(\omega)$ must be known, experimental data are used. The experiments usually involve the determination of a (directly from the attenuation of light after passing through a layer of specified thickness) and the simultaneous determination of n (from the phase shift or from the reflection coefficient R). The polarization ratios in reflected light are often determined. The equations expressing n and k in terms of the reflection characteristics are rather cumbersome. It is convenient to use specially calculated tables and graphs [178a].

These measurements have been carried out for a large number of different substances and the results may be found in reference books. Later, we consider actual data for water and other substances important in the optics of the ocean but first we look at some general ideas which give guidelines for dealing with the chaos of experimental data.

First, we note that the quantities $n(\omega)$ and $k(\omega)$ are not independent. They are related to each other by integral relationships. These were found in 1927 by G. Kramers and R. Kronig and are called the dispersion relationships. These relationships are a direct consequence of the causality principle.

An elegant example giving an elementary illustration of the heart of the matter is given in [1, chapter 2]. Imagine that at time $t = 0$, a source of electromagnetic radiation is switched on. The field, equal to 0 at $t < 0$ and becoming constant at $t > 0$, is the function $E(t)$ shown in Fig. 3.1a. It is known that this single-step function $E(t)$ can be regarded as the sum of an infinite number of plane waves existing from $t = -\infty$ to $t = +\infty$; the frequencies of these waves cover the entire spectrum from $-\infty$ to $+\infty$:

$$E(t) = \tfrac{1}{2} + (1/\pi) \int_{-\infty}^{\infty} (e^{i\omega t}/2i\omega)\, d\omega = \tfrac{1}{2} + (1/\pi) \int_{0}^{\infty} (\sin\ \omega t/\omega)\, d\omega \qquad (3.8)$$

As the result of adding 1/2 and waves of the form $[1/(2\pi i\omega)]e^{i\omega t}$, we find: $E(t)$

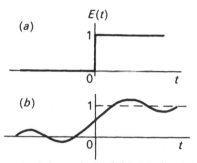

Fig. 3.1. (a) step function, actual change in the field over time (at time $t = 0$ the field source was switched on); (b) 'truncated' step function, components are omitted with frequencies greater than a certain frequency ω_0 (observation of the field source is through an imaginary filter which absorbs all frequencies $\omega > \omega_0$).

$= 0(t < 0)$, $E(t) = 1/2(t = 0)$, $E(t) = 1$ $(t > 0)$. We now assume that when $t > 0$ we are looking at a field source, not yet switched on, through a layer of the substance which does not allow a certain frequency band $\Delta\omega$ to pass through it. It is clear that if we remove certain terms from the sum which previously gave zero for $t < 0$, then we no longer obtain zero for all values of $t < 0$ (Fig. 3.1b). This means that if we look at a lamp through a suitable colour filter then we will see light before the lamp is switched on. This absurd conclusion follows from the assumption that it is possible to have a substance (filter) which completely absorbs a band of frequencies $\Delta\omega$, without any effect on all the other waves. In order not to violate the causality principle, it is necessary for our filter to also have phase shifts in all other waves so that their sum (3.8) is always automatically equal to zero for $t < 0$. This means that the function $n(\omega)$ which determines the phase shift and the function $k(\omega)$ which determines amplitude of the transmitted wave must be closely related.

The dispersion relationships take the form (the integral being understood in the sense of the principal value):

$$n(\omega_0) = 1 - (2/\pi) \int_0^\infty k(\omega)\omega \cdot d\omega/(\omega^2 - \omega_0^2)$$

$$k(\omega_0) = (2\omega_0/\pi) \int_0^\infty n(\omega) \, d\omega/(\omega^2 - \omega_0^2) \tag{3.9}$$

A rigorous derivation of these equations is given in the book [49, §62]. Equation (3.9) is often used in optics for the practical determination of $n(\omega)$ and $k(\omega)$ for actual substances. Using it is not a trivial matter as it is necessary to know the functions $n(\omega)$ or $k(\omega)$ over the entire range of frequencies from zero to infinity. In addition, since Equation (3.9) is a consequence of the causality principle, it must also apply to those cases when the nature of the physical fields is not known to us. Therefore, it is widely used in studying the interaction of elementary particles. No matter how complex the nature of nuclear forces unknown to us it must not be possible to see light earlier than the switching on of a lamp. It is interesting, as A. Angot [1] noted, that before the dispersion relationships and their connection to the causality principle were discovered, the arguments concerning the lamp given above forced many physicists to deny the reality of the Fourier integral.

Apart from the dispersion relationships, the important laws governing the functions $n(\omega)$ and $k(\omega)$ are established in the electronic theory of matter. Let us first examine the dielectrics of the type of water whose molecules possess a constant dipole moment. In a constant or slowly changing field; the molecules are oriented along the field. This leads to high values of $n:n = 9$ ($n^2 = \varepsilon = 81$). On increasing the frequency, the heavy water molecules gradually lag behind the external field and when $\omega \approx 10^{12}$ per second ($\lambda \approx 1$ mm), the orientation polarizability of water disappears. However, the ionic and electronic polarizability remain. In addition, when the wave frequency is close to the eigen frequency of the ionic or electronic vibrations in the molecule, a resonance occurs, the amplitudes of the vibrations sharply increase and the polarization of the substance P sharply increases. In the general case:

$$P \sim \sum_i f_i/(\omega_i^2 - \omega^2 - 2i\gamma_i\omega) \qquad (3.10)$$

Here, ω_i, γ_i and f_i are the eigen frequencies, the damping coefficients and the so-called oscillator forces for these frequencies. The summation is carried out over all 'oscillators' inside the molecule and over all molecules in unit volume of matter. On further increase in the frequency ω, the ionic polarizability first disappears and then also the electronic polarizability. For $\omega \rightarrow$ infinity, when for all ω_i, the frequency $\omega \gg \omega_i$, the polarization P tends to zero as ω^{-2} and the dielectric constant $\varepsilon(\omega) \rightarrow 1$—the polarization processes simply do not have time to take place. In this case, the electrons in the substance may be regarded as free. For them it is easy to see that:

$$\varepsilon(\omega) = 1 - 4\pi Ne^2/(m\omega^2) \qquad (3.11)$$

where N is the number of electrons in unit volume; e and m are the charge and mass of the electron, respectively.

This universal equation holds for the far ultraviolet. It can be equally applied to dielectrics and to conducting bodies such as metals.

3.2 Optical constants of pure water, pure ocean water and dissolved organic substances

3.2.1. Pure water

The starting point for understanding the absorption spectrum of a substance is the absorption spectrum of the isolated molecule. However, when condensed matter is formed from the individual molecules, the spectrum of the molecules is strongly distorted due to intermolecular interactions. These interactions have an effect on the structure of the energy levels of the system. Furthermore, fresh absorption bands arise in the spectrum associated with the formation of short-range order in the liquid, the formation of associates, etc. In the majority of liquids, the function $a(\lambda)$ differs appreciably from $a(\lambda)$ for the isolated molecule. Nevertheless, such a comparison is still useful to us.

The situation is considerably complicated in water by the fact that it is a closely packed system and the interaction between the molecules is extremely strong. In order to estimate the extent of this interaction, we note that the energy produced by the hydrogen bond relative to one molecule will be of the order of 8×10^{-20} J for water. This is 20 times higher than the energy of thermal motion kT which at room temperature is $\approx 4 \times 10^{-21}$ J. Due to the strong interaction between molecules, liquid water may be regarded as a macrocrystal consisting of hydrogen and oxygen atoms.

A curious confirmation of this is provided by the thermal capacity. We remember that by the beginning of the 19th century, Dulong and Petit had already established an empirical law according to which the atomic thermal capacity of crystals is the same for all elements and is approximately $25 \, \text{J}/(^\circ\text{C} \times \text{mole})$. We may add that the modern theory of thermal capacity of the solid state developed by Debye which regards thermal motion as a system of elastic waves in a crystal also gives a value of $25 \, \text{J}/(^\circ\text{C} \times \text{mole})$ for the atomic thermal capacity. From this point of view, the thermal capacity of 1 mole of water, if it is regarded as a crystal consisting of 2 hydrogen moles and 1 oxygen mole, must be $75 \, \text{J}/(^\circ\text{C} \times \text{mole})$. This exactly corresponds to the experimental value for liquid water. This means that on heating, water behaves as a molecular crystal, the energy of the individual atoms of which are equally distributed over the degrees of freedom.

Nevertheless, we start with the spectrum of the isolated H_2O molecule, ie the spectrum of water vapour. We see that there is a qualitative agreement between the absorption bands of liquid water and the absorption bands of water vapour. An analysis of the spectrum of the H_2O molecule may be found in courses on molecular spectroscopy. We make a few comments here on this subject. It is known that the energy E which the molecule has is a sum of E_e, the energy of the electrons in the molecule, E_v, the vibration of the individual atoms relative to the centre of the molecule, E_r, rotation of the molecule as a whole, and E_t, the translational motion of the molecule as a whole:

$$E = E_e + E_v + E_r + E_t \qquad (3.12)$$

According to this, electronic, vibrational and rotational transitions should be observed in the spectrum of the molecule. The energy of translational motion is not quantized and accordingly changes in it are not associated with absorption or radiation. However, the data of spectroscopy show that pure electronic and vibrational spectra are not observed. Electron transitions are accompanied by vibrational and rotational transitions and vibrational transitions are accompanied by rotational transitions. In real spectra, we are always dealing with electronic–vibrational–rotational transitions or with vibrational–rotational transitions or with purely rotational transitions. In spectroscopy, the first transitions are called electronic transitions, the second are called vibrational transitions and the third rotational transitions.

In the water molecule, electronic transitions lead to strong absorption bands situated in the far ultraviolet region (for $\lambda < 18.6$ nm) and the rotational transitions cover a broad spectral region from $\lambda = 8\,\mu$m to several centimetres. The entire intervening region, the near ultraviolet, the visible spectrum and the near infrared is taken up by the vibrational–rotational spectrum of the molecule. It is very complex. The vibrational spectrum of the water molecule has three fundamental frequencies with centres situated at wave numbers: $\tilde{v}_1 = 3657\,\text{cm}^{-1}$, $\tilde{v}_2 = 1595\,\text{cm}^{-1}$, $\tilde{v}_3 = 3756\,\text{cm}^{-1}$ which correspond to the wavelengths 2·734, 6·270 and 2.662 μm. The type of molecular vibrations corresponding to these fundamental frequencies is shown in Fig. 3.2. In the first and third vibrations, there is compression (or extension) of the chemical bonds in the molecule, symmetric in the first and asymmetric in the third. In the second vibration, there is deformation of the shape of the molecule; the angle at the oxygen atoms fluctuates. This deformation requires minimum energy and therefore has the lowest frequency. Approximately, $\tilde{v}_1 = \tilde{v}_3 = 2\tilde{v}_2$ and at low resolution we observe two strong absorption bands in H_2O: the main band in the region $\lambda \approx 2.7\,\mu$m and a weaker band in the region 6·3 μm. The harmonics of these fundamental frequencies as well as the component frequencies produce an entire series of very weak bands in the visible region, in the interval 543–847 nm and strong bands in the near and central infrared region from 944 nm and beyond.

The spectrum of liquid water is considerably simpler than the spectrum of water vapour; much of the fine structure disappears. Figure 3.3 shows the results of measuring the absorption coefficient of liquid water and the equivalent

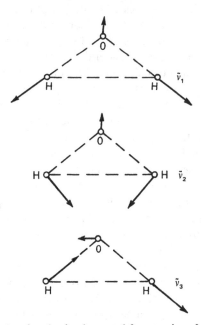

Fig. 3.2. Types of vibration for the fundamental frequencies of the water molecule.

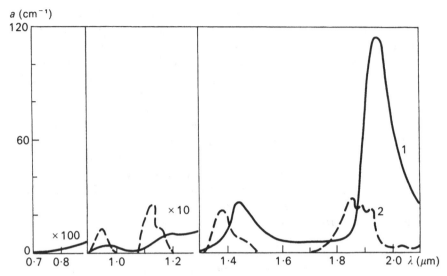

Fig. 3.3. Comparison of the absorption coefficient of liquid water (1) and steam (2). Over the section 0·9–1·3 μm the scale is enlarged by a factor of 10 and over the section 0·7–0·9 μm by a factor of 100.

amount of water vapour (water at room temperature, vapour at 127°C) in the spectral region 0.7–2.1 μm. For the data to be reliably compared, the measurements were made with the same spectral apparatus [143]. Due to the strong interaction between the molecules, the fine rotational structure in water is broadened. Bands with simple structure remain and are shifted in the direction of larger λ; they are strongly broadened and overlap each other. Absorption in liquid water is continuous and considerably higher than in vapour. We have already noted that an exact calculation of all of this transformation requires the determination of the position of the energy levels of the condensed system which at present is impossible. The general picture is as follows. The ultraviolet absorption is associated with electron transitions, the infrared with vibrational transitions. Due to intramolecular and intermolecular interactions, a large number of vibrational harmonics are produced, which overlap and lead to absorption in the visible region of the spectrum.

We now turn to the experimental data. For transparent liquids, the values of n are most simply determined by interferometry from the phase advance of the wave. The value of k is found from the coefficient a using Equation (3.7). In order to determine a in pure water, it is possible to use the measurement of the attenuation coefficient c, since in this case, the scattering coefficient b may be calculated with good accuracy (Section 3.5). In the visible region, the attenuation in water is small and the measurement of c is not easy. The great difficulty is to purify the water (free from dissolved substances and specks of dust) and to maintain its purity during the measurements. In [41], the results of different measurements of c in the spectral region 250–600 nm are compared starting from the first measurements of Ashkinass (1895). These values with those for the interval 600–800 nm (according to [48]) are shown in Fig. 3.4. The wide scatter of the data of different authors is worth noting. This scattering may be caused by: (1) the difference in the degree of purity of the water of different authors; (2) the difference in methodology, ie measurement errors. Apart from laboratory data, Fig. 3.4 also gives measurement data of $c(\lambda)$ in extremely transparent waters to the north-west of the island of Rarotonga in the Pacific Ocean. If it is assumed that c for pure ocean water cannot be smaller than c for pure water, then it is possible to determine the most probable curve $c(\lambda)$ for pure water. This curve is shown in Fig. 3.4 as a solid line. The optical characteristics corresponding to it are given in Table 3.1 according to [110].

However, recently, data on pure water have been revised in [148, 172, 178]. We look at the last of these studies [178]. For measurements of the absorption coefficient a, the authors use the optico-acoustic effect already observed at the time by J. Tyndall and widely used in the studies of M.L. Veingerov. On absorption of a pulsed beam of light in water, acoustic waves are produced whose

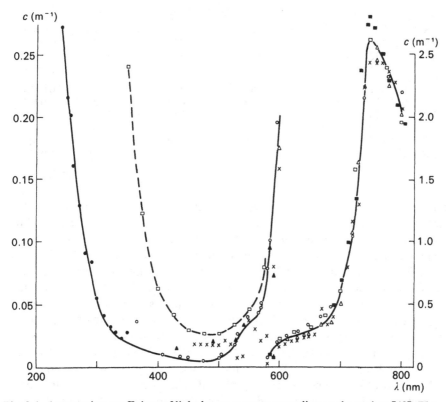

Fig. 3.4. Attenuation coefficient of light by pure water according to the review [48]. The different symbols refer to the data of different authors (for references see [48]). The scale on the left refers to the spectral interval 250–600 nm and on the right to 580–800 nm.

Table 3.1. Optical properties of pure water ($T = 20°C$, $p = 10^5$ Pa)

λ(nm)	n	$k \times 10^{10}$	$b \times 10^3$ m^{-1}	$c \times 10^3$ m^{-1}	$a \times 10^3$ m^{-1}	$\Lambda = b/c$
250	1·337	3·78 1	32·0	220	190	0·15
300	1·359	9·55	15.0	55	40	0·27
320	1·354	5·09	12.0	32	20	0·38
350	1·349	3.34	8·2	20	12	0·41
400	1·343	1·91	4·8	11	6	0·44
420	1·342	1·67	4·0	9	5	0·54
440	1·340	1·40	3·2	7	4	0·46
460	1·339	7.32 − 1	2·7	5	2	0·54
480	1·337	1·15	2·2	5	3	0·44
500	1·336	2·39	1·9	8	6	0·24
520	1·336	5·79	1·6	16	14	0·10
530	1·335	9·28	1·5	23	22	0·065
540	1·335	1·25 1	1.4	30	29	0·047
550	1·334	1·53 1	1·3	36	35	0·036
560	1·334	1·74 1	1·2	40	39	0·030
580	1·333	3·42 1	1·1	75	74	0·015
600	1·333	9·55 1	0·93	200	200	0·0046
620	1·332	1·18 2	0.82	240	240	0·0034
640	1·332	1·38 2	0·72	270	270	0·0027
660	1·331	1·63 2	0·64	310	310	0·0021
680	1·331	2·06 2	0·56	380	380	0·0015
700	1·330	3·34 2	0·50	600	600	0·0008
740	1·329	1·32 3	0·40	2250	2250	0·0002
750	1·329	1·56 3	0·39	2620	2620	0·0001
760	1·329	1·55 3	0.35	2560	2560	0·0001
800	1·328	1·29 3	0.29	2020	2020	0·0001

amplitude is a measure of the intensity of absorption of the light. A detailed analysis of the method, given in [178], shows that it ensures an absolute measurement accuracy of approximately $\pm 10\%$. This is the highest accuracy achieved to date. The data of different authors are compared in Table 3.2 and in Fig. 3.5. We see that in the visible region which is important for ocean optics there is considerable divergence of the data. In the red and infrared regions, where the absorption is significant, the data of different authors show significantly better agreement. The new method is free of many of the errors typical of measurements of a from c. This is an absolute method, in which the absorption is directly determined from the thermal effect. It is not necessary to introduce corrections for the scattering by particles, the displacement of the axis and loss of collimation of the beam, for absorption and reflection at windows, etc. These corrections may be overestimated in some cases and underestimated in other cases. The data for a from [178] together with k and c (in the calculation of c, we took b from Table 3.1) are shown in Table 3.3. Comparing Table 3.3 with Table

Table 3.2. Absorption coefficient for pure water according to various sources $(a \times 10^3\,\mathrm{m}^{-1})^a$

Source	λ(nm)						
	400	450	500	550	600	650	700
[151]	14	7	9	40	190	280	450
[140]	41	21	35	70	170	230	393
[177]	58	33	—	—	272	351	648
[150]	100	20	25	35	150	250	600
[142]	76 ± 6	22 ± 5	24 ± 10	36 ± 13	170 ± 50	280 ± 60	560 ± 120
[41, 48]	6	3	6	35	200	290	600
[148]	—	—	22	—	—	—	—
[172]	—	37	45	76	260	340	—
[178]	—	23 ± 2	23.3 + 2	57 ± 6	205 ± 20	324 ± 30	590 ± 60

^{a} The purification of the water sample was very carefully carried out in [151] (this is especially important for the blue and ultraviolet regions where contamination is critical). The data of [140] and [177] are given in Jerlov's monograph [25]. The review [150] is widely used in geophysical studies. The monograph [142] is considered to be the most careful review of the physical properties of pure water (we give the average values and variations according to the data of different authors considered in [142]). The data in [178] are given with the errors indicated by the authors.

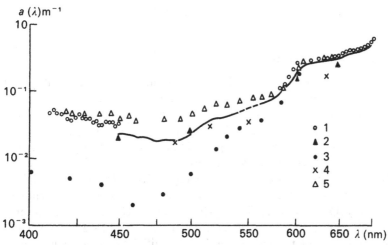

Fig. 3.5. Light absorption coefficient for pure water, corrected data from [178]. The values in the dotted section have been determined by extrapolation. For the sake of comparison, data from the studies: 1—[177], 2—[150], 3—[41], 4—[148] and 5—[172] are also shown.

Table 3.3. Optical properties of pure water ($T = 20°C$, $p = 10^5$ Pa) according to new measurements

λ (nm)	$a \times 10^3 \, m^{-1}$	$c \times 10^3 \, m^{-1}$	$k \times 10^{10}$
440	25·8	29·0	9·03
460	21·0	23·7	7·69
480	18·6	20·8	7·10
500	23.3	25·2	9·27
520	38·0	39·6	1·571
530	41·0	42·5	1·731
540	46·0	47·4	1·981
550	52·0	53·3	2·281
560	60·8	62·0	2·711
580	100	101	4·611
600	212	213	1·012
620	288	289	1·422
640	310	311	1·582
660	366	367	1·922
680	426	427	2·312
700	590	590	3·292

3.1, we can see that at present there is a large spread of values for the optical characteristics of pure water.

The optical constants of water were published in [27] over a wide spectral range from 0.1×10^{-10} to 1 m. In the region 2–50 μm, the authors found n and k from the measurements of reflection spectra. The values of n and k outside this spectral region were determined from the dispersion relationships (Equation (3.9)).

3.2.2. Pure ocean water

All the above are for pure water. For pure ocean water, ie pure water with the addition of inorganic salts, k and n both change. We first examine the effect of salts on absorption. The main change occurs in the ultraviolet with a small change in the infrared region. The ions of inorganic salts have electron absorption spectra situated in the ultraviolet region. They make a small contribution to the visible and near ultraviolet region but this increases appreciably with decreasing λ. This contribution has been investigated especially in the the work of J. Lenoble [157]. She determined the absorption of aqueous solutions of eight salts (NaCl, $MgCl_2$, Na_2SO_4, $CaCl_2$, KCl, $NaHCO_3$, NaBr, $SrCl_2$), whose ions form the main ionic composition of seawater. Then, using Beer's law and assuming additivity, she calculated the absorption of artificial seawater with a chloride content of $19^0/\!_{00}$ made up from the individual components. Lenoble's data are given in Table 3.4 and Fig. 3.6.

It can be seen from Fig. 3.6 that the data of [157] are somewhat higher than the

Table 3.4. Absorption of ultraviolet light by artificial seawater

λ (nm)	$a \times 10^2 \, \text{m}^{-1}$	λ (nm)	$a \times 10^2 \, \text{m}^{-1}$	λ (nm)	$a \times 10^2 \, \text{m}^{-1}$
390	2·1	340	6·9	290	25
380	2·1	330	7·3	280	35
370	5·1	320	11·5	270	45
360	5·3	310	14	260	47
350	6·0	300	19	250	51

data for pure ocean water. It is possible that this is the contribution of particles remaining in the sample after filtration (during measurement in [157]). For smaller values of λ, the absorption of seawater increases sharply and becomes significantly larger than a of the solution of inorganic salts. Armstrong and Boalch [132] ascribed this difference to the effect of organic matter. They also noted the important role of nitrates particularly in deep water (the concentration of nitrates increases with depth, see [28]). Ogura and Hanya [163] made a detailed study of absorption in the ultraviolet region. They investigated samples taken from the western part of the Pacific Ocean and from the Sagami Gulf to depths of ~ 5000 m. The samples were filtered through a filter of pore diameter

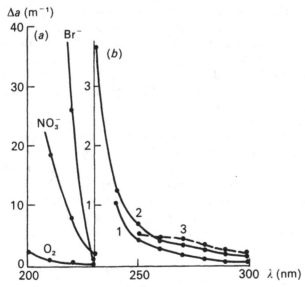

Fig.3.6. Increase in the absorption coefficient caused by dissolved salts.
(a) the effect of bromine-containing compounds nitrates and oxygen [typical concentrations for sea water: Br^-, 65 mg/l, NO_3^-, 10μ mole/l, O_2, $5^o/oo$ (with respect to volume)].
(b) filtered Mediterranean water: 1—surface treated with ultraviolet light (in order to destroy the dissolved organic matter); 2—deep water (low in organic matter). Data are also given here for the solution of salts—artificial sea water according to [157] (3).

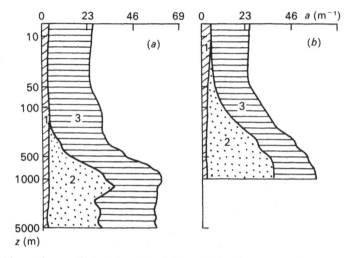

Fig. 3.7. Absorption coefficient for ultraviolet radiation in ocean water:
(a) station in the Pacific Ocean to the south of Japan (28°00′ latitude north, 135°55′
longitude east); (b) Sagami Gulf (34°26′ latitude, 139°40′ longitude east). 1, organic matter;
nitrates (NO_3^-). 3, bromides (Br^-).

$0.45 \mu m$ and the attenuation was measured in the region 210–300 nm. The
contribution of the scattering by small particles remaining in the sample will be
negligibly small in the 210 nm region. However, its proportion increases with
increasing λ and may amount to 15–30 % at $\lambda = 300$ nm. For an accurate
estimate, it is necessary to have data on the suspension remaining in the sample.
According to their data, in this region, the absorption of organic materials in
surface waters is 15–20 %, decreasing with depth and the absorption of nitrates,
which is negligibly small at the surface, increases with depth to 60 %. The
remaining part is associated with bromides, the contribution of which is
approximately the same at all depths. The increase in a compared to distilled
water for $\lambda = 220$ nm is shown in Fig. 3.7 and in Table 3.5 for the data of [163].

Copin-Montegut and Saliot (according to [28]) carried out careful investigations of absorption in the ultraviolet region. They found that dissolved oxygen
absorbs in the spectral region 200–230 nm, in addition to bromides and nitrates.
Their measurement data are given in Fig. 3.6. The concentration of nitrates is
given in micromoles of nitrogen and the concentration of oxygen in millilitres. All

Table 3.5. Contributions to the ultraviolet absorption by ocean water (Δa m^{-1})

	Organic matter	Br$^-$	NO$_3^-$	Σ
Surface waters	5	18–25	—	23–30
Deep waters	2·5	18–25	≤ 35	≤ 60

values are based on one litre of seawater. Figure 3.6 shows the additional absorption of light (compared to pure water) for water samples from the Mediterranean Sea (salinity $37\cdot8^0/oo$) filtered through a fine filter (in order to remove the suspensions) and illuminated with a mercury lamp (in order to eliminate absorption by the dissolved organic matter). For $\lambda = 250$ nm, a value was obtained of $0\cdot45$ m^{-1} and for 300 nm a value of $\sim0\cdot1$ m^{-1}; the absorption did not change with depth. These values do not contradict the measurement data for natural ocean water (the minimum values of $c(\lambda)$ are of the order of 1 m^{-1} for 250 nm and ~0.3 m^{-1} for 300 nm according to [48]). The curves in Fig. 3.6. make it possible to estimate a of water in the ultraviolet region and for any other concentrations of Br$^-$, NO$_3$$^-$ and O$_2$. Using the data on the distribution of these substances, as given in Chapter 2, it is possible to obtain the distribution of $a(\lambda)$ in the world's oceans.

The optical properties of pure ocean water in the infrared region have been studied in [144]. The author believes that the presence of the dissociated ions MgSO$_4$, MgCl$_2$, CaCl$_2$ and NaCl leads to two effects; a small increase in n and a shift in the position of the absorption bands (this is expressed in a reduction of k). The main changes occur in the region 9–15 μm and are associated with the absorption band of sulphates. For water with salinity $S = 34\cdot3^0/oo$ and chloride content Cl $= 19^0/oo$ it is recommended in [144] that the following corrections to n and k are made: (1) in the visible and infrared region up to $\lambda = 9\,\mu$m, the correction Δn_0 to the value of n is equal to 6×10^{-3}; $\Delta k_0 = 0$; (2) in the infrared region 9–15 μm, the corrections Δn_0 and Δk_0 are given in Table 3.6 (they depend on the wavelength).

For any other water with salinity differing from standard water, the author of [144] recommends linear interpolation, ie it is assumed that:

$$\Delta n = \Delta n_0 S/S_0$$
$$\Delta k = \Delta k_0 S/S_0 \tag{3.13}$$

The corrections to n and k are defined by the following equations:

$$n = n_w + \Delta n; \; k = k_w - \Delta k \tag{3.14}$$

where n_w and k_w are the optical constants of pure water. In a more recent study [169], the authors confirmed the results of [144].

There are no exact data on the variation in the absorption coefficient for pure ocean water $k(\lambda)$ with salinity in the visible region. The absorption coefficient is very small and it is usually assumed to be the same as for pure water. The refractive index $n(\lambda)$ increases slightly with increasing salinity S, for $\lambda = 0.59\,\mu$m ($T = 25°C$) by approximately 0.7% for a change in S from 1 to 35$^0/oo$. This fact is used in laboratory interference refractometers. On changing S by 0.001$^0/oo$ at the 10 cm shoulder, the increment to the optical path will be 0.003λ which is not difficult to measure. In situ optical salinometers have not yet been described. The refractive index $n(\lambda)$ has been measured by a number of authors. There are many data available, some of which have been determined with great accuracy [25, 28, 74]. We examine this question in detail in Section 3.5. Here we

Table 3.6. Corrections to the optical constants of ocean water for salinity

$\lambda(\mu m)$	$\Delta n_0 \times 10^3$	$\Delta k_0 \times 10^3$	$\lambda(\mu m)$	$\Delta n_0 \times 10^3$	$\Delta k_0 \times 10^3$
9·0	7	1	12·2	6	10
9·2	7	1	12·4	6	10
9·4	8	1	12·6	6	10
9·6	8	1	12·8	5	10
9·8	9	1	13·0	4	11
10·0	9	1	13·2	3	11
10·2	9	2	13·4	2	10
10·4	9	2	13·6	2	10
10·6	9	2	13·8	2	9
10·8	9	3	14·0	1	8
11·0	9	3	14·2	1	8
11·2	9	4	14·4	1	7
11·4	8	4	14·6	1	7
11·6	8	5	14·8	1	7
11·8	7	7	15·0	0	6
12·0	7	9			

Table 3.7. Values of $n^* = (n - 1\cdot3) \times 10^5$ for pure seawater ($T = 20°C$, $p = 10^5$ Pa, $S = 35°/oo$)

$\lambda \times 10^{10}$ m	n^*	$\lambda \times 10^{10}$ m	n^*	$\lambda \times 10^{10}$ m	n^*
4047	4951	4880	4356	5770	3982
4358	4692	5017	4285	5791	3976
4579	4536	5083	4252	5893	3943
4678	4473	5145	4224	6328	3813
4800	4401	5461	4092	6438	3781

present Table 3.7. It gives the values of $n(\lambda)$ for standard seawater in the visible region.

With the help of the increment table Δn (Table 3.11), n can be easily estimated in the visible region for any value of S, p and T.

The optical constants of pure ocean water over the wide spectral region 0.35–15 μm are given in Table 3.8 and in Fig. 3.8. More detailed data are given in [24].

3.2.3 Dissolved organic substances

Up till now we have discussed pure water and pure seawater. We now consider dissolved organic matter, more exactly that part of it which is called yellow substance. The absorption coefficient $a_y(\lambda)$ is shown in Fig. 3.9. On changing wavelength from $\lambda = 600$ nm to $\lambda = 250$ nm for example, the coefficient a_y

Table 3.8. Optical constants of pure ocean water ($T = 20°C$, $p = 10^5$ Pa, $s = 35°/oo$)

$\lambda\,(\mu m)$	n	k	$\lambda\,(\mu m)$	n	k
0.35	1.356	3.3 − 10	4.00	1.354	3.7 − 3
0.40	1.350	1.9 − 10	4.50	1.338	1.1 − 2
0.45	1.346	8.4 − 10	5.00	1.336	1.4 − 2
0.50	1.343	9.3 − 10	5.50	1.313	1.5 − 2
0.55	1.341	2.3 − 9	6.00	1.276	1.0 − 1
0.60	1.339	1.0 − 8	6.50	1.336	3.6 − 2
2.50	1.246	1.7 − 3	7.00	1.318	3.2 − 2
2.86	1.180	2.2 − 1	7.50	1.307	3.3 − 2
3.25	1.471	7.6 − 2	8.00	1.292	3.5 − 2
3.30	1.450	5.2 − 2	9.00	1.259	4.0 − 2
3.35	1.437	3.8 − 2	10.00	1.210	5.5 − 2
3.40	1.426	2.9 − 2	11.00	1.167	1.1 − 1
3.45	1.413	1.8 − 2	12.00	1.125	2.0 − 1
3.50	1.404	1.2 − 2	13.00	1.144	3.1 − 1
3.55	1.396	6.4 − 3	14.00	1.218	4.0 − 1
3.60	1.390	5.0 − 3	15.00	1.282	4.3 − 1

increases by a factor of 200. This leads to the yellowish colour of some seawater, for example the Baltic Sea, which contains a large amount of yellow substance. The absorption coefficient of yellow substance is approximated by the equation:

$$a_y(\lambda) = C\,e^{-\mu\lambda} \tag{3.15}$$

since in the semilogarithmic scale it is represented by a straight line in [25]. The coefficient μ for this straight line is approximately equal to $15 \times 10^{-3}\,\text{nm}^{-1}$. The

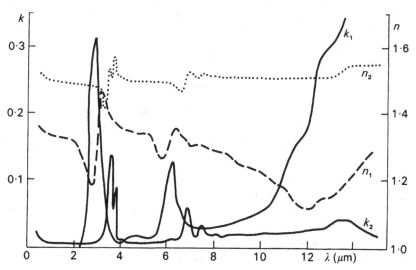

Fig. 3.8. Optical constants of ocean water (n_1, k_1) and oil (n_2, k_2) over a wide spectral range.

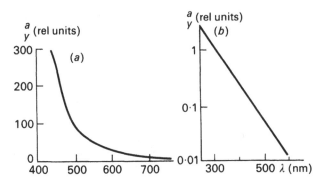

Fig. 3.9. Absorption coefficient for yellow substance. (a) standard scale; (b) semilogarithmic scale.

measurements carried out in the voyages of IOAN showed that the coefficient μ is equal to approximately $15 \times 10^{-3}\,\text{nm}^{-1}$ for surface waters and $\sim 17 \times 10^{-3}\,\text{nm}^{-1}$ for deep waters [48]. A close dependence is recommended in [149]:

$$a_y(\lambda) = a_y(\lambda_0)e^{-0.0140(\lambda - \lambda_0)} \tag{3.16}$$

where λ_0 is the fixed wavelength in the region between 280 and 250 nm. The dispersion of the index in the exponent is equal to 0·0025.

There are different points of view in the literature relating to the stability of the absorption characteristics of light by yellow substance. Thus, Jerlov in the first edition of [25, p. 78] assumed that as a whole, the function $a_y(\lambda)$ has a universal character. A. Ivanoff [28] points out that the amount of yellow substance in water fluctuates strongly and consequently the absolute value of absorption associated with yellow substance, and to a smaller extent the spectral functions [28, Fig. 15.11] appreciably vary also. In [149], it is pointed out that for $\lambda_0 = 450\,\text{nm}$, $a_y = 0.212\,\text{m}^{-1}$ at a concentration of yellow substance of $C = 1\,\text{mg/l}$. N. Højeyslev in [149] thus recommends the following relationship for determining the concentration of yellow substance (in mg/l) from the coefficient $a_y(\lambda)$:

$$C = 4{\cdot}72 a_y(\lambda)e^{0.0140(\lambda - 450)} \tag{3.17}$$

Qualitatively, it is not difficult to understand the basic behaviour associated with yellow substance, the sharp increase in absorption at shorter wavelengths. The ultraviolet region which we are dealing with concerns the electron absorption spectra. For solutions of organic molecules in water, the electronic absorption spectra are broadly blurred bands. They have an asymmetric shape: they are steeper on the long-wave side than on the short-wave side. The absorption of yellow substance which we observe in the ultraviolet region is the long-wave part of the electronic absorption band.

In order to establish the relationship between the absorption spectra and molecular structure, it is necessary to consider how the electronic spectrum is formed in complex molecules [22]. Carbon in organic compounds forms two types of bonds: single bonds or σ bonds and conjugated bonds or π bonds. In single bonds, the electrons forming the bond are localized between the bonding atoms. In conjugated bonds, the electrons forming the bond belong to the entire interacting system as a whole. An approximate method for calculating the absorption spectrum of complex molecules, the molecular orbital method, is based upon consideration of the π electrons only. All the remaining part of the molecule, all the atomic nuclei and all the remaining electrons are considered as a framework or skeleton in the field of which the π electrons move freely. In the simplest organic compound, in the benzene molecule C_6H_6 for example, this simplification makes it possible to consider six delocalized electrons instead of the system of 12 nuclei and 42 electrons. The picture is very similar to the system of absorption bands arising in the solid state.

In order to calculate the energy spectrum of the system of π electrons, imagine that we are dealing with a linear chain of length L. Since the electronic waves at the ends of the chain must have nodes, an integer number n of electronic half waves $\lambda/2 = L/n (n = 1, 2, ...)$ must fit over a length L. Accordingly, electrons can only exist in the system with velocities $v_n = h/(m\lambda) = hn/(2mL)$ and the energy spectrum is determined by the formula $E_n = h^2 n^2/(8mL^2)$. Let the system consist of N atoms with the length of each bond 1. Obviously, $L = (N - 1)1$. Each atom provides one π electron. According to Pauli's principle, they occupy $N/2$ levels in the optical band. On absorption, there is a transition from the upper occupied level to the nearest free level. Thus, the transition frequency v is determined from the equation:

$$hv = h^2[(N/2 + 1)^2 - (N/2)^2]/[8ml^2(N - 1)^2] \approx [(h^2/8ml^2)](1/N)$$

It can be seen from this equation that the smaller the system (the smaller N), the higher the frequency of the absorbed light. It is apparent that the frequency of short sections of conjugated bonds is significantly higher than for long sections if only because a long conjugated chain is unstable. Under the effect of various perturbations, thermal and other, it will bend and break, becoming several short pieces. In addition, the smaller the pieces, the more stable they will be and the more there will be of them. The shorter the chain, the higher the frequency of the virtual vibrator which must be ascribed to it. Since the absorption intensity is proportional to the number of vibrators capable of absorbing light, the coefficient $a(v)$ increases with frequency in complex molecules.

3.3. Suspended particles

The optical constants of suspended particles are determined by the nature of the particles. Terrigenous suspensions are a mixture of different minerals. Their composition is roughly known. A.P. Vinogradov, A.P. Lisitsyn, T. Sasaki et al. have carried out studies on them. For references to the original work see [123]. It is assumed that the most common mineral substances are those given in Table 3.9 where n_0 and n are the absolute and relative refractive indices of these substances in the visible region of the spectrum. The average relative refractive

Table 3.9. Refractive indices of minerals in marine suspensions

Index	Hydromica	Kaolinite	Montmorillonite	Chloride	Byeleyite	Mountain leather
n_0	1.58	1.56	1.51	1.57	1.55	1.55
n	1.19	1.17	1.135	1.18	1.165	1.165

index n corresponding to Table 3.9 is equal to 1.17. The composition of minerals may differ somewhat from that given in the table and the average refractive index lies in the range 1.15–1.20. Comparison of the measured optical characteristics with model characteristics gives values for n in the range 1.15–1.25 in different cases. It is shown in Chapter 6 that for terrigenous particles in the visible region of the spectrum, best agreement of the calculated volume scattering function (at moderate and large angles) with the observations gives $n = 1.15$.

Biogenic suspensions are produced directly in the sea and consist of living organisms (microbes, plankton) as well as organic detritus (fragments of skeletons, residues of plasma organic matter). There are very few direct measurements of the refractive index of biological particles. It has been determined by immersion refractometry that the absolute refractive index of the spherical bacteria *Stafilococcus aurens* is equal to 1.40. It is usually assumed that the relative refractive index of organic particles in the sea is less than 1.05. Absorption by these particles in the visible region is observed in the absorption bands of phytoplankton pigments.

As we saw in Chapter 2, living phytoplankton cells make up aproximately 10% of the ocean suspension. Algae cells have a complex organization. They contain intracellular formations called chloroplasts. Chloroplasts consist of proteins, lipids and pigments. They are the centres of photosynthesis in the cell. Their structure depends on the actual organism. Thus, the chloroplast of the single-cellular algae *Chlorella vulgaris*, studied by F.Ya. Sid'ko *et al.* [79], consists of four and eight unequal layers with 16–32 monomer layers of chlorophyll. Apart from chlorophyll, plankton cells contain other pigments, carotenoids, phycobilins, etc. The pigment composition in the cell depends on its species, age, conditions under which it is developing, etc. Two factors simplify this in the general complex picture: (1) the optical properties of different pigments do not differ greatly. This can be seen from Fig. 5.3 where the optical density spectra of three different pure cultures of plankton and a natural population from the waters near Woods Hole are given (according to [179]). All the curves have an absorption minimum in the green region, a strong absorption in the blue region (due to chlorophylls and carotenoids) and a strong absorption in the red region (due to chlorophyll *a*). The similarity in the optical properties of different pigments is related to the similarity of their chemical structures. In addition, the pigments of the green leaf are structurally similar to the blood pigment, haem. Both are metal derivatives of cyclic pyrrol pigments. In chlorophyll, the centre of the pyrrol ring is occupied by magnesium and in haem by iron. (2) There is a close relationship between the total content of all pigments in the upper 100 m layer of the ocean and the content of chlorophyll a. This means that the concentration of pigments as a whole may be characterized by the concentration of chlorophyll a (Fig. 5.4). The first approximation in the modelling of cells is to regard them as homogeneous spheres in which pigments and non-absorbing substances are uniformly mixed. According to the data of [79], the chlorophyll fraction of dry matter in the cells is a few per cent ($<4\%$) and the average absorption coefficient a per cell is equal to $(5.5–9) \times 10^4 \, \mathrm{m}^{-1}$ for $\lambda = 435 \, \mathrm{nm}$ and $(4.5–6) \times 10^4 \, \mathrm{m}^{-1}$ for $\lambda = 680 \, \mathrm{nm}$. This gives a value of $k = (2.6–3) \times 10^{-3}$ for the imaginary part of the refractive index of the cell substance for $\lambda = 435$ and $680 \, \mathrm{nm}$. For the algae *Chlorella pyrenoidosa* at $\lambda = 675 \, \mathrm{nm}$, the value given for $a = 33 \times 10^4 \, \mathrm{m}^{-1}$ is six times higher in [156]. The authors of [156] point out that due to the absorption the scattering maximum at the cells is shifted in the long-wave direction [156, Fig. 6]. This is the phenomenon of anomalous dispersion already noted by R. Wood. It should be noted that a distinction has to be made between the refractive index of the matter of the suspended particles and the refractive index of the suspension itself (this is given in Fig. 5.3).

The values of n for the suspended particles given above are obtained from samples. A method for determining n *in situ* is suggested in [181]. The authors made use of the fact established in a number of studies of a close correlation between the total coefficient of hydrosol scattering b_h and the scattering coefficient $b_h(45°)$.

$$\bar{b}_h(45°)/\bar{b}_h = \text{constant}$$

The bar signifies averaging over the distribution curve. Using Equation (4.85) for the scattering cross-section of an individual particle and assuming a gamma distribution with $\mu = 0$, the authors of [181] obtained for the hydrosol scattering coefficient:

$$b_h(\lambda) = c[\bar{a}^2 + (\delta^2 - 1)/(\delta^2 + 1)]$$

$$\delta = (2\pi/\lambda)\bar{a} | n_1 - n_0 |$$

Here, $\bar{a}$ is the average size of the suspended particles; n_1 and n_0 are the refractive indices of suspended particles and ocean water, respectively.

Measuring $\bar{b}_h(\lambda, 45°)$ at different depths in the ocean for two wavelengths $\lambda_1 = 436$ and $\lambda_2 = 546 \, \mathrm{nm}$ and assuming that:

$$\bar{b}_h(\lambda_1, 45°)/\bar{b}_h(\lambda_2, 45°) = \bar{b}_h(\lambda_1)/\bar{b}_h(\lambda_2)$$

$| n_1 - n_0 |$ can be easily found from the preceding equations if certain values are assumed for $\bar{a}$. In Fig. 3.10, we give a graph for the quantity $| n_1 - n_0 |$ plotted in [181]. Measurements of $\bar{b}_h(45°)$ were carried out in the ocean trench at two stations in the Pacific Ocean. For both stations (separated by a distance of 210 km), the vertical profile of $| n_1 - n_0 |$ was approximately the same.

It follows that marine suspensions comprise particles with $n = 1\cdot02, 1\cdot05, 1\cdot10,$ $1\cdot15, 1\cdot20$ and $1\cdot25$. It can be assumed that the first two values ($1\cdot02$ and $1\cdot05$) refer to biogenic particles and the following four (1.10, 1.15, 1.20 and 1.25) to terrigenous particles. As far as the absorption coefficient is concerned, it is usually assumed that $k = 0$ for 'dead' particles. Of course, this is a crude model. We have already noted in Chapter 2 that up to 45% of yellow substance is absorbed by small suspended particles (Fig. 5.5). This means that it is necessary to ascribe $k \neq 0$ to these particles (of any origin). However, an accurate value for k of 'dead' particles is not known.

In recent years, there has been great interest in the optical constants of oil. At first, oil enters the sea in the form of a film which then breaks up into individual aggregates. The data for the ultraviolet, visible and infrared regions are given in [24]. Almost all oils absorb strongly in the ultraviolet part of the spectrum. In the visible region, k is considerably reduced and takes a value of the order of 10^{-3}. The refractive index changes significantly less, from 1.57–1.67 to 1.48–1.52. The typical spectral variation of $n(\lambda)$ and $k(\lambda)$ is given in Fig. 3.8.

In the infrared region 2.5–25 μm, oil has two strong absorption bands, intense in the region 3.3–3.6 μm and less intense at 6.7–7.1 μm, and bands of average intensity in the region 11.1–16.7 μm. In comparison to seawater, there are two distinguishing characteristics: (1) the refractive index of oil n is practically constant ~ 1.55 over the entire range examined; (2) the two first strongest absorption bands of oil are close to the absorption bands of water (in water the centres of the strong bands lie at $3\cdot2$ and $6\cdot1$ μm). A simple explanation of these characteristics is given in [24]. The fact is that the regions of strong absorption are regions of the eigen frequencies of molecular vibration ν. The

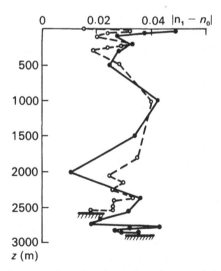

Fig. 3.10. Difference between the refractive index of suspended particles (n_1) and ocean water (n_0) with depth for two stations in the vicinity of the Ecuador coast.

vibrational frequency of the molecules v is defined by the formula:

$$v = [1/(2\pi c)]\sqrt{f/\mu} \tag{3.18}$$

where f is the bond strength and μ is the reduced mass.

It is known that the strengths of different bonds in molecules vary relatively little. Therefore, the eigen frequencies of the molecules are essentially determined by the reduced mass μ:

$$\mu = m_1 \cdot m_2/(m_1 + m_2) \tag{3.19}$$

where m_1 and m_2 are the masses of the particles involved in the vibrating bond. In water, we are dealing with vibrations of the OH bond, and in oil, of the CH bond. In both cases, since $m_1 \ll m_2$, then the reduced mass $\mu_1 \approx m_1$, ie in both cases, the hydrogen atom vibrates around the heavier atom (oxygen or carbon). This is why the eigen frequencies of oil and water are similar. This is also why the wavelengths which determine the absorption minima of the light and heavy isotopes of water have a ratio of $\sqrt{2}$ (Section 2.1.1).

In conclusion, we add that in the upper layers of water there is a certain amount of bubbles. The situation with bubbles is very simple. The absolute refractive indices of air are practically equal to unity so that for bubbles the relative indices will be:

$$m_{\text{bub}} = n/(n^2 + k^2) + ik/(n^2 + k^2) \tag{3.20}$$

where n and k are the optical constants of pure seawater.

3.4. Theory of the molecular scattering of light by ocean water

3.4.1. Scattering due to fluctuations in density and temperature

In contrast to absorption, the molecular scattering of light by seawater may be calculated theoretically. Furthermore, this calculation may be based on classical electrodynamics and thermodynamics. The reason for scattering is the optical heterogeneity of the medium arising from the thermal motion of the molecules. If the molecules are isotropic, the variations in the refractive index of the substance, n, will be caused by local fluctuations in density and temperature of the body; if they are anisotropic (as in the case of water molecules), there will also be fluctuations in orientation. In solutions, an additional reason for heterogeneities will be local fluctuations in the concentration of the solution.

The idea of fluctuations as the reason for the scattering of light was first put forward by M. Smoluchowsky, who suggested it as an explanation of the phenomenon of critical opalescence. In every element of volume, the fluctuations are considered as random and independent of each other. This idea formed the basis for the quantitative theory of light scattering developed by A. Einstein.

Einstein found a formula for the scattering coefficient $b(\gamma)$ for a pure liquid and for a solution [128]. Fluctuations in the orientation of the molecules were taken into account by J. Cabannes [135] who made a certain correction to Einstein's formula.

Let us first consider the scattering of light in pure water caused by fluctuations in density and temperature. We examine a small element of volume v from the illuminated volume V. As a result of the heterogeneities, let the permittivity ε in the element v deviate by $\Delta\varepsilon$ from its average value ε_0. We assume that the element v is spherical with radius a and that a is very much less than λ, so that the scattering of light by the element v obeys Rayleigh's formula for small particles (Chapter 4). On the other hand, we assume that a is very much greater than the size of the molecules, so that the properties of v will be described phenomenologically. We remember that a sphere placed in a constant field E_0 will acquire a dipole moment:

$$P = [a^3\Delta\varepsilon/(\Delta\varepsilon + 3\varepsilon_0)]E_0 \approx [a^3/(3\varepsilon_0)]\Delta\varepsilon E_0$$

In the electric field of the light wave $E = E_0 e^{i(\omega t - kr)}$, which will lead to the formation of an additional dipole moment:

$$\Delta P = (\Delta\varepsilon/4\pi)(v/\varepsilon_0)E_0 e^{i(\omega t - kr)}$$

Assuming, that the point of observation M is in the wave zone at a distance r from the centre of the dipole, we find that for the secondary field (see [90, p. 82]):

$$\Delta E' = [(\Delta\ddot{P}, R_0)R_0 - \Delta\ddot{P}]/(c^2 r\varepsilon_0)$$

It is necessary to substitute the dipole moment here, not at the point in time under consideration t but at the time $t' = t - r/c_a$ (taking into account the delay). Here, c_a is the velocity of light in the substance and R_0 is the unit vector from the centre of the dipole to the point of observation. Noting that $\Delta\ddot{P} = -\omega^2\Delta P$, we obtain $(c_a^2\varepsilon_0 = c^2)$:

$$\Delta E' = -[\omega^2 v/(c^2 r)][(E_0, R_0)R_0 - E_0](\Delta\varepsilon/4\pi)e^{i(\omega t - kr)}$$

We introduce the spherical system of coordinates (r, γ, ϕ) at point M. The projection of the expression in square brackets on the spherical axis ϕ will be $\sin\phi$ (see [90]). This means that the amplitude, the ϕ-component of the secondary wave produced by the element v, will be:

$$A = [\pi\Delta\varepsilon/(\lambda_0^2 r)]vE_0 \sin\phi$$

For the intensity of light scattered at an angle γ, $I(\gamma)$ we find:

$$I(\gamma) = I_0[\pi^2\Delta\varepsilon^2/(\lambda_0^4 r^2)]v^2 \sin^2\phi \qquad (3.21)$$

Let the illuminated volume V consist of N identical elements v $(V = Nv)$. Assuming that the different elements contained in the area V scatter light incoherently, ie that there is no correlation between the fluctuations of ε in the different elements, we find for the total intensity of the scattered light (adding intensities and not fields):

$$I(\gamma) = I_0[\pi^2/(\lambda_0^4 r^2)]V\overline{\Delta\varepsilon^2}v \sin^2 \phi \qquad (3.22)$$

The observed intensity is the result of averaging Equation (3.22) over the time of observation. We emphasize that the wavelength in a vacuum λ_0 and not the wavelength in the substance λ stands for the scattered field and intensity in Equations (3.21)–(3.22). Equation (3.22) describes the scattering of a linearly polarized wave. In order to obtain the equation for the scattering of natural light it is necessary to substitute $\sin^2 \phi$ by $(1 + \cos^2 \gamma)/2$ as usual (see [90, p. 84]). Furthermore, we remember that the scattering coefficient in the given direction $b(\gamma)^*$ is determined by the equation (Table 1.1):

$$b(\gamma) = I(\gamma)r^2/(I_0 V)$$

Finally, we now find:

$$b(\gamma) = b(90°)(1 + \cos^2 \gamma)$$

$$b(90°) = [\pi^2/(2\lambda_0^4)]v\overline{\Delta\varepsilon^2} \qquad (3.23)$$

Equation (3.23) completes the 'electrodynamic' part of the derivation.

We now obtain an expression for $v\overline{\Delta\varepsilon^2}$. In a pure liquid, $\varepsilon = f(\rho, T)$ and the fluctuations of ε in Equation (3.23) may be expressed in terms of the fluctuations of density and temperature:

$$\Delta\varepsilon = (\partial\varepsilon/\partial\rho)_T\Delta\rho + (\partial\varepsilon/\partial T)_\rho\Delta T$$

The density and temperature of the body are independent thermodynamic variables. Therefore, their fluctuations are also independent. This means that $\overline{\Delta\rho\Delta T} = 0$ and:

$$\overline{\Delta\varepsilon^2} = (\partial\varepsilon/\partial\rho)_T^2\overline{\Delta\rho^2} + (\partial\varepsilon/\partial T)_\rho^2\overline{\Delta T^2} \qquad (3.24)$$

For the density fluctuation $\overline{\Delta\rho^2}$ and temperature fluctuation $\overline{\Delta T^2}$ of the small volume v, the following equations are derived in thermodynamics (see [19, p. 66]; [84, p. 437]):

$$v(\overline{\Delta\rho^2}/\rho^2) = kT\beta_T$$
$$v(\overline{\Delta T^2}/T^2) = kT\beta_T(\gamma - 1)/(\alpha^2 T^2) \qquad (3.25)$$

Here, $\beta_T = -[(1/v)(\partial v/\partial\rho)]_T$ is the coefficient of isothermal compressibility; $\gamma = c_p/c_v$ is the ratio of the thermal capacities and $\alpha = [(1/v)(\partial v/\partial T]_p$ is the coefficient of thermal expansion of the body. It is clear that when the volume v is reduced, the relative fluctuations in density and temperature in it must increase. It follows from Equation (3.25) that this obeys a $v^{-1/2}$ law. Thus, we have:

$$v\overline{\Delta\varepsilon^2} = kT\beta_T[(\rho \ \partial\varepsilon/\partial\rho)_T^2 + (\gamma - 1)[(1/\alpha)(\partial\varepsilon/\partial T)]_\rho^2] \qquad (3.25')$$

---

* In optics (see [87, 19], the quantity $\sigma(\gamma)$ is usually designated by $R(\gamma)$ and is called the scattering coefficient of the liquid.

The second term on the right-hand side of Equation (3.25') describes the deviation of ε caused by local fluctuations in temperature at constant density of the substance ρ. For practically all liquids it is significantly smaller than the first term and it is usually neglected. An exact value of the error produced by this has been estimated differently by different authors. Thus, according to the estimate in [84, p. 44], the second term is 1.6 % of the first term and according to the data of [19, p. 66] it does not amount to more than 0.5 %. Apparently, with an error not exceeding 2 %, the final equations for the scattering of linearly polarized or natural light will be:

$$b(\gamma) = (\pi^2/\lambda_0^4)kT\beta_T(\rho\ \partial\varepsilon/\partial\rho)_T^2 \sin^2\phi$$
$$b(\gamma) = (\pi^2/\lambda_0^4)kT\beta_T(\rho\ \partial\varepsilon/\partial\rho)_T^2(1 + \cos^2\gamma)/2 \tag{3.26}$$

In order to calculate $b(\gamma)$ from Equation (3.26), it is necessary to know the quantities occurring in them: the coefficient of isothermal compressibility β_T and $(\rho\ \partial\varepsilon/\partial\rho)_T$. For the determination of β_T, the adiabatic compressibility β_S is usually used which can be easily determined from the velocity of 'adiabatic' sound C [28] and the ratio of the specific thermal capacities γ. The quantities β_T, β_S, C and γ are related by the equations:

$$C = \sqrt{1/(\beta_S\rho)}$$
$$\beta_S = 1/(\rho C^2)$$
$$\beta_T/\beta_S = c_p/c_v = \gamma$$

In normal sound and ultrasound, the compressions take place adiabatically and in very long wavelength sounds, isothermally so that in the equation for C, β_T should replace β_S.

For liquids (in contrast to gases), the ratio of the specific thermal capacities γ is close to unity. Exact values of β_T for seawater were recently determined in [137]. For a salinity of $S = 36^0/\text{oo}$, $T = 20°C$ and $p = 0, 2 \times 10^7$ and 10^8 Pa, β_T will be equal to 42.60, 40.59 and 37.82 $\times 10^{-11}$ Pa^{-1}. More detailed data on the coefficient β_T are given in [28, chapters 17; 34].

The main problem is the determination of the second quantity $(\rho\ \partial\varepsilon/\partial\rho)_T = (2n\rho\ \partial n/\partial\rho)_T$. The simplest would be to find $(\partial n/\partial\rho)_T$ from experimental data but direct measurements of $n = f(\rho)$ are very difficult; there are in fact no such data. A way out is to use a theoretical relationship between n and ρ. This approach was adopted by Einstein in [128]. He used the Lorentz-Lorenz relationship:

$$[(\varepsilon - 1)/(\varepsilon + 2)](1/\rho) = [(n^2 - 1)/(n^2 + 2)](1/\rho) = \text{constant} \tag{3.27}$$

We can easily find from Equation (3.27) that:

$$(\rho\ \partial\varepsilon/\partial\rho)_T = (n^2 - 1)(n^2 + 2)/3 \tag{3.28}$$

On substituting Equation (3.28) in (3.26) we obtain:

$$b(90°) = [\pi^2/(18\lambda_0^4)](n^2 - 1)^2(n^2 + 2)^2\beta_T kT \tag{3.29}$$

However, it has been found that the values of $b(90°)$ calculated from Equation (3.29) appreciably exceed the data of direct measurements [19, 84]. Therefore, instead of Equation (3.27), some authors suggested using the Laplace-Maxwell relationship.

$$(\varepsilon - 1)/\rho = (n^2 - 1)/\rho = \text{constant} \tag{3.30}$$

which gives for:

$$(\rho \; \partial\varepsilon/\partial\rho)_T = n^2 - 1 \tag{3.31}$$

On substituting Equation (3.31) in (3.26), we find:

$$b(90°) = [\pi^2/(2\lambda_0^4)](n^2 - 1)^2\beta_T kT \tag{3.32}$$

Equation (3.32) in particular was recommended in the first edition of Jerlov's monograph [25, p. 40]. It differs from Equation (3.29) by the factor $\eta = (n^2 + 2)^2/9$. For seawater in the visible region, the quantity $\eta \approx 1\cdot6$, ie the values of $b(\gamma)$ determined from Equation (3.32) are 60% smaller than those determined from Equation (3.29). Comparison with the measurement data shows that Equation (3.32) is also unsatisfactory; opposite to Equation (3.29) it leads to extremely low values. This queston was analysed by M.F. Vuks [19] who made measurements on a large number of different liquids. He recommends the following semi-empirical equation:

$$(\rho \; \partial\varepsilon/\partial\rho)_T = (n^2 - 1) \cdot 3n^2/(2n^2 + 1) = \phi(n) \tag{3.33}$$

which leads to values between those of Equations (3.29) and (3.32). Substituting in Equation (3.26) we finally obtain:

$$b(\gamma) = \pi^2\phi^2(n)\beta_T kT(1 + \cos^2 \gamma)/2\lambda_0^4 \tag{3.34}$$

The application of the semi-empirical Equation (3.33) is not the only method of calculating $b(90°)$. If we use the thermodynamic relationship:

$$(\rho \; \partial\varepsilon/\partial\rho)_T = (\rho \; \partial\rho/\partial T)_T(\partial\varepsilon/\partial p)_T = (1/\beta_T)(2n \; \partial n/\partial p)_T$$

then we can substitute $\partial n/\partial p$ instead of $\partial\varepsilon/\partial p$ in Equation (3.26). It is not complicated to measure the piezooptical coefficient of the liquid $-[(1/n)(\partial n/\partial p)]_T$ and there is a significant amount of data available. Thus, we arrive at the equation:

$$b(90°) = [\pi^2/(2\lambda_0^4)](1/\beta_T)(2n \; \partial n/\partial p)_T^2 kT \tag{3.35}$$

This equation is convenient for calculating b for pure water. For this it is necessary to have direct experimental data on β_T, n and $\partial n/\partial p$.

We now point out some important circumstances associated with Equations (3.34) and (3.35).

1. It is often maintained that the spectral dependence of molecular scattering is determined by the factor λ^{-4}. This is not accurate. In fact, the spectral variation of $b(\lambda)$ depends on the factor $\phi^2(n)/\lambda^4$. Using the data on n from

Table 3.8, it is not difficult to show that on increasing λ from 360 to 800 nm, the quantity $\phi^2(n)$ decreases by 13 %. This factor may be taken into account if it is assumed that $b \sim \lambda^{-r}$ where r is equal not to 4 but to 4·17. A. Morel in [164] recommends $r = 4·32$. However, direct calculation of the experimental values of $b(\lambda)$, which he gives for seawater, gives $r = 4·22$ (according to his data, $r = 4·32$ for pure water). The values of b of other authors show that r varies from 4·05 to 4·35 and depends on the spectral interval considered.

2. We look at the physical meaning of Equation (3.26). The density $\rho = Nm$, where N is the number of molecules in unit volume and m is the mass of the molecule. The fluctuations in density are caused by fluctuations in N. Thus, $(\rho \partial \varepsilon / \partial \rho)^2 = N^2 (\partial \varepsilon / \partial N)^2$. Equation (3.26) was obtained from general thermodynamic considerations and can thus be applied to any states of matter: gaseous, liquid, solid. The simplest situation is in gases. Let us consider this case. Here, the permittivity ε is related to the dipole moment of the molecule χ by the simple equation $\varepsilon = 1 + 4\pi\chi N = 1 + 4\pi\mu; \mu = \chi N$ is the dipole moment of unit volume. Since $\partial\varepsilon/\partial N = 4\pi\chi$, then for $b(\gamma)$ we find:

$$b(\gamma) = NkT\beta_T N\chi^2 (2\pi/\lambda_0)^4 (1 + \cos^2 \gamma)/2$$

It is natural to assume that the quantity χ does not fluctuate for thermal motion. We also note that $\overline{\Delta N^2} = N$. We then find that:

$$\overline{\Delta\mu^2} = \overline{\Delta(N\chi)^2} = N\chi^2$$

Thus, we obtain:

$$b(\gamma) = NkT\beta_T \overline{\Delta\mu^2} (2\pi/\lambda_0)^4 (1 + \cos^2 \gamma)/2$$

For gases, the quantity NkT is equal to the pressure p, the volume $v \sim T/p$ and $\beta_T = -[(1/v)(\partial v/\partial p)]_T = 1/p$. Thus, we simply have:

$$b(\gamma) = \overline{\Delta\mu^2} (\omega^2/c^2)^2 (1 + \cos^2 \gamma)/2$$

Consequently, for gases the coefficient $b(\gamma)$ is simply equal to the fluctuations of μ associated with the fluctuations of N and the square of the quantity ω^2/c^2 ($2\pi/\lambda_0 = \omega/c$), determining the amplitude of the electromagnetic field produced by the vibrating dipole. The factor $(1 + \cos^2 \gamma)/2$ determines the angular structure of the light scattered by the dipole on illuminating it with a natural beam of light.

3. The molecular scattering of water is anomalously small in comparison with other liquids (see [19]). This is caused by its small coefficient of isothermal compressibility β_T and the comparatively small refractive index n. This is one of the anomalies of water associated with its closely packed molecular structure.

3.4.2. The effect of fluctuations in molecule orientation

We now examine the effect of fluctuations in orientation of the water molecule. We designate the components of light intensity scattered perpendicular and parallel to the scattering plane by i_1 and i_2, respectively. To characterize the polarization characteristic of light resulting from scattering, we introduce either the degree of polarization of the scattered light:

$$p(\gamma) = [i_1(\gamma) - i_2(\gamma)]/[i_1(\gamma) + i_2(\gamma)] \qquad (3.36)$$

or the depolarization factor or Cabannes' depolarization coefficient [135]:

$$\Delta = i_2(90°)/i_1(90°) \qquad (3.37)$$

It is obvious that:

$$p(90°) = (1 - \Delta)/(1 + \Delta)$$

On the scattering of natural light by a small isotropic sphere (Rayleigh scattering), for example, we have (see [90, p. 85]): $i_1 = 1$, $i_2 = \cos^2 \gamma$. Hence $p_{Ray} = \sin^2 \gamma/(1 + \cos^2 \gamma)$. When $\gamma = 90°$, then $p = 1$ and $\Delta = 0$, ie the scattered light is completely polarized.

The first experiments on molecular scattering of light in gases showed that in reality $\Delta \neq 0$, ie that light scattered at an angle of $\gamma = 90°$ is not completely polarized. Since gases were involved, the reason for this was the anisotropy of the molecules. If the molecules are modelled as ellipsoids, then $\Delta \neq 0$ and the ratio of the axes of the ellipsoid may be selected for the observed value of Δ. The anisotropy of the molecules must lead not only to depolarization of the scattered beam but also to an increase in the total amount of scattered light since the scattering due to fluctuations in orientation is added to the scattering due to fluctuations in density (concentration). Measurements in liquids also showed that $\Delta \neq 0$. In particular, for pure water at $T = 20°C$ the value of Δ is close to 0·09 for $\lambda = 436\,nm$ [84]. The theory of scattering in liquids consisting of anisotropic molecules was developed in [135]. Although the intervening calculations are rather cumbersome and will not be presented here, the final equations are simple.

We formulate them as in [19, 84]:

$$b_1(90°) = b(90°)(6 + 6\Delta)/(6 - 7\Delta)$$

$$b_1(\gamma) = b_1(90°)[1 + p_1(90°) \cos^2 \gamma] \qquad (3.38)$$

Here, $b(\gamma)$ is the 'isotropic' scattering coefficient [from Equation (3.26)]; $b_1(\gamma)$ is the scattering coefficient allowing for the anisotropy of the molecules. The total scattering coefficient b_1, the volume scattering function $\beta_1(\gamma)$ and the degree of polarization $p_1(\gamma)$ will now be:

$$b_1 = 2\pi \int_0^\pi b_1(\gamma) \sin \gamma \, d\gamma = b\eta, \quad \eta = (6 + 9\Delta)/(6 - \Delta) \approx 1 + (\tfrac{5}{3})\Delta$$

$$b_1(\gamma) = b_{\text{Rayleigh}}(\gamma) f(\gamma, \Delta)$$

$$f(\gamma, \Delta) = \left(\frac{1 + \Delta}{1 - (\tfrac{7}{6})\Delta}\right)\left(\frac{1 - (\tfrac{1}{6})\Delta}{1 + (\tfrac{3}{2})\Delta}\right)\left(\frac{1 + p_1(90°)\cos^2 \gamma}{1 + \cos^2 \gamma}\right) \approx$$

$$(1 + \Delta/2)\left(\frac{1 + p_1(90°)\cos^2 \gamma}{1 + \cos^2 \gamma}\right) \tag{3.39}$$

$$p_1(\gamma) = \frac{p_1(90°)\sin^2 \gamma}{1 + p_1(90°)\cos^2 \gamma} = p_{\text{Rayleigh}}(\gamma)\phi(\gamma, \Delta)$$

$$\phi(\gamma, \Delta) = p_1(90°)\frac{1 + \cos^2 \gamma}{1 + p_1(90°)\cos^2 \gamma}$$

Thus, if the anisotropy of the molecules and the fluctuations in orientation connected with it are taken into account: (1) there is an increase in the total scattering by an amount η; (2) the form of the volume scattering function is changed by the factor $f(\gamma, \Delta)$; 3) the degree of polarization of the scattered light is reduced by the factor $\phi(\gamma, \Delta)$.

For pure ocean water, assuming $\Delta = 0.09$, we find: $\eta = 1.15$, $p_1(90°) = 0.835$ and $f(0°)/f(90°) = 0.92$. This means that because of the anisotropy of the molecules: (1) the total scattering increases by approximately 15%; (2) the volume scattering function is somewhat less drawn out than the Rayleigh function, for example forwards by 8%; (3) for all scattering angles γ, the degree of polarization of the scattered light is less than that for Rayleigh scattering, for example by 16.5% for $\gamma = 90°$. Table 3.10 gives values of the functions $f(\gamma)$ and $\phi(\gamma)$ for $\Delta = 0.09$. The graphs of these functions are given in Fig. 3.11.

Table 3.10. Values of $f(\gamma)$ and $\phi(\gamma)$ for $\Delta = 0.09$

$\gamma°$	$f(\gamma) \times 10$	$\phi(\gamma) \times 10$	$\gamma°$	$f(\gamma) \times 10$	$\phi(\gamma) \times 10$
0	9.70	9.10	50	10.1	8.77
10	9.71	9.09	60	10.2	8.63
20	9.75	9.05	70	10.4	8.50
30	9.82	8.98	80	10.5	8.39
40	9.92	8.89	90	10.6	8.35

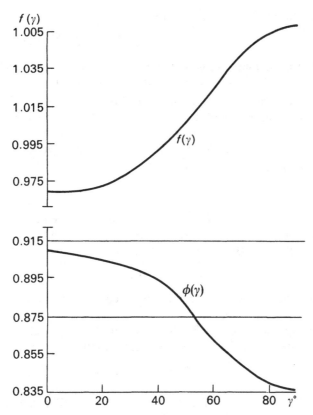

Fig. 3.11. Correction for the anisotropy of water molecules for the volume scattering function $f(\gamma)$ and for the degree of polarization of the scattered light $\phi(\gamma)$.

3.4.3. The effect of dissolved substances

Up till now we have dealt only with the scattering of light by pure water. We now turn to pure seawater, ie we take into account the dissolved inorganic salts. They have a two-fold effect: first, they change the refractive index, the density and β_T of water. This may be taken into account by substituting the values of these quantities for pure water by their values for salt water in Equation (3.26). This is the effect of an average (uniform) concentration of salts. However, apart from this, there is the effect of fluctuations in the concentration of the salts. These fluctuations take place independently of the fluctuations considered previously and simply must be added to them. This effect was also studied in [128]. For the sake of simplicity, we limit ourselves to a two-component solution which is known to give a good description of actual seawater (Chapter 2). Let the concentration of dissolved substance be C (in grams per cubic centimetre).

Turning to Equation (3.24) for $\overline{\Delta\varepsilon^2}$, we now obtain:

$$\overline{\Delta\varepsilon^2} = (\partial\varepsilon/\partial\rho)^2_{T,C}\overline{\Delta\rho^2} + (\partial\varepsilon/\partial T)^2_{\rho,C}\overline{\Delta T^2} + (\partial\varepsilon/\partial C)^2_{T,\rho}\overline{\Delta C^2} \qquad (3.40)$$

Thus, in solutions, the fluctuation $\Delta\varepsilon$ is the sum of density, temperature and concentration terms. As in the case of homogeneous liquids, the term with $\overline{\Delta T^2}$ is neglected and the first term is determined from Equation (3.33) or (3.35) (using the values characteristic of the solution and not of pure water). As far as the third term is concerned, in a similar manner to Equation (3.25), an equation is derived in thermodynamics for the fluctuation of the concentration of a solution $\overline{C^2}$ in a small volume v (see [84, p. 438]):

$$v\overline{\Delta C^2}/C^2 = kT/[C(\partial p/\partial T)_{T,\rho}] \qquad (3.41)$$

Here, p is the osmotic pressure of the dissolved substance (see [28, §19.2]). For a weak solution such as ocean water, the osmotic pressure p is determined by the van't Hoff Law for ideal solutions:

$$p = v(C/\mu)N_A kT \qquad (3.42)$$

Here, v is the dissociation coefficient for the dissolved substance; μ is its molecular mass; N_A is the Avogadro number.

According to Equation (3.40) we find for b_C the additional part of the scattering coefficient associated with fluctuations in concentration:

$$b_C(90°) = H(\mu/v)C; \quad H = (2\pi^2/\lambda_0^4)(n^2/N_A)(\partial n/\partial C)^2_{p,T}{}^* \qquad (3.43)$$

The main ions in seawater, Cl^-, Na^+, are isotropic. The fluctuations in concentration consequently should lead to an 'isotropic' equation for $b_C(\gamma)$ and b_C:

$$b_C(\gamma) = b_C(90°)(1 + \cos^2\gamma) \qquad (3.44)$$

and correspondingly:

$$b_C = (16\pi/3)b_C(90°) \qquad (3.44')$$

Salts dissolved in water should therefore lead to a reduction in the depolarization coefficient Δ of the solution as a whole, which is confirmed in practice [168]. However, it is less than expected from simple addition of water and salts. It is found that isotropic salt ions polarize the water molecules with their electrostatic field so that the complexes of water molecules and isotropic ions in solutions have a greater anisotropy than that of pure water molecules [174]. As a whole, it can be assumed that the anisotropy correction for a solution leads to the same Cabannes' factor as for the pure liquid [see Equation (3.38)]. Thus, for the total scattering coefficient of the solution b_2 we find:

$$b_2(90°) = [b_1(90°) + b_C(90°)](6 + 6\Delta^p)/(6 - 7\Delta^p) \qquad (3.45)$$

where Δ^p is the depolarization coefficient of the solution. Generally speaking,

---

* In [110], N_A is shown in the numerator. This is a printing error.

it is different from the coefficient Δ for pure water. However, by virtue of the above-mentioned opposing effects, the total effect depends on the nature of the ions. Thus, for an aqueous solution of magnesium chloride, investigated in [174], Δ^p is larger than Δ and linearly increases with the concentration of the solution (this is the effect of the strong field of Mg^{2+} ions). For the solution of NaCl in water, Δ^p and Δ agree according to the data of [174]. Therefore, in calculations on the scattering by pure ocean water, it is usually assumed that $\Delta^p = \Delta = 0.09$. This is the approach we adopt in Section 3.5. If it is found in future experiments that $\Delta^p \neq \Delta$, then in the calculations of Section 3.5 it will be necessary to make an insignificant correction, changing only the Cabannes' correction.

We now estimate the magnitude of both effects of dissolving salts. Lochet investigated the first effect [158]. It was found that for seawater, the factor:

$$F = (\beta_T/\beta_{T,0})[(\rho n \, \delta n/\delta \rho)/(\rho_0 n_0 \, \partial n_0/\partial \rho_0)]^2 \qquad (3.46)$$

is close to unity. The subscript zero here denotes quantities relating to pure water. This happens because as the concentration increases, the ratio $\beta_T/\beta_{T,0}$ decreases and the expression in square brackets increases by approximately the same amount. The second fluctuating effect is of greater importance. It follows from Equations (3.45) and (3.43) that the ratio μ of the total scattering coefficient of the solution $b_2(90°)$ to the scattering coefficient of pure water $b_1(90°)$ grows linearly with the concentration of the solution C:

$$\mu = b_2(90°)/b_1(90°) = 1 + \alpha C. \qquad (3.47)$$

According to the data of [164], for the solution NaCl in water, the coefficient $\alpha = 5.6$, for artificial seawater $\alpha = 6.9$, for natural water $\alpha = 8.0$ (the concentration C is expressed here in grams per gram of solution). It follows from Equation (3.47), for example, that for natural seawater with $C = 3.5 \times 10^{-2}$ g/g, $\mu = 1.28$ and for a solution of NaCl with the same concentration, $\mu = 1.20$. This means that for a solution of NaCl, the value of b will be 20% higher and for natural seawater will be 28% higher than for pure water. Thus, in order to obtain the total scattering coefficient for ocean water, we have to multiply b determined for pure water from Equation (3.38) by the value of μ determined from Equation (3.47).

3.4.4. Fine structure of the molecular scattering of light

Measurements show that in the molecular scattering of monochromatic light, additional lines appear in the scattered beam. This phenomenon is called the fine structure of molecular scattering of light. It is of key importance in understanding the mechanism of the scattering of light by liquids. Recently, the fine structure has been applied to the optics of the ocean. We briefly dwell on the physics of the phenomenon.

For the sake of simplicity, we just limit ourselves to the picture of the phenomenon associated with fluctuations in density. Already in his first study [128], A. Einstein represented the field of density fluctuation in a medium in the form of a superimposition of three-dimensional density waves:

$$\rho(x, y, z) = \bar{\rho} + \Delta\rho(x, y, z)$$

$$\Delta\rho(\mathbf{r}) = \sum_q \Delta\rho_q e^{iqr}$$

The wave vector $\mathbf{q}$ of these waves has the components q_x, q_y and q_z; they take values corresponding to standing waves which may exist inside the volume V. The scattering of light at such a lattice of density waves is very similar to the scattering of X-ray waves in crystals. It is well known that the observed scattering pattern corresponds to reflection from a three-dimensional lattice of condensation planes and scattering maxima will be observed in directions corresponding to the Bragg-Wulf condition:

$$2\Lambda \sin(\gamma/2) = \lambda$$

Here, Λ is the density wavelength; λ is the wavelength of light in the liquid; γ is the scattering angle (Fig. 3.12).

It is known that a standing wave may be regarded as a combination of two progressive waves of the same frequency propagating in opposite directions. In accordance with Doppler's principle, on reflection of light from density waves moving with the velocity of sound C, there is a change in the wavelength $\Delta\lambda$:

$$\Delta\lambda/\lambda = \pm 2(Cn/c) \sin(\gamma/2) \tag{3.48}$$

Because we are dealing with two waves one moving towards us and another away from us, we observe a doublet. This doublet is called the Mandelshtam-Brillouin (M-B) doublet because they theoretically predicted its existence. The scattering of light at thermal density waves was experimentally investigated by E.F. Gross. For water at 20°C, the shift in the doublet line from the centre for observation at a scattering angle $\gamma = 90°$ was equal to $\Delta\lambda = 4{\cdot}5 \times 10^{-3}$ nm for $\lambda = 435{\cdot}8$ nm. Calculation using Equation (3.48) gives $\Delta\lambda = 4 \times 10^{-3}$ nm.

Careful observations, however, showed that in liquids, as distinct from crystals, scattered light is actually a triplet—a central unshifted Rayleigh line is observed

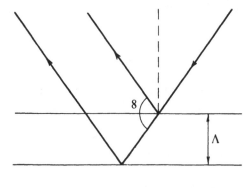

Fig. 3.12. Reflection of light from a density wave lattice.

together with the M-B doublet. The appearance of a central line is caused by the special characteristics of fluctuations in the density of the liquid. According to the 'hole' theory of Ya. I. Frenkel, there are two types of density fluctuation in the liquid: adiabatic fluctuations which are propagated at the velocity of sound and are responsible for the M-B doublet and isobaric fluctuations which are propagated very slowly and in practice do not give a Doppler shift. These isobaric fluctuations produce the central line.

In 1934, L. Landau and G. Plachek showed that the ratio of the intensity of the central line I_c to the total intensity of the shifted lines to $2I_{M-B}$ equals:

$$I_c(2I_{M-B}) = (c_p - c_v)/c_v = \gamma - 1 \tag{3.49}$$

For water, $\gamma = c_p/x_v = 1.01$. This means, that in water, the central line is fifty times weaker than the doublet line. Investigations into the scattering fine structure of water were carried out in [162]. The spectrum obtained in this study for $T = 20°C$ is shown in Fig. 3.13. The central line is very weak. On lowering the temperature, the value of γ for water decreases and at 4°C becomes equal to unity. According to (3.49), this means that for this $I_c = 0$, ie the central line disappears.

It is convenient to use the fine structure of molecular scattering in laser sounding for the determination of the vertical structure of suspended materials in the ocean. It is known that in methods of lidar sounding in turbid media, the main difficulty is the problem of isolating the vertical profile of the back scattering coefficient and the attenuation coefficient. Since the coefficient of molecular scattering is known, then taking the scattered light on the doublet lines, we obtain a simple method for determining the vertical profile $c(z)$. Since the central line and the doublet line are very close, the coefficient of $c(z)$ is the same for both. For the vertical profile, of the attenuation coefficient $c(z)$, we find: $c(z) = -(\frac{1}{2})[d\ln F(z)/dz]$. Here, $F(z)$ is equal to $I(z)z^2$, where $I(z)$ or $I(t)$ is the dependence of the intensity of the return signal on the sounding distance z (or on the time t, $t = 2z/c$). Assuming the velocity of sound $C \approx 1.5 \times 10^3$ m/second for a displacement $\Delta\lambda/\lambda$ for back scattering, we find: $|\Delta\lambda/\lambda| = n \times 10^{-5}$. In order to implement this procedure, it is necessary to displace the radiation receiver by $\Delta\lambda$ and cover the central line.

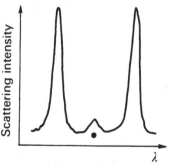

Fig. 3.13.   Fine structure of molecular scattering of light by water at 20°C.

The scattering of light by hydrosol particles in ocean water occurs without a change in frequency, ie the intensity of the central line increases. Therefore, by measuring the ratio (3.49) for actual seawater, it is possible to determine directly the ratio between hydrosol and molecular scattering in the ocean. N. Jerlov points out [25] that this method was successfully tested on samples of seawater, although measurements using it in the sea have not been described.

We add in conclusion that the study of the fine structure of the molecular scattering of light has in recent years become a powerful tool for studying the structure of liquids. A large number of interesting details may be found in the monographs of I.L. Fabelinsky [84] and M.F. Vuks [19].

3.5. Data on the molecular scattering of light by ocean water

3.5.1. Empirical formula for the refractive index of light in ocean water

The theory developed in Section 3.4 was used by a number of authors to calculate the coefficient of molecular scattering of light by seawater. The results of these calculations are given in [25, 28, 110, 164]. In these studies, the molecular scattering of light was considered as a small part of the total scattering and therefore its variation was not considered. In reality, the contribution of molecular scattering to medium and large angles is significant. Therefore, in [112, 113], an attempt was made to calculate the coefficient $b_m(\gamma)$ for the entire range of conditions under which seawater is found in the ocean. We discuss these calculations briefly.

The difficulty was that in order to determine $b_m(\gamma)$ from Equation (3.45), it is necessary to have the values of $n(\lambda)$ and in particular the derivatives of $n(\lambda)$ with respect to p, S and T over the entire range investigated. Generally speaking, the determination of $n(\lambda)$ is the object of experiment. Although quite a large number of such measurements have been made up till now [28, 74, 142] and a number of values of n have been determined with great accuracy, the available data are still inadequate. If we regard the refractive index n in the visible region as a function of p, S and T, then in particular there are few data on the dependence of n on the pressure p, ie on the depth z. There is a small amount of data in [28] and [142]. Therefore, based on the available publications, using the method of least squares, an empirical formula was devised in [113] which represents n uniformly as a function of four variables: the wave number $\tilde{\nu} = 1/\lambda,^*$ p, S and T over the entire range of variation of these variables encountered in the ocean. The range of variation of the arguments and the increment Δn for each argument are shown in Table 3.11.

---

* The use of $\tilde{\nu}$ and not λ is connected with the fact that the visible range is the range of normal dispersion of water in which formulae of the Cauchy-Sellmeier type [8] work well.

Table 3.11. Variation in the refractive index

$\tilde{v}\ \mu m^{-1}$	1/0.76–1/0.40	1/0.6328	1/0.5876	1/0.6563
$\rho \times 10^{-5}$ (Pa)	1	1–1000	1	1
$S(^0/oo)$	0	35	0–40	0
$T(^\circ C)$	20	15	15	0–40
$\Delta n \times 10^4$	130	140	75	−33

Table 3.11 gives the interval of variation of the argument and the values for which the other arguments are fixed inside a four-dimensional parallelepipid W. The total variation of n is equal to $\delta n = n(1/0.40, 1100, 40, 0) - (1/0.76, 1, 0, 40) = 300 \times 10^{-4}$. There are a large number of gaps inside W where the values of n are unknown. The use of linear or quadratic interpolation in these empty spaces is cumbersome in view of the four-dimensionality of W and does not give the necessary accuracy. The situatiton is particularly bad in the neighbourhood of the boundaries of the region W.

For gases, when the dependence of n on the thermodynamic variables (p, T) is studied, use is usually made of the fact that the variable $R = [(n^2 - 1)/(n^2 + 2)]$ $(1/\rho)$, the so-called specific refraction (ρ is the density of the medium) is a constant. This statement is called the Lorentz-Lorenz law [see Equation (3.27)]. Within the accuracy of the constant factor, the quantity R is proportional to the polarizability of the individual molecule [8]. The independence of R on p and T simply means that the polarizability of the individual molecules does not depend on pressure and temperature (of course, the polarizability and refraction depend on $\tilde{v}$). Thus, for gases, we can assume to a certain approximation that n is a function of only two variables $n = n(\tilde{v}, \rho)$ where $\rho = \rho(p, T)$.

In general, this situation also applies to liquids although the accuracy with which R is constant for them is appreciably less than for gases. Of course, the situation is much worse in cases when we are dealing not with pure liquids but with solutions such as seawater. If we compare the relative variation of the specific refraction R with the variation of n on varying p, S and T in the region W, then it is found that the use of R has some advantage for p and T [$\Delta R/R \approx (1/4)(\Delta n/n)$] but for the salinity S the variation of R is appreciably higher than for n. Therefore, in [113], an empirical formula was directly selected for $n = n(\tilde{v}, p, S, T)$ in terms of all four variables. The accuracy which the formula for $n = n(\tilde{v}, p, S, T)$ must have is determined by the necessity for calculating the derivatives. It is known that the derivative $\Delta y/\Delta x$ is extremely sensitive to errors in the function y. An estimate shows that in order to ensure the necessary accuracy of calculation of $b_m(\gamma)$, it is necessary to know the function $n(\tilde{v}, p, S, T)$ with an error of not more than 3×10^{-4}. Generally speaking, this is a low accuracy. It is only 1 % of the total variation of n in the region W. However, it must be ensured at all points in the region W. This makes it possible to obtain complete data on $b_m(\gamma)$.

An analysis of the initial experimental data showed that some authors publish absolute values n_{abs} and others publish relative values $n = n_{abs}/n_{air}$. The quantity $(n_{air} - 1) \approx 3 \times 10^{-4}$; its value varies as a function of v, p, T. Since 3×10^{-4} is the accuracy which we wish to ensure for n, it is necessary to make a distinction between n_{abs} and n. In [113], an interpolation formula is constructed for n.

Approximate formulae for the four variables were formed from combinations of simpler expressions describing the dependence of $n = n(\tilde{v}, p, S, T)$ on one or two variables. Here, it is convenient to replace the arguments which are true to scale by the dimensionless quantities $\tilde{v}$, $\tilde{p}$, $\tilde{S}$, $\tilde{T}$, each of which varies between -1 and $+1$. In order not to complicate the notation, we designate the dimensional and dimensionless wave number in the same way: $\tilde{v}$. Transformation of the region W into a four-dimensional cube W is accomplished by the formula:

$$\tilde{x} = \eta(x - a) - 1 \tag{3.50}$$

where x is any of the variables; $\eta = 2/(b - a)$; a and b are the lower and upper limits for the range of variation of x.

The investigated functions were represented in the form of polynomials in terms of one or two arguments with indeterminate coefficients. They have the form:

$$n = n_0 + \Delta n$$

$$\Delta n = \sum_{l=0} q_i \tilde{x}^i \tag{3.51}$$

$$\Delta n = \sum_{i=0} \sum_{j=0} q_{il} \tilde{x}^i \tilde{y}^j$$

where n_0 is a constant convenient for the calculations; Δn is the variable part of n; $\tilde{x}$ and $\tilde{y}$ are any arguments selected from the set $\tilde{v}$, $\tilde{p}$, $\tilde{S}$, $\tilde{T}$.

In defining the coefficients q_i, approximating functions were sought in the form of an expansion in orthogonal polynomials. This method is analogous to the method of least squares but is more stable in the calculation of errors. After simplifying terms in such expansions, we arrive at equations of the same type as Equation (3.51).

One-dimensional approximations were first considered for each of the arguments taken separately. It was assumed that:

$$n = n_0 + \Delta n, \, n_0 = 1.33250 \tag{3.52}$$

The results for one-dimensional approximations of Δn are given in Table 3.12. This gives information on the sections of W in the 1st to the 4th row, values of the parameters in Equation (3.50) for the given variable in the 5th and 6th rows, the coefficients q_l, q_k, q_j, q_i (the subscripts refer to the different variables) in the 8th to 11th row and the maximum deviations of the calculated values of Δn from the measured values in the 12th row. References for the studies from which the points were selected are given in [113]. The points were selected from the available experimental data in order to cover the area under study as uniformly as possible.

Table 3.12. Coefficients for one-dimensional approximations

Row number	2	3 $\Delta n(\tilde{v})$	4 $\Delta n(\tilde{p})$	5 $\Delta n(\tilde{S})$	6 $\Delta n(\tilde{T})$
		Column Number			
1	$1/\tilde{v}$ μm	–	0.6328	0.5876	0.5893
2	$p \times 10^{-5}$ Pa	1	–	1	1
3	$S^0/_{00}$	0	35	–	0
4	T°C	20	20	20	–
5	η	1.688 889	1.8198 -3	5 -2	5 -2
6	a	1.315 789	1	0	0
7	l, k, j, i	q_l	q_k	q_j	q_i
8	0	2.8437 -3	1.29702 -2	4.1856 -3	4.921 -4
9	1	6.8654 -3	7.0097 -3	3.6900 -3	-1.7633 -3
10	2	7.610 -4	-3.606 -4	–	-7.121 -4
11	3	2.081 -4	–	–	8.80 -5
12	$\Delta_{l,k,j,i}$	3·2 -6	2·2 -6	9·4 -6	7·1 -6

It follows from Table 3.12 that over one-dimensional sections, Δn may be approximated: in terms of $\tilde{S}$ by a polynomial of the first degree; in terms of $\tilde{p}$ by a polynomial of the second degree; in terms of $\tilde{T}$ and $\tilde{v}$ by polynomials of the third degree. The maximum deviations of such one-dimensional functions do not exceed $\pm 10^{-5}$.

In constructing the two-dimensional sections, linear combinations of the formulae used in the one-dimensional approximations were used with the corresponding arguments but with indefinite coefficients. Here, the structure of the dependences obtained in the one-dimensional cases was retained. The coefficients q_{ki} were determined in the same way as for the one-dimensional sections. It was found that this gives approximately the same accuracy as that of the initial data.

Two versions of a total approximating function are given in [113] by combining one-dimensional and two-dimensional sections. In combining one-dimensional sections, successively passing to two, three and four variables and using common points to the different sections of the W region, the simple equation was obtained:

$$n(\tilde{v}, \tilde{p}, \tilde{S}, \tilde{T}) = n(\tilde{v}) + n(\tilde{p}) + n(\tilde{S}) + n(\tilde{T}) - c. \qquad (3.53)$$

where $c = 4.00413$ and $n(\tilde{v})$, $n(\tilde{p})$, $n(\tilde{S})$ and $n(\tilde{T})$ are one-dimensional polynomials specified in Table 3.12.

An estimate of the accuracy of Equation (3.53) was carried out on a sample of values of n from 250 points selected at random from the available material. Based on the results of the verification, a graph was plotted of the distribution probability density of the deviations Δn (Fig. 3.14). The variable $\eta = \Delta N/(N\Delta n)$ is plotted along the ordinate where N is the total number of points taken for the verification and ΔN is the number of points where the deviation of n falls in the given interval Δn. It can be seen from Fig 3.14 that with a confidence level of 0.95. the absolute error of the calculated values of the refractive index is less than $\varepsilon = \pm 6 \times 10^{-4}$; the curve 1 is asymmetric with respect to the mode and the mode itself is shifted by 10^{-4} with respect to zero. Omitting terms in Equation (3.53) whose contribution has an insignificant effect on the accuracy, we find the simple expression for n:

$$n = n_0 + 10^{-4}[(69 + 8\tilde{v})\tilde{v} + (70 - 4\tilde{p})\tilde{p} + 37\tilde{S} - (18 + 7\tilde{T})\tilde{T}] \quad (3.54)$$

where $n_0 = 1.34636$.

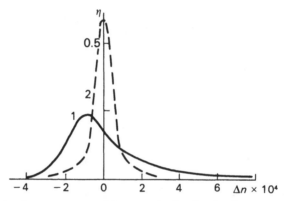

Fig. 3.14. Distribution of the probability density for the absolute error of the refractive index based on superimposition of : one-dimensional (1) and two-dimensional (2) sections.

The combination of two-dimensional sections was carried out using the same procedure as for one-dimensional sections. An expression was obtained similar to Equation (3.53) but containing a linear superimposition of two-dimensional sections. The final equation contains 40 empirical coefficients.

The accuracy of this equation was checked in the same way as Equation (3.53). The result is represented in the form of a distribution curve for the variable Δn (curve 2 in Fig. 3.14). For the complete equation, the curve of the error distribution is significantly narrower than for Equation (3.53); it is practically symmetrical with respect to zero. The absolute error when calculating n according to the complete equation does not exceed 2×10^{-4} with a probability of 0.95. This equation ensures the necessary accuracy for calculating $b(v, p, S, T)$.

3.5.2. Molecular scattering of light by ocean water

According to the general theory developed in Section 3.4, it is necessary to know β_T, Δ and $\rho\, \partial\varepsilon/\partial\rho$ for calculating the coefficient of molecular scattering of light in seawater. We have already discussed the quantities β_T and Δ. For a direct calculation of $\rho\, \partial\varepsilon/\partial\rho$, apart from the data on $n(\tilde{v}, p, S, T)$, it is necessary to have data on the derivative $\partial n/\partial\rho$. A convenient method of numerical differentiation was suggested in the C. Lanczos monograph [50]. Using this method with certain natural assumptions concerning the function $f(x)$, then in practice the calculation of its derivative reduces to integration of the tabulated function. We represent $f(x + t)$ in the form of a Taylor series in terms of t:

$$f(x + t) = f(x) + tf'(x) + \frac{t^2}{2} f''(x) + \frac{t^3}{6} f'''(x) + \ldots$$

Multiplying both sides of this equation by t and integrating over the interval $[-\varepsilon, +\varepsilon]$:

$$\frac{3}{2} \frac{1}{\varepsilon^3} \int_{-\varepsilon}^{+\varepsilon} tf(x + t)\, dt = f'(x) + \frac{\varepsilon^2}{10} f'''(x) + \ldots = f'(x) + \delta(x, \varepsilon)$$

Hence, we find:

$$f'(x) = \frac{3}{2\varepsilon^3} \int_{-\varepsilon}^{+\varepsilon} tf(x + t)\, dt \tag{3.55}$$

The error in this equation $\delta(x, t)$ is of the order of ε^2 and for a small third derivative it is not large. An accurate estimate of the error made in this way is given in [50].

We now turn to the equation for the molecular scattering coefficient [Equation (3.35)]. Using Equation (3.55) for the calculation of the derivative, it is possible to calculate the coefficient of molecular scattering and also the values for all other characteristics. Some of the results of these calculations are given in Table 3.13.

We now look at some consequences of these calculations.

Molecular scattering of light is usually considered as a relatively small quantity and a certain standard is adopted for it, varying little for different conditions of water in the world's oceans. This is approximately true in so far as we are dealing with the total scattering coefficient.

However, it is interesting that the role of molecular scattering becomes significant when we come to an analysis of the values of the scattering coefficient in a given direction $b(\gamma)$. The proportion of molecular scattering $\eta = b_m(\gamma)/b(\gamma)$ may reach quite appreciable values for medium and large scattering angles.

Furthermore, it increases sharply on passing to short waves. Different authors give somewhat different estimates for η. Thus, in [112] it is stated that in pure

Table 3.13. Molecular scattering coefficient of light by ocean water ($b \times 10^3 \, \mathrm{M}^{-1}$), $\lambda_0 = 546 \, \mathrm{nm}$

$T(°C)$	$S(‰)$					$p \times 10^{-5} \, \mathrm{Pa}$
	0	20	30	35	40	
0	1.45	1.74	1.88	1.95	2.02	1
10	1.48	1.80	1.95	2.03	2.10	
20	1.49	1.83	1.99	2.07	2.14	
40	1.50	1.86	2.04	2.13	2.22	
0	1.43	1.72	1.86	1.94	2.01	50
10	1.47	1.78	1.94	2.01	2.09	
20	1.48	1.81	1.98	2.06	2.13	
40	1.50	1.86	2.04	2.12	2.21	
0	1.40	1.70	1.85	1.92	1.99	100
10	1.43	1.77	1.92	2.00	2.07	
20	1.47	1.80	1.96	2.04	2.12	
40	1.49	1.85	2.03	2.12	2.21	
0	1.33	1.64	1.79	1.87	1.94	250
10	1.40	1.72	1.88	1.95	2.03	
20	1.43	1.77	1.93	2.01	2.09	
40	1.48	1.84	2.02	2.10	2.19	
0	1.24	1.56	1.72	1.79	1.87	500
10	1.32	1.65	1.81	1.89	1.96	
20	1.37	1.71	1.87	1.95	2.03	
40	1.45	1.80	1.98	2.07	2.15	
0	1.10	1.43	1.59	1.67	1.74	1000
10	1.19	1.52	1.68	1.76	1.83	
20	1.25	1.59	1.75	1.83	1.90	
40	1.36	1.71	1.89	1.97	2.06	

ocean waters, molecular scattering makes up only 7% of the total scattering but at angles larger than 90° it produces approximately two thirds of the total intensity. In particular, for an angle of 90° it is 0.70 and for an angle of 135° equal to 0.83 of the total scattering. In [110], it is maintained that at angles in the vicinity of 135° in pure waters, the value of η is more than 40% and that even in such turbid waters as in the coastal region of Peru it is not less than 5%. In [25] when $\gamma = 90°$, the value of η is estimated as $\sim 70\%$ for deep waters and $\sim 13\%$ for surface waters (for $\lambda_0 = 546 \, \mathrm{nm}$). In [28], it is maintained that the contribution of total molecular scattering at $\lambda_0 = 546 \, \mathrm{nm}$ is a little more than 10% for purest ocean waters and that it is usually less than 5% for waters in open regions and negligibly small for coastal waters. However, when $\gamma = 135°$, the proportion of $b_m(\gamma)$ to $b(\gamma)$ is more than 10 % even in turbid coastal waters and exceeds 80 % in transparent waters. Thus, all authors assume that molecular scattering determines light scattered sideways and backwards by a volume element. It is a key factor in remote sensing of the ocean using passive as well as active methods and for a number of other operations. In [112], a graph of $\eta(\gamma)$ is shown for different depths z_m. We reproduce it in Fig. 3.15. For $b(\gamma, z)$, data are taken from direct measurements in the summer of 1973 in the Indian Ocean in the vicinity of Madagascar in the zone of the southern tropical convergence. We see that the ratio η reaches a maximum not near the bottom but in intermediate layers of the ocean ($z = 2000$ m). The value of $\eta(\gamma, z)$ has a minimum not in the surface layer ($z = 0$ m) but somewhat lower above the pycnocline ($z = 100$ m). This is connected to the fact that in the vicinity of the pycnocline, the concentration of suspended material and correspondingly $b(\gamma)$ increases sharply.

It is not only important that molecular scattering has a high value in the range of medium and large scattering angles but it is also important that it varies appreciably in the ocean. The data on the variation of the molecular scattering coefficient are given in Table 3.14. For $\lambda_0 = 546$ nm, the average percentage of variation of b_m in the intervals Δp, ΔS and ΔT with respect to the average value of b_m is specified here.

It can be seen from Table 3.14 that as the pressure increases, b_m decreases (in Table 3.14 this is denoted by the minus sign in brackets after the number) and as T and S increase, the value of b_m increases. We see that the variation of b_m is not small and it should not be regarded as an invariant quantity.

Using the data on $b(v, p, S, T)$, it is possible to estimate the spatial variation of molecular scattering in the ocean. The macro variation of b over the ocean may be estimated from the latitude curves of T and S (Fig. 2.1). The results of such a calculation are shown in Fig. 3.16a. It can be seen that an appreciable latitudinal variation can be traced for molecular scattering. The characteristics

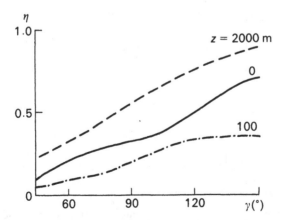

Fig. 3.15. Proportion of molecular scattering η of the total scattering for different depths z (m) and scattering angles $\gamma(°)$. $\lambda_0 = 546$ nm, $T = 25°C$, $S = 35°/oo$.

Table 3.14. Variations in the molecular scattering coefficient

$p \times 10^{-5}$ Pa	1–1000	1	1
S (°/∞)	35	0–40	35
T(°C)	5	5	0–40
Δb_m(%)	15(−)	35(+)	10(+)

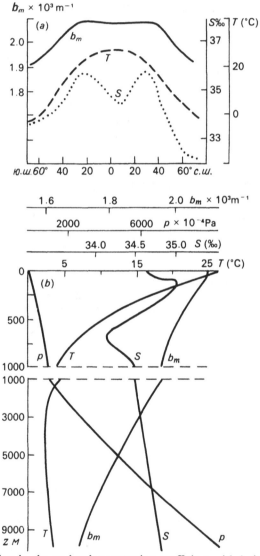

Fig. 3.16. Variation in the molecular scattering coefficient with latitude (average over the ocean) (a) and with depth (b) (August, region of the Phillipine Trench). $\lambda_0 = 546$ nm, $\Delta = 0.09$.

of molecular scattering in ocean water vary with respect to the average latitudinal values of b_m in the range $\pm 5\%$. As an illustration of the variation of b_m with depth, we can examine Fig. 3.16b where a profile is given of the variation of the molecular scattering coefficient with depth z in the region of the Phillipine Trench. The profile is constructed from the data on $p(z)$, $T(z)$ and $S(z)$ for August.

It can be seen from Fig. 3.16 that on varying the depth from 0 to 1000 m, b_m varies from 8% with respect to its average value in this interval of z and on changing the depth from 0 to 10 km, the coefficient b_m varies by 21% with respect to its average value over the entire depth considered. Such a large variation in the characteristic of molecular scattering with depth is the result of the fact that in this case both effects, an increase in pressure and a fall in temperature, act in the same direction.

Chapter 4

Scattering by suspended particles

4.1. Light scattering by a sphere. Scattering coefficients and the volume scattering function

4.1.1. Components of the scattered field. Partial waves

The scattering of light by suspended particles is the most important optical phenomenon occurring in the sea. As we saw in Chapter 2, particles in seawater have a complex irregular form. There is no rigorous theory of light scattering by such particles. The majority of authors in their theoretical analysis make use of the model of equivalent spheres, ie they consider the particles as homogeneous spheres.

An exact solution of the problem of diffraction of a plane electromagnetic wave by a homogeneous sphere was found independently by A. Love (1899) and G. Mie (1908). It is as follows (see [87, 90]).

We choose the origin of coordinates O at the centre of a sphere and position the axes as shown in Fig. 4.1. We denote the radius of the sphere as a. The

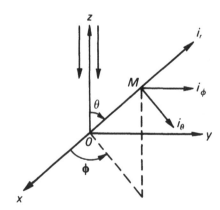

Fig. 4.1. Position of the axes for the diffraction at a sphere. Point O is the centre of the sphere, M is the point of observation. The unit vectors $\mathbf{i}_r$, $\mathbf{i}_\theta$, $\mathbf{i}_\phi$ are the unit vectors of the spherical system of coordinates (r, θ, ϕ) constructed at point M.

z-axis is directed at an infinitely distant source (a linearly polarized plane wave is propagated in the negative direction of the z-axis), the x-axis is the direction of electric vibrations and the y-axis is the direction of magnetic vibrations. The electric and magnetic fields in the incident wave at time $t = 0$ are expressed in the form:

$$E^0 = E_0 e^{ikz}; \; E_{ox} = E_0; \qquad E_{oy} = E_{oz} = 0;$$

$$H^0 = H_0 e^{ikl}; \; H_{oy} = -m_a E_0; \quad H_{ox} = H_{oz} = 0 \qquad (4.1)$$

where $k = m_a k_0$ is the wave vector of the incident radiation in the external medium (in which the sphere is immersed) and k_0 is the wave vector in vacuum, $|k_0| = \omega/c$.

The incident field excites an internal field in the sphere. At the surface of the sphere, the external and internal fields must satisfy the conditions of continuity of tangential components:

$$E_\theta^a = E_\theta^i; \; H_\theta^a = H_\theta^i;$$

$$E_\theta^a = E_\theta^i; \; H_\theta^a = H_\theta^i. \qquad (4.2)$$

The superscript i denotes variables relating to the internal area of the sphere and the superscript a to the external medium. It is known tht the system of boundary conditions [Equation (4.2)] cannot be satisfied unless it is assumed that in the external space, apart from the incident field $\mathbf{E}^0$, there is another additional, diffracted field $\mathbf{E}$. This is a perturbation which the sphere introduces to the external medium. Thus, the total external field is $\mathbf{E}^a = \mathbf{E}^0 + \mathbf{E}$. The spherical components of this field are also on the left-hand side of Equation (4.2).

It is known that Maxwell's equations for the components of the electric and magnetic fields can be reduced to the wave equations:

$$\Delta \mathbf{E} + k^2 \mathbf{E} = 0; \; \Delta \mathbf{H} + k^2 \mathbf{H} = 0 \qquad (4.3)$$

In contrast to Equation (3.1), here we use the usual designation k for the wave number. In writing Equations (4.1) and (4.3), we assumed that the electric and magnetic fields are pure periodic functions of time of the form $e^{i\omega t}$, where ω is the frequency of the field. This factor common to both sides of the equations has been omitted everywhere. The full Maxwell's equations and not Equation (4.3) make it possible of course to consider the field with any time dependence. In the theory of electric circuits our restriction considerably simplifies the problem and corresponds to the elimination of transient conditions for the switching on and switching off of currents. We limit ourselves only to forced, sinusoidal vibrations and assume that the transitional process has already died down. This approach cannot be used, for example, for the case in which the light wave front arrives at the particle and a complex non-steady state problem arises but it correctly describes the phenomenon when the front has moved far away beyond the particle and the steady state conditions have been established.

We represent an electromagnetic field as the superimposition of two types of vibrations. We call the first type electric and assume that in these vibrations the radial component of the magnetic field is equal to zero:

$$H_r = 0; \; E_r \neq 0 \tag{4.4}$$

The second type are magnetic vibrations. For these, the radial component of the electric field is equal to zero everywhere and the magnetic component is non-zero:

$$E_r = 0; \; H_r \neq 0 \tag{4.5}$$

With this formulation, the problem of finding six unknown functions $E(r)$ and $H(r)$ reduces to the problem of finding two auxiliary scalar functions: the electric and magnetic potentials $U_1(r)$ and $U_2(r)$. By means of these potentials, the components of the fields are calculated by simple differentiation. The idea of an electromagnetic field as the superimposition of two partial fields [Equations (4.4) and (4.5)] was already raised in the studies of H. Hertz and J. Rayleigh. In electrodynamics, it is proved that the solution obtained in this way is complete.

In a similar manner to the fields E and H, the potentials $U_1(r)$ and $U_2(r)$ are also solutions of the wave equation.

For the tangential components of the fields to be continuous [Equations (4.2)], it is necessary that the following four variables are continuous at the surface of the sphere:

$$m^2 r U_1, \; \partial(rU_1)/\partial r$$
$$rU_2, \quad \partial(rU_2)/\partial r \tag{4.6}$$

At infinity, the functions $U_1(r)$ and $U_2(r)$ must satisfy the radiation principle. The radiation principle is a consequence of the fact that we impose the steady state solution on the problem (for more details see [90, p. 17]). Thus, we have to find two independent solutions of the wave equation with independent boundary conditions. We obtain the solutions using the Fourier method. We seek a particular solution for the potentials in the form:

$$U = f(r)Y(\theta, \phi) \tag{4.7}$$

It is easy to prove that the radial functions $f(r)$ are Bessel functions with half-integral index and $Y(\theta, \phi)$ are spherical functions. Only functions of the first kind $J_{l+1/2}(x)$ may be used inside the sphere and only Hankel functions of the second kind $H^2_{l+1/2}(x)$ may be used outside the sphere since only they give waves diverging from the source of diffraction.

Thus, particular solutions of the wave equation inside the sphere will be of the type $[\psi_l(kr)/kr]Y_l(\theta, \phi)$ and outside the sphere of the type $[\zeta_l(kr)/(kr)]Y_l(\theta, \phi)$ where:

$$\psi_l(x) = \sqrt{\pi x/2}\, J_{l+1/2}(x),$$
$$\zeta_l(x) = \sqrt{\pi x/2}\, H^{(2)}_{l+1/2}(x), \tag{4.8}$$

$Y_l(\theta, \phi)$ is the lth spherical function of the angles θ and ϕ.

A complete solution should obviously be represented in the form of a superimposition of particular solutions with undetermined coefficients. These coefficients may be calculated from the boundary conditions given in Equations (4.2) and (4.6). The incident plane wave is also included in the boundary conditions. Therefore, its potentials U_1^0 and U_2^0 should also be represented in the form of a superimposition of waves of the type given in Equation (4.7). This may be easily done using the remarkable Heine-Rayleigh equation:

$$e^{ikr\cos\theta} = \sum_{l=0}^{\infty} i^l (2l+1) \frac{\psi_l(kr)}{kr} P_l(\cos\theta) \qquad (4.9)$$

This equation represents a plane wave in the form of an infinite sum of spherical waves emanating from the origin of coordinates. In Equation (4.9), $P_l(x)$ is the Legendre polynomial of the lth order of x. Using Equation (4.9), we find for U_1^0 and U_2^0 that:

$$U_1^0 = \frac{1}{k^2 r} \sum_{l=1}^{\infty} i^{l-1} \frac{2l+1}{l(l+1)} \psi_l(kr) P_l^{(1)}(\cos\theta)\cos\phi$$

$$U_2^0 = -\frac{m_a}{k^2 r} \sum_{l=1}^{\infty} i^{l-1} \frac{2l+1}{l(l+1)} \psi_l(kr) P_l^{(1)}(\cos\theta)\sin\phi \qquad (4.10)$$

where:

$$P_l^{(1)}(\cos\theta) = \sin\theta \cdot dP_l(\cos\theta)/(d\cos\theta) \qquad (4.11)$$

is the first associated Legendre polynomial. For two unknown amplitudes of the electric vibrations in the fields $\mathbf{E}^i$ and $\mathbf{E}$ (for the lth spherical harmonic) it is possible to write two linear equations from Equation (4.6). These equations are easy to solve and it is easy to produce explicit expressions for the potential U_1 and similarly for U_2. Having determined the expression for the potential, by simple differentiation we find expressions for the components of the electric and magnetic fields outside the sphere $\mathbf{E}$ as well as inside the sphere $\mathbf{E}^i$.

We limit ourselves to the external field. The spherical components of the electric field in the diffracted wave will be:

$$E_\phi = E_0 \frac{\sin\phi}{kr} e^{-ikr} E_1; \quad E_\theta = E_0 \frac{\cos\phi}{kr} e^{-ikr} E_2$$

$$E_1 = -\sum_{l=1}^{\infty} [c_l(\rho, m)Q_l(\cos\theta) + b_l(\rho, m) S_1(\cos\theta)]$$

$$E_2 = \sum_{l=1}^{\infty} [c_l(\rho, m)S_l(\cos\theta) + b_l(\rho, m)Q_l(\cos\theta)] \qquad (4.12)$$

$$E_r = E_0 \frac{i\cos\phi}{(kr)^2} e^{-ikr} \sum_{l=1}^{\infty} c_l(\rho, m)l(l+1)P_l^{(1)}(\cos\theta)$$

Here, the asymptotic expressions have been written for the components of the field for the case $kr \gg 1$, ie under the assumption that the distance to the point of observation $r \gg \lambda$. The following notation has also been introduced:

$$c_l(\rho, m) = i^{2l+1} \frac{2l+1}{l(l+1)} \frac{\psi_l(\rho)\psi_l'(m\rho) - m\psi_l'(\rho)\psi_l(m\rho)}{\zeta_l(\rho)\psi_l'(m\rho) - m\zeta_l'(\rho)\psi_l(m\rho)}$$

$$b_l(\rho, m) = i^{2l+3} \frac{2l+3}{l(l+1)} \frac{\psi_l'(\rho)\psi_l(m\rho) - m\psi_l(\rho)\psi_l'(m\rho)}{\zeta_l'(\rho)\psi_l(m\rho) - m\zeta_l(\rho)\psi_l'(m\rho)}$$

(4.13)

The variables $c_l(\rho, m)$ and $b_l(\rho, m)$ are expressed in terms of the cylindrical functions $\psi_l(x)$ and $\zeta_l(x)$ with half-integral index. They depend on the two parameters ρ and m:

$$\rho = ka = k_0 m_a a = (2\pi a/\lambda_0) m_a a$$

$$m = m_i/m_a$$

(4.14)

The quantity ρ is called the diffraction parameter or the Mie parameter; m is the relative complex refractive index equal to the ratio of the refractive index of the substance of the particle m_i to the refractive index of the external medium m_a in which the particle is immersed. The prime denotes a derivative with respect to the argument under the function sign.

The spatial distribution of the diffracted field is determined by the angular functions $Q_l(\cos \theta)$ and $S_l(\cos \theta)$:

$$Q_l(\cos \theta) = P_l^{(1)}(\cos \theta)/\sin \theta = dP_l(x)/dx, \; x = \cos \theta$$

$$S_l(\cos \theta) = dP_l^{(1)}(\cos \theta)/d\theta = -\sqrt{1-x^2} \; dP_l^{(1)}(x)/dx$$

(4.15)

Equations similar to Equations (4.12) to (4.15) may also be written for the components of the magnetic field **H** in the diffracted wave—H_ϕ, H_γ and H_r. We note that at a sufficiently large distance from the particle, in the so-called wave zone, we can ignore the components E_r and H_r in comparison with the components with respect to ϕ and γ. If we restore the previously omitted factor $e^{i\omega t}$, the full components of the electric field according to Equation (4.12) will have the form:

$$E_\phi = E_\phi^0 \; e^{i(\omega t - kr)}; \; E_\phi^0 = E_0[\sin \phi/(kr)]E_1$$

$$E_0 = E_\theta^0 \; e^{i(\omega t - kr)}; \; E_\theta^0 = E_0[\cos \phi/(kr)]E_2$$

(4.16)

Equation (4.16) represents the components of the field of scattered light. According to Equation (4.12), the fields $\mathbf{E}_\phi$ and $\mathbf{E}_\gamma$ represent infinite series in terms of elementary solutions of the wave equation. Individual terms of these series are called partial waves. The amplitudes of these waves are determined by the quantities $c_l(\rho, m)$ and $b_l(\rho, m)$; since they originate from the solutions of Equations (4.4) and (4.5), they are called the amplitudes of electric and magnetic vibrations, respectively.

The physical picture of the phenomenon is very simple. Under the influence of the incident field, the particle is polarized and becomes a source of an infinite number of secondary waves. These waves are propagated in all directions. Their amplitudes and phases are determined by the parameters ρ and m and by the order of the wave l. The angular structure of the wave is determined by the functions Q_l and S_l. It does not depend on ρ and m and is only determined by the order l. Observed scattered light is the result of interference of all coherent waves arriving at the receiver.

4.1.2. Convergence. Two methods of notation

Since the solution of the problem is obtained in the form of an infinite series, an important question arises concerning the convergence of these series. It can be proved that all series obtained are absolutely and uniformly convergent (for more details see [90]). However, in practice they converge slowly and the number of terms which have to be considered for summation in Equation (4.12) depends on ρ.

For small values of ρ, only the coefficients c_l and b_l with small numbers have any appreciable importance. The field associated with the coefficient c_1 corresponds to the so-called Rayleigh scattering. On increasing the dimensions of the particles, the intensity of excitation of the subsequent partial waves increases. All the partial electric and magnetic waves further away begin to come into play and in addition the intensities of the waves with small numbers oscillate. P. Debye showed that for large values of ρ the number of partial waves which has to be taken into account is of the order of ρ. The intensity of excitation of the following waves falls off rapidly since the number ρ determines the order at which we can truncate the infinite series in practice.

More detailed calculations using Equation (4.12) show that at large values of ρ, in order to obtain satisfactory accuracy, the value of l^*, to which it is necessary to summate in Equation (4.12) is approximately 1.2ρ. The graph in Fig. 4.2 shows the behaviour of the modulus of the amplitudes of electric vibrations $|c_l| = f(l)$ as l approaches ρ. It refers to the case $\rho = 60$ and $m = 1.3300$ (typical cloudy drop with radius $a = 6.25\,\mu m$ and $\lambda = 0.656\,\mu m$). The amplitudes of magnetic vibrations behave in a similar manner. Analytical analysis of the structure of the functions $c_l(l)$ and $b_l(l)$ is given in [90]. The curve in Fig. 4.2 gives a good illustration of the basic result of this analysis: when $l < \rho$, the amplitudes c_l and b_l behave as $(2l + 1)/[2l(l + 1)]$ with an accuracy up to the oscillations and when $l > \rho$ decay exponentially.

In several papers on the scattering of light by a particle, the system of coordinates of Fig. 4.1. (as used by Mie in the book by V.V. Shulejkin [127] and in many other studies) is replaced by another system in which the z-axis runs in the direction of propagation of the incident wave. It is obtained from Fig. 4.1 by rotating the entire system of axes by $180°$ about the y-axis (Fig. 4.3). In this way, the new x-axis points in the opposite direction. In this system, the electric field of the wave given in Equation (4.1) initially will be:

$$E_{ox} = -E_0; \quad E_{oy} = E_{oz} = 0 \tag{4.17}$$

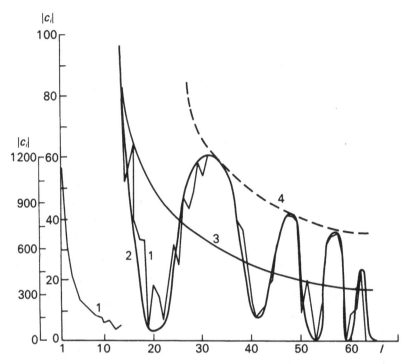

Fig. 4.2. Modulus of the amplitude of electric vibrations $|c_l|$ as a function of l for a particle with $\rho = 60$, $m = 1.3300$.

The graph shows values for $|c_l| \times 10^3$ in two scales; for $l > 13$, the scale on the ordinate axis is enlarged by a factor of 20. 1, exact curve; 2, curve with the 'roughness' smoothed out; 3, average curve (for $l \gg 1$, ie $\approx 1/l$); 4, fitted to the maxima (for $l \gg 1$, ie $\approx 2/l$).

The book by V.V. Shulejkin [127] adopts notations for the special functions of the problem (radial and angular) introduced by Mie. Mie's notation differs from those used earlier by Love (and following him Rayleigh, Ray and others) and from the modern notation which was introduced by Debye. The relationship between the different systems of coordinates and notations is given in [81, vol. 2; 90].

We introduce the scattering angle γ. This angle is calculated from the propagation direction of the light and not from the direction to the light source, as in the case of θ. It is related to θ by the equation:

$$\gamma = \pi - \theta \tag{4.18}$$

In addition, the azimuthal angle in the new system of coordinates, ϕ', is added to the previous angle ϕ:

$$\phi' = \pi - \phi \tag{4.19}$$

The detailed transformation of the fields on changing from the axes of Fig. 4.1

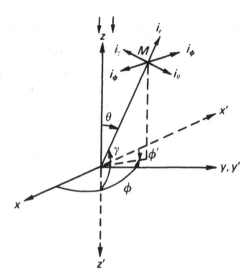

Fig. 4.3. The two systems of coordinates used in the problem of diffraction at a sphere (old system x, y, z and new system x', y', z') and the corresponding unit vectors of the spherical systems constructed at the point of observation $M(i_r, i_\theta, i_\phi, \text{and } i_r, i_y, i_\phi')$.

to the axes of Fig. 4.3 is given in [81]. It is shown here that in the new system, the fields E_ϕ, and E_y will be:

$$E_{\phi'} = E_0 \frac{\sin \phi'}{kr} e^{-ikr} E_1; \; E_y = -E_0 \frac{\cos \phi'}{kr} e^{-ikr} E_2 \tag{4.20}$$

$$E_1 = i \sum_{l=1}^{\infty} [c_l^* Q_l(\cos \gamma) + b_l^* S_l(\cos \gamma)] = i\tilde{S}_1(\gamma)$$

$$E_2 = i \sum_{l=1}^{\infty} [c_l^* S_l(\cos \gamma) + b_l^* Q_l(\cos \gamma)] = i\tilde{S}_2(\gamma) \tag{4.21}$$

Here, $\tilde{S}_1(\gamma)$ and $\tilde{S}_2(\gamma)$ denote the sums on the right-hand side of Equation (4.21). The quantities c_l^* and b_l^* are related to c_l and b_l as well as to a_l and b_l by the equations used by van de Hulst [87]:

$$c_l^* = i^{2l+3} c_l = a_l(2l + 1)/[l(l + 1)]$$
$$b_l^* = i^{2l+1} b_l = \bar{b}_l(2l + 1)/[l(l + 1)] \tag{4.22}$$

The quantity $\bar{b}_l$ denotes the amplitude b_l used in [87]. The bar is used to distinguish it from the value of b_l in Equation (4.13). Equations (4.17) to (4.22) establish the relationship between both notations used in the theory of scattering. In the following, provided it does not cause any confusion, the prime placed above the angle ϕ in the system of Fig. 4.3 is omitted.

4.1.3. Intensity of scattered light. Scattering coefficients and volume scattering function. Characteristics of scattering asymmetry

We now look at calculations of the intensity of scattered light. On the basis of Equation (4.1) for the fields E^0 and H^0, we obtain for the intensity of the incident radiation I_0 in the absence of absorption in the external medium:

$$I_0 = cm_a E_0^2/(8\pi) \tag{4.22'}$$

For the intensity of the ϕ component in the scattered light, we obtain $(E_\gamma = H_\phi = 0)$:

$$I_\phi = I_1 = I_0 i_1 a^2 \sin^2 \phi \ (r^2\rho^2)$$

$$i_1 = \left| \sum_{l=1}^{\infty} (c_l^* Q_l + b_l^* S_l) \right|^2 \tag{4.23}$$

In a similar manner for the γ component $(E_\phi = H_\gamma = 0)$, we have:

$$I_\gamma = I_2 = I_0 i_2 a^2 \cos^2 \phi/(r^2\rho^2)$$

$$i_2 = \left| \sum_{l=1}^{\infty} (c_l^* S_l + b_l^* Q_l) \right|^2 \tag{4.24}$$

Equations (4.23) and (4.24) are very important. They give the intensity of the scattered light for the case when the incident beam is linearly polarized. The total intensity of the scattered light:

$$I = I_1 + I_2 = I_0(a^2/r^2)(i_1 \sin^2 \phi + i_2 \cos^2 \phi)/\rho^2 \tag{4.25}$$

and the scattering coefficient b in the given direction based on one particle will be:

$$b(\gamma, \rho) = I(\gamma)r^2/I_0 = a^2(i_1 \sin^2 \phi + i_2 \cos^2 \phi)/\rho^2 \tag{4.26}$$

For the scattering of natural light, the azimuthal angle ϕ takes quite different values. The average value $\overline{\sin^2 \phi} = \overline{\cos^2 \phi} = 1/2$ and thus:

$$I_1 = I_0(a^2/r^2)[i_1/(2\rho^2)]; \ I_2 = I_0(a^2/r^2)[i_2/(2\rho^2)]$$

$$I = I_0(a^2/r^2)(i_1 + i_2)/(2\rho^2); \ b(\gamma, \rho) = a^2(i_1 + i_2)/(2\rho^2) \tag{4.27}$$

If there are N particles per unit volume, then relative to unit volume we find that:

$$b(\gamma, \rho) = N\pi a^2[i_1(\gamma, \rho)\sin^2 \phi + i_2(\gamma, \rho)\cos^2 \phi]/(\pi\rho^2)$$

$$b(\gamma, \rho) = N\pi a^2[i_1(\gamma, \rho) + i_2(\gamma, \rho)]/(2\pi\rho^2) \tag{4.28}$$

Knowing the equations for the fields, it is possible to obtain an expression for the scattering coefficient k_s, the attentuation coefficient k and the absorption coefficient k_a of light for a spherical particle. They may be found by integrating the expressions for the fields over the entire solid angle. Omitting the intermediate stages, we give the final result:

$$k_s = \pi a^2 K_s, \quad K_s = (2/\rho^2) \sum_{l=1}^{\infty} (|c_l^*|^2 + |b_l^*|^2)$$

$$k = \pi a^2 K, \quad K = (2/\rho^2)\operatorname{Re} \sum_{l=1}^{\infty} l(l+1)(c_l^* + b_l^*) \tag{4.29}$$

$$k_a = k - k_s, \quad k_a = \pi a^2 K_a, \quad K_a = K - K_s$$

Here, K_s, K and K_a are dimensionless cross-sections corresponding to scattering, attenuation and absorption, respectively. If there are N cm^{-3} particles per unit volume, then the scattering coefficient, attenuation coefficient, absorption coefficient and photon survival probability (denoted b, c, a and Λ in Chapter 1) will be given by:

scattering coefficient: $Nk_s = N\pi a^2 K_s$

attenuation coefficient: $Nk = N\pi a^2 K$

absorption coefficient: $Nk_a = N\pi a^2 K_a$
$$\tag{4.30}$$
photon survival probability: $K_s/(K_s + K_a)$

Using Equation (1.8) and Equation (4.26) for $b(\gamma)$, we find for the volume scattering function of linearly polarized and natural light, respectively that:

$$\beta(\gamma) = 4\pi b(\gamma)/b = [4/(\rho^2 K_s)](i_1 \sin^2 \phi + i_2 \cos^2 \phi)$$

$$\beta(\gamma) = [4/(\rho^2 K_s)][(i_1 + i_2)/2] \tag{4.31}$$

The volume scattering function $\beta(\gamma)$ gives the scattering probability distribution for different angles γ. If the polarization is ignored, then $\beta(\gamma)$ contains full information on the angular structure of the scattering. Often such complete information is not necessary and it is sufficient to know the degree of anisotropy of the scattering. This is described with the help of a few characteristics. The most common are the asymmetry coefficient and factor (Table 1.1).

In order to calculate b_b, it is not necessary to calculate first the volume scattering function and then integrate it over the hemisphere. Using the property of orthogonality of the Legendre polynomials, an equation was obtained in [92], directly expressing b_b in terms of c_l^* and b_l^* (relative to one particle):

$$b_b = (\tfrac{1}{2})b - \pi a^2(C'B' + C''B'' + \sum_{l>m} \operatorname{Re} \eta_{l,m}) \tag{4.32}$$

Where:

$$C = C' + iB' = -i_i \sum_{i=1}^{\infty} f_i c_i^*$$

$$B = B' + iB'' = i \sum_{l=1}^{\infty} f_l b_l^*$$

$$\eta_{l,m} = (c_l^* \tilde{c}_m^* + b_l^* \tilde{b}_m^*)x_{l,m} \tag{4.33}$$

where the auxiliary variables of the column vectors f_l and the matrix $x_{l,m}$ will be:

$$f_l = \left\{ \frac{(-1)^{k+1}(2k+1)!!}{2^k k!} 2k+1, \ x_{l,m} = \frac{(m+1)(l+1)}{l(l+1)-m(m+1)} (mf_m f_{l-1} - lf_l f_{m-1}) \right\}$$

(4.34)

In calculating β_0 in terms of the amplitudes c_l^* and b_l^*, it is convenient to know the values of the auxiliary variables in Equations (4.32) and (4.33) of the column vector f_l and the matrix $x_{l,m}$. Seven figure tables for these values up to $l = 100$ are published in [102]. We note that the values of f_l differ here from the analogous values in [102] by the factor $(-1)^l$. This means that when the tables of f_l and $x_{l,m}$ are used in the equations written here, it is necessary to change the sign of all numbers specified in [102]. The sign $\sim$ above c_m and b_m indicates a complex conjugate number.

For the asymmetry characteristic, the distribution moments are often also used:

$$v_k = (\tfrac{1}{2}) \int_0^\pi \beta(\gamma) \cos^k \gamma \sin \gamma \, d\gamma$$

(4.34′)

By virtue of the normalization condition given in Equation (1.8), $v_0 = 1$ for any $\beta(\lambda)$. The first moment is denoted in a particular manner:

$$v_1 = g(\rho, m) = (\tfrac{1}{2}) \int_0^\pi \beta(\gamma) \cos \gamma \sin \gamma \, d\gamma$$

(4.35)

The variable $g(\rho, m)$ is called the asymmetry factor of the scattering function. It is the average cosine of the scattering angle $g = \cos \gamma$. This variable characterizes the exchange of momentum between the electromagnetic field and the particles and is simply related to the cross-section of the light pressure of the light beam on the particle.

It is known that if we have a certain volume filled with an electromagnetic field with energy E, the momentum p, the velocity of light c and E are related by $p = E/c$.

If an obstacle is encountered in the paths of these waves, then, interacting with it, the field will transfer its momentum to it. The exchange of momentum is observed in the form of the force of light pressure. From considerations of the dimensionality, we obtain for this force:

$$F = [(I_0 \pi a^2)/c] K_p(\lambda, m, a)$$

where K_p is the dimensionless cross-section of the light pressure. A general equation for K_p was obtained by Debye. It takes the form:

$$K_p = K - q$$

$$q = -\frac{4}{\rho^2} \operatorname{Re} \sum_{l=1}^\infty \frac{l(l+1)}{2l+1} \left[\tilde{c}_l b_l + \frac{l(l+2)^2}{2l+3} (\tilde{c}_l c_{l+1} + \tilde{b}_l b_{l+1}) \right]$$

The total energy of the field 'captured' by the particles in unit time will be

$I_0 \pi a^2 K$. It is obvious that the total momentum transferred by the field to the particle will be $(I_0 \pi a^2/c)K$. The momentum, 'reradiated' by the particles, ie returned by the particles to the field will be $(I_0 \pi a^2/c) \times K_s \overline{\cos \gamma}$. Hence, we find that:

$$K_p = K - K_s \overline{\cos \gamma} = K - gK_s$$

Thus:

$$g = (K - K_p)/K_s = 1 + (K_a - K_p)/K_s \qquad (4.36)$$

In the absence of absorption, we obtain:

$$g = 1 - K_p/K \qquad (4.36')$$

There are a number of tables for the pressure cross-section K_p. Using Equation (4.36) they can be used to determine the asymmetry coefficient g. In particular, for models of the marine suspension, values of K_p are given in [123]. We examine them in detail below.

4.2 Polarized light scattering

4.2.1. Methods for describing the state of polarization. The Stokes parameters

In electrodynamics, it is proved that in the general case, a plane monochromatic wave will be elliptically polarized. In such a wave, the end of the electric field vector describes a helix wound around an elliptical cylinder. A natural method for describing the state of polarization of such a wave involves a geometric definition of the direction of vibrations of the electric field.

Let the wave be propagated in a direction perpendicular to the plane of Fig. 4.4, away from us (in the direction of the z-axis not shown in Fig. 4.4). We

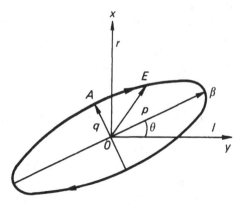

Fig. 4.4. Characteristics describing the state of polarization of a plane wave.

denote the angle between the major axis of the ellipse described by the end of
the electric vector and the y-axis by θ. The plane yOz is called the reference
plane. This plane is always selected so that the propagating ray lies in it.

We denote the unit vector along the x-axis by $\mathbf{r}$ and along the y-axis by $\mathbf{l}$.
The electric field of the propagating wave may be written in the form:

$$\mathbf{E} = E_r\mathbf{r} + E_l\mathbf{l} \tag{4.37}$$

where E_r and E_l are complex amplitudes:

$$E_r = a_r\, e^{i(\tau - \varepsilon_r)}$$

$$E_l = a_l\, e^{i(\tau - \varepsilon_l)}, \quad \tau = \omega t - kz \tag{4.37'}$$

where a_r and a_l are positive amplitudes of corresponding vibrations and ε_r
and ε_l are their phase shifts. In an optical experiment, the measured variables
will be a_r, a_l and $\delta = \varepsilon_l - \varepsilon_r$, the difference between the phases of the vibrations
about the axes l and r.

Together with the vectors $\mathbf{r}$ and $\mathbf{l}$, we introduce the unit vectors $\mathbf{p}$ and $\mathbf{q}$ along
the major and minor axes of the ellipse. The geometrical relationships character-
izing the elliptical wave can be seen from Fig. 4.4.

The relationship between the vectors l and r, p and q is given by the equations:

$$\mathbf{p} = \cos\theta \cdot \mathbf{1} + \sin\theta \cdot \mathbf{r}$$

$$\mathbf{q} = -\sin\theta \cdot \mathbf{1} + \cos\theta \cdot \mathbf{r}$$

We represent the expression for the field $\mathbf{E}$ in vector form as:

$$\mathbf{E} = a\cos\beta\sin\psi \cdot \mathbf{p} + a\sin\beta\cos\psi \cdot \mathbf{q}$$

$$\psi = \omega t - kz + \zeta \tag{4.38}$$

where ζ is a suitable phase angle; $|a\cos\beta|$ and $|a\sin\beta|$ are the lengths of the
major and minor axes of the ellipse, respectively; the quantity a^2 characterizes
the intensity of the beam. The ratio of the axes of the ellipse, the so-called degree
of ellipticity of the beam. is given by $|\tan\beta|$.

It can be seen from Equations (4.37) and (4.38) that it is possible to describe
an elliptically polarized wave with the help of different groups of four parameters.
These are ReE_r, ImE_r, ReE_l and ImE_l or a_r, a_l, ε_r, and ε_l or a, β, ζ and θ. These
parameters are different in nature. In the first case, they are four amplitudes of
the field, in the second case, they are two amplitudes and two angles and in the
third case they are one amplitude and three angles. Each of these groups can
be easily expressed in terms of another.

However, the application of any of the above-mentioned groups of parameters
to the characterization of the polarized state of the radiation is inconvenient
because these parameters are very different. In particular, considerable difficulties
arise on addition of the beams. In order to obtain the state of polarization of
the combined beam, it is necessary to construct a geometrical picture of the
position of the vibrations in the component beams and to add them vectorially,
taking into account the phase difference. Such calculations were made in early

studies on the polarization of underwater light fields. A complex calculation procedure was obtained. This cumbersome system can be avoided if the state of polarization of the light beam is described with the help of the Stokes parameters. These are four homogeneous quadratic variables which are defined by the following equations:

$$S_1 = I = E_l E_l^* + E_r E_r^*$$
$$S_2 = Q = E_l E_l^* - E_r E_r^*$$
$$S_3 = U = E_l E_r^* + E_r E_l^*$$
$$S_4 = W = i(E_l E_r^* - E_r E_l^*)$$

(4.39)

We use two notations found in the literature on hydrooptics for the Stokes parameters.

With an accuracy to a constant factor, all these variables have the dimensionality of the intensity of light, ie in a similar manner to the Pointing vector, they represent an energy flux across $1\,cm^2$. The Stokes parameters contain full information on the intensity, degree and shape of polarization of the beam which may be obtained with the help of an energy receiver. These are four real numbers which may be constructed from quadratic combinations of the two complex numbers E_r and E_l. For a plane wave all the Stokes parameters can be easily represented in terms of the geometrical characteristics:

$$S_1 = a_l^2 + a_r^2 = a^2$$
$$S_2 = a_l^2 - a_r^2 = a^2 \cos(2\beta)\cos(2\theta)$$
$$S_3 = 2a_l a_r \cos \delta = a^2 \cos(2\beta)\sin(2\theta)$$
$$S_4 = 2a_l a_r \sin \delta = a^2 \sin(2\beta)$$

(4.40)

In this case there are only three independent parameters since the following relationship applies:

$$S_1^2 = S_2^2 + S_3^2 + S_4^2$$

With the help of Equation (4.40) it is not difficult to use the Stokes parameters for a plane wave to determine the quantities defining the direction of vibrations of E in a system of axes l and r or p and q. The Stokes parameters form a complete set of variables observed in optics. Stokes formulated this statement in the form of the principle of optical equivalence: 'two beams with the same parameters are indistinguishable'. In [89] it is pointed out that this statement is inaccurate and that in certain interference experiments such beams can be distinguished. In any case, it is possible to do this with the help of apparatus in which the higher field correlators are measured, for example in the interferometers of Brown and Twiss (they are used to measure field correlators of the fourth order). The Stokes parameters are especially convenient for the description of incoherent beams. It may be proved that when such beams are added, the Stokes parameters of the combined beam are equal to the sums of the

Stokes parameters of the individual beams:

$$S_1 = \sum_i S_{1i}; \ S_2 = \sum_i S_{2i}; \ S_3 = \sum_i S_{3i}; \ S_4 = \sum_i S_{4i} \qquad (4.40')$$

The first Stokes parameter S_1 is the intensity of the light. The normalized Stokes parameters $S_i = S_i/S_1$ are often used since they are dimensionless quantities $(1, S_2/S_1, S_3/S_1, S_4 S_1)$. Their meaning can easily be seen from Equation (4.40). Thus, for example, if $a_r = 0$, we are dealing with horizontally polarized light and its normalized Stokes parameters are equal to $(1, 1, 0, 0)$. If $a_1 = a_r$ and $\delta = 0$, the light is polarized at an angle of $45°$ $(1, 0, 1, 0)$, etc. For non-polarized (natural) light, $S_2 = S_3 = S_4 = 0$. In general, it is not difficult to determine all the parameters of an actual beam with the help of an analyser and a quarter-wave plate. A summary of the values of Stokes parameters for different forms of polarized light may be found in [89]. It should be borne in mind that the directions of rotation used in [89] and in this text are opposite. We follow the definition adopted in the optics of turbid medium, ie we call polarization right-handed if the field is rotated clockwise for the observer looking in the direction of propagation, into the 'back' of the beam. See [8, p. 53] for a discussion of the confusion in the definition of right-handed rotation.

4.2.2. The scattering matrix

In scattering, there is a linear transformation of the vector parameter of the incident beam S_0 to the vector parameter of the scattered beam S:

$$S_i = F_{ik} S_{0k}$$

In the general case, the transformation matrix of any four-dimensional vector into another four-dimensional vector must contain 16 components. However, it may be proved that in the case of scattering by any (stable) particle, there are nine relationships between the 16 elements of the matrix so that the number of independent elements is only seven. Nevertheless, for scattering by a statistical ensemble of particles the matrix may contain all 16 independent components. If the particles or their ensemble possess symmetry elements, the number of independent parameters is reduced and some of the F_{ik} are equal to zero. This question was investigated by F. Perrin and developed by van de Hulst [87] and G.V. Rosenberg [75].

In the simplest case of a homogeneous sphere, 10 elements of the matrix F_{ik} are equal to zero. The remaining six depend only on three independent variables: i_1, i_2 and the phase shift $\delta = \sigma_1 - \sigma_2$ [see Equation (4.47)]. In order to show this, we select the first two Stokes vector parameters in the form:

$$I_l = E_l E_l^*; \ I_r = E_r E_r^* \qquad (4.41)$$

The variables I_l and I_r are related to I and Q in Equation (4.39) by the obvious relationships:

$$I = I_l + I_r; \ Q = I_l - I_r \qquad (4.41')$$

The fields E_r and E_l are complex components of the electric field vector of the wave perpendicular and parallel to the scattering plane.

In the notation of Equation (4.41), we have for the components of the Stokes vector parameter of the scattered field [see Equation (4.16)]:

$$I_r = E_1 E_1^* \sin^2 \phi \cdot E_0^2/(k^2 r^2)$$

$$I_l = E_2 E_2^* \cos^2 \phi \cdot E_0^2/(k^2 r^2)$$

$$U = -(E_1 E_2^* + E_2 E_1^*)\sin \phi \cdot E_0^2 \cos \phi/(k^2 r^2)$$

$$W = i(E_2 E_1^* - E_1 E_2)\sin \phi \cos \phi \cdot E_0^2/(k^2 r^2) \tag{4.42}$$

In the incident wave:

$$E_{0r} = -E_0 \sin \phi; \ E_{0l} = E_0 \cos \phi \tag{4.43}$$

and the components of its Stokes vector parameter will be:

$$I_{0_r} = \sin^2 \phi \cdot E_0^2; \ I_{0_l} = \cos^2 \phi \cdot E_0^2$$

$$U_0 = -2 \sin \phi \cos \phi \cdot E_0^2; \ W_0 = 0 \tag{4.44}$$

We introduce the notation [see Equation (4.21)]:

$$i_1 = E_1 E_1^*; \ i_2 = E_2 E_2^*$$

$$V = E_1 E_2^* + E_2 E_1^*; \ \eta = -i(E_1 E_2^* - E_2 E_1^*) \tag{4.45}$$

If, in addition, we introduce the phases σ_1 and σ_2 of the complex numbers E_1 and E_2, then:

$$E_1 = \sqrt{i_1} \ e^{i\sigma_1}; \ E_2 = \sqrt{i_2} \ e^{i\sigma_2} \tag{4.46}$$

It is not difficult to show that the components of the Stokes vector parameter in the scattered field are related to its components in the incident beam (these components are designated by subscript zero) by the following equations:

$$I_r = I_{0r} i_1/(k^2 r^2); \ I_l = I_{0l} i_2/(k^2 r^2)$$

$$U = (U_0 \cos \delta - W_0 \sin \delta)\sqrt{i_1 i_2}/(k^2 r^2) \tag{4.47}$$

$$W = (U_0 \sin \delta - W_0 \cos \delta)\sqrt{i_1 i_2}/(k^2 r^2)$$

where:

$$\delta = \sigma_1 - \sigma_2 \tag{4.48}$$

Thus, the scattering matrix F_{ik} takes the form:

$$F_{ik} = (1//k^2 r^2) \begin{pmatrix} i_1 & 0 & 0 & 0 \\ 0 & i_2 & 0 & 0 \\ 0 & 0 & \sqrt{i_1 i_2} \cos \delta & -\sqrt{i_1 i_2} \sin \delta \\ 0 & 0 & \sqrt{i_1 i_2} \sin \delta & -\sqrt{i_1 i_2} \cos \delta \end{pmatrix} \tag{4.49}$$

The variables i_1 and i_2 are defined by Equation (4.45) and the phase shift δ by Equation (4.48).

The components of the scattering matrix F_{ik} are expressed by the following equations:

$$F_{11} = i_1/(k^2 r^2); \; F_{12} = F_{13} = F_{11} = 0$$

$$F_{22} = i_2/(k^2 r^2); \; F_{21} = F_{23} = F_{21} = 0$$

$$F_{33} = V/(2k^2 r^2); \; F_{31} = -\eta/(2k^2 r^2); \; F_{31} = F_{32} = 0 \qquad (4.50)$$

$$F_{43} = \eta/(2k^2 r^2); \; F_{44} = -V/(2k^2 r^2); \; F_{41} = F_{42} = 0$$

Equation (4.50) makes it possible to calculate S of the scattered beam from S_0 of the incident beam and describes the transformation of the beams with any polarization state.

We present equations for the determination of the intensity I, degree of polarization p and degree of ellipticity q of the scattered light beam using the variables given in Equation (4.45):

For scattering of a linearly polarized beam:

$$I_r = I_0 i_1 \sin^2 \phi/(k^2 r^2)$$

$$I_l = I_0 i_2 \cos^2 \phi/(k^2 r^2)$$

$$I = I_0 (i_1 \sin^2 \phi + i_2 \cos^2 \phi)/(k^2 r^2)$$

$$p = (i_1 \sin^2 \phi - i_2 \cos^2 \phi)/(i_1 \sin^2 \phi + i_2 \cos^2 \phi)$$

$$q = (\sin \phi \cos \phi \cdot \eta)/(i_1 \sin^2 \phi + i_2 \cos^2 \phi) \qquad (4.51)$$

For the scattering of a natural beam:

$$I_r = I_0 i_1/(2k^2 r^2); \; I_l = I_0 i_2/(2k^2 r^2)$$

$$I = I_0 (i_1 + i_2)/(2k^2 r^2)$$

$$p = (i_1 - i_2)/(i_1 + i_2); \; q = 0 \qquad (4.52)$$

Of greatest practical interest is the calculation of the intensity and degree of polarization for the scattering of a natural beam. As can be seen, the calculation of these basic characteristics with the help of the variables given in Equation (4.45) is especially simple.

4.3.　Limiting cases for small and large particles

4.3.1. Small particles

The equations given above represent primary hydrooptical characteristics of the particles of marine suspensions as an infinite sum of terms corresponding to individual partial waves. The calculation of the fields using the exact equations is rather complicated. For example, for a typical biological particle with radius 10 μm and for light with a wavelength of $\lambda = 0.546\,\mu$m, the diffraction parameter $\rho = 115$, and 140 terms have to be retained in the sums. In addition, since the resulting field is the result of interference of 140 partial waves, it is necessary to calculate every wave with great accuracy since the individual terms almost completely cancel each other out. At present, such calculations are carried out on a computer and the results are presented in the form of tables of scattered fields or the radiance of scattered light.

In the theory of scattering of light by particles, great importance is attached to the limiting cases when the general formulae may be significantly simplified. These cases depend on the values of m and ρ, the basic parameters of the problem. A general analysis of the different limiting cases has been carried out by van de Hulst [87]. We restrict ourselves to those which are important for ocean optics.

We note that since Equation (4.12) represents a general solution of the problem of diffraction by a sphere, all limiting cases should be simply obtained from it. This can be done for the cases considered in this section.

Let us consider the limiting case of small particles whose relative refractive indices m are also small so that the two conditions are satisfied:

$$\text{(a) } \rho \ll 1$$
$$\text{(b) } |m\rho| \ll 1$$

(4.53)

If these two conditions are satisfied, we are dealing with the famous Rayleigh scattering which was first investigated by Rayleigh in 1871. It is very similar to the molecular scattering of light by pure water which was described in Chapter 3. In the optics of the sea, those conditions practically coincide. However, generally speaking, they are independent and if condition (b) is not satisfied then both the scattering function and all other variables describing the scattering will significantly differ from the Rayleigh variables. If both conditions are met, then expanding the cylindrical functions in Equation (4.13) for c_1 and b_1, in a series with respect to a small argument, we obtain (see [90]):

$$c_1 = \rho^3(m^2 - 1)/(m^2 + 2); \quad c_2 = \rho^5(m^2 - 1)/[18(2m^2 + 3]$$
$$b_1 = -\rho^3(m^2 - 1)/30$$

(4.54)

Since $\rho \ll 1$, then the coefficient c_1 is significantly larger than all other coefficients. Limiting ourselves only to this first term and noting that $Q_1(\theta) = 1$ and $S_1(\theta) = -\cos \theta$, we find that:

$$E_\phi = -[(\sin \phi \cdot E_0 \, e^{-ikr})/r]k^2a^3[m^2 - 1)/(m^2 + 2)]$$
$$E_\theta = (\cos \phi \cos \theta \cdot E_0 \, e^{-ikr}/r)k^2a^3[(m^2 - 1)/(m^2 + 2)]$$

(4.55)

This expression coincides with the equation for the field of a vibrating dipole at a large distance from it (in the wave zone). This dipole is excited in the sphere by a constant field (along the sphere) which varies periodically with the frequency ω. We used Equation (4.55) in Chapter 3.

As the size of the particle increases, apart from the electric dipole radiation it is necessary to take into account the dipole magnetic radiation, the quadrupole electrical radiation and subsequent partial waves. The distribution of the electric lines of force for the first eight partial waves was given by G. Mil. They are given in the books by M. Born and E. Wolf [8] and V.V. Shulejkin [127]. An elegant explanation of these lines of force patterns was given by Ya.I. Frenkel in 1946 and is presented in [90].

We now look at the volume scattering function. For linearly polarized light we obtain from Equation (4.55):

$$I(\gamma, \phi) = I_0 \frac{16\pi^4 a^6}{\lambda^4 r^2} \left| \frac{m^2 - 1}{m^2 + 2} \right|^2 (1 - \sin^2 \gamma \cos^2 \phi) \qquad (4.56)$$

The volume scattering function of natural light may be obtained from Equation (4.56) by averaging over ϕ. Performing this averaging we obtain:

$$I_1 = I_0 \frac{9\pi^2 v^2}{\lambda^4 r^2} \left| \frac{m^2 - 1}{m^2 + 2} \right|^2 \frac{1}{2}$$

$$I_2 = I_0 \frac{9\pi^2 v^2}{\lambda^4 r^2} \left| \frac{m^2 - 1}{m^2 + 2} \right|^2 \frac{\cos^2 \gamma}{2}$$

$$(4.57)$$

The complete function will be:

$$I(\gamma) = I_0 \frac{9\pi^2 v^2}{\lambda^4 r^2} \left| \frac{m^2 - 1}{m^2 + 2} \right|^2 \frac{1 + \cos \gamma}{2} \qquad (4.58)$$

Hence for the scattering coefficient in the given direction $b(\gamma)$ and for the total b for Rayleigh scattering relative to one particle, we have:

$$b(\gamma) = b(90°)(1 + \cos^2 \gamma)$$

$$b(90°) = \frac{8\pi^4 a^6}{\lambda^4} \left| \frac{m^2 - 1}{m^2 + 2} \right|^2 \qquad (4.59)$$

$$b = \frac{16\pi}{3} b(90°)$$

Equation (4.59) relates to the scattering coefficient of a small transparent particle. We note that it follows that for Rayleigh scattering, one of the fundamental laws of colloidal optics, Beer's law for attenuation, is not satisfied. According to this law, the attenuation coefficient $c \sim \mu$ where μ is the density of the attenuating substance. It can be seen from Equation (4.59) that $c \sim \mu^2/N$, where N is the number of particles in 1 cm³. If the absorption coefficient of the substance of the particle $k \neq 0$, then absorption of light by the particles will take place, in addition to scattering. The total attenuation coefficient c for real values of ρ will be equal to:

$$c = a^* + b = \pi a^2 4\rho Im \left[-q \left(1 + \frac{3}{5} \rho^2 q - i \frac{2\rho^3}{3} q + \cdots \right) \right.$$

$$\left. + \frac{\rho^2}{30} q(m^2 + 2) - \frac{\rho^2}{6} q \left(\frac{m^2 + 2}{2m^2 + 3} \right) + \cdots \right]^* ; \quad q = \frac{m^2 - 1}{m^2 + 2} \qquad 4.60)$$

*This a^* is the absorption coefficient.

If the particle does not absorb ($k = 0$), then $c = b$ and Equation (4.60) turns into Equation (4.59). If k is not very small, then:

$$c = \pi a^2 \cdot 4\rho \, Im \frac{1 - m^2}{m^2 + 2} = \frac{36\pi nk}{|m^2 + 2|^2} \frac{v}{\lambda} \tag{4.61}$$

In this case, the attenuation ceofficient $c \gg b$ and is practically equal to the absorption coefficient a^*. It is proportional to the volume v of the particle and not to v^2 as is the scattering coefficient. In the presence of absorption, Beer's law will be satisfied. In addition, the dependence of c on λ will also change: instead of λ^{-4} we will have λ^{-1}. We are speaking of that part of the dependence which is associated with diffraction. The total spectral variation of b and c also contains the functions $n(\lambda)$ and $k(\lambda)$ in accordance with Equations (4.59) and (4.61).

4.3.2. Large particles

We now turn to another limiting case, that in which $\rho \gg 1$. We have already noted that in this case all amplitudes of the partial waves up to $l \sim \rho$ are approximately of the same order and consequently they all have to be considered. This means, for example, that for phytoplankton with $a = 100 \, \mu m$, it is necessary to retain 1000 terms in the summation of Equation (4.12). The calculations are only possible using special computers and are very complex. Obviously, it is necessary to find some approximate methods for this limiting case. Large particles, in terms of what interests us, do not differ from small spherical lenses which have been calculated using the methods of geometrical optics. Naturally, attempts have been made to apply these methods to the analysis of the scattering of light by spheres with $\rho \gg 1$.

The basis for this procedure has been given in [90, chapter 7, paragraph 2]. It has been rigorously shown there that if the diffraction equations are taken to the limit $\rho = \infty$, then the intensity of light scattered by the 'internal' beam is exactly that which results from the equations of geometrical optics.

The internal beam undergoes multiple reflection and refraction. Here, beams emanating from the particles will be observed as scattered light. Their intensity may be calculated from the Fresnel equations. The entire pattern of reflected and refracted beams has been examined in detail in [90]. We note some of the important features. We refer to Fig. 4.5. Let a parallel beam of light fall on a spherical particle in direction Ox. We take any one of the planes passing through the centre of the particle and containing direction x. We will call all rays emanating from the particle derivatives of the ray which initially fell on the particle. The reflected ray will be the first derivative, the ray refracted twice, the second derivative, and so on. It can easily be seen that the angle of rotation $\gamma^{(k)}$ of the derivative ray of the kth order is defined by the equation:

$$\gamma^{(k_1)} = (k - 2)\pi + 2[\phi - (k - 1)\psi], \; k \geqslant 2 \tag{4.62}$$

Here, ϕ and ψ are the angles of incidence and refraction of the rays, respectively.

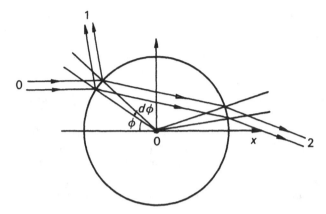

Fig. 4.5. Formation of derivative rays when a plane beam of rays falls on a drop. The numbers 0, 1 and 2 refer to the incident ray and the first and second derivative rays, respectively.

For the reflected beam, the scattering angle is:

$$\gamma^{(1)} = \pi - 2\phi \tag{4.63}$$

Its intensity at an infinitely distant observation sphere of radius R will be:

$$I_{s,p}^{(1)} = \frac{a^2}{4R^2} I^0 r_{s,p} \tag{4.64}$$

We distinguish rays with different states of polarization as s and p rays. These rays have different coefficients of reflection $r_{s,p}$ and refraction $d_{s,p}$. According to Equation (4.64), the intensity of the reflected light is determined by a geometric factor, the factor $a^2/4$ (by 'turning out' the initially plane beam) and by the reflection coefficient $r_{s,p}$. All derivative rays with $k > 1$ undergo $(k - 2)$ reflections and two refractions. Let dF^0 be the flux initially incident on a strip of the particle with angular width $d\phi$ and area $ds_0 = a^2 \sin 2\phi \cdot d\phi$. The proportion of it which emanates from the particle in rays of the kth order will be:

$$dF^0 \cdot r^{k-2} d^2 = I_0 \cos \phi \cdot ds_0 \cdot r^{k-2} d^2$$

The spatial distribution of the intensity of the kth derivative rays is determined by the equation:

$$I^{(k)} = dF^0 \cdot r^{k-2} d^2 / dS^{(k)}$$

Here, $dS^{(k)}$ is the area of the part of the observation sphere receiving the kth derivative rays initially incident on the strip $d\phi$. Obviously, we have:

$$dS^{(k)} = 2\pi R^2 \sin \gamma^{(k)} \, d\gamma^{(k)}$$

$$d\gamma^{(k)} = \frac{d\gamma^{(k)}}{d\phi} \, d\phi = 2\left(1 - \frac{k-1}{n} \frac{\cos \phi}{\cos \psi}\right) d\phi \tag{4.64'}$$

For the intensity $I^{(k)}$, we obtain:

$$I^{(k)} = \frac{dF^0 \cdot r^{k-2} \, d^2}{2\pi \sin \gamma^{(k)} |d\gamma^{(k)}/d\phi| \, d\phi} = \frac{a^2}{4R^2} I^0 r^{k-2} \, d^2 |\Theta^{(k)}| \qquad (4.65)$$

The quantity $4/(a^2 |\Theta^{(k)}|)$ geometrically defines the 'turn out' of the rays in the kth order. The total intensity of light in [90] is determined by the sum of the intensity of all derivative rays scattered in a given direction. This requires some explanation. All rays are coherent and one should therefore add the fields and not the intensities In fact, since we are considering the case $\rho = \infty$, due to the different degrees of 'roughness' the interference term between fields of different orders disappears and the addition of intensities is correct. It is the same reason why we do not observe interference in thick plates in optics.

We introduce the quantities $\beta_{s,p}^{(k)}$ defined by the equation:

$$\beta_{s,p}^{(k)} = \frac{I_{s,p}^{(k)}}{I_{s,p}^0} \frac{4R^2}{a^2} = r_{s,p}^{(k)} \, d_{s,p}^2 |\Theta^{(k)}(\gamma)| \qquad (4.66)$$

With the help of these quantities, the geometric optical coefficient $b_{go}(\gamma)$, the function $\beta_{go}(\gamma)$ and the degree of polarization of the scattered light $p_{go}(\gamma)$ for irradiation by a natural light beam will be:

$$\beta_{s,p}(\gamma) = \sum_{k=1}^{\infty} \beta_{s,p}^{(k)}(\gamma); \; \beta_{go}(\gamma) = \tfrac{1}{2}[\beta_s(\gamma) + \beta_p(\gamma)]$$

$$b_{go}(\gamma) = \frac{a^2}{4} \beta_{go}(\gamma); \; p_{go}(\gamma) = \frac{\beta_s(\gamma) - \beta_p(\gamma)}{\beta_s(\gamma) + \beta_p(\gamma)} \qquad (4.67)$$

The values of $\beta_{go}(\gamma)$ and $p_{go}(\gamma)$ for typical particles of marine suspensions are given in Table 4.1 (only two orders have been considered). The graph of the scattering function and the degree of polarization for typical particles of marine suspensions are shown in Fig. 4.6. It can be seen from Table 4.1 that when $k = 0$, ie the particle is transparent and m is close to 1, the geometrical optical function is elongated forwards strongly.

At some scattering angles, $\beta(\gamma)$ becomes infinite. These are directions for which the denominator in $\Theta^{(k)}$ becomes zero. These directions are determined from Equation (4.64'):

$$1 - \frac{k-1}{n} \frac{\cos \phi}{\cos \psi} = 0$$

which may be written in the form:

$$\cos \phi = \sqrt{\frac{n^2 - 1}{k^2 - 2k}} \qquad (4.68)$$

It agrees with the known condition determining the position of rainbows. The scattering angle $\gamma_{\text{rain}}^{(k)}$ at which a rainbow in the kth order is observed and the degree of polarization of light in rainbows p_k can be easily determined from

Table 4.1. The geometric optical scattering function $\beta_{go}(\gamma)$ and the degree of polarization of scattered light $p_{go}(\gamma)$ in the the geometric optical approximation

	m = 1·02		m = 1·05		m = 1·15	
γ	$\beta_{go}(\gamma)$	$P_{go}(\gamma)$	$\beta_{go}(\gamma)$	$p_{go}(\gamma)$	$\beta_{go}(\gamma)$	$f_{go}(\gamma)$
0	2.451 3	0	4.417 2	0	5.921 1	0
5	7.558 1	0	1.364 2	−1.833 −3	4.904 1	−1.121 −3
10	5.756	−6.167 −3	2.488 1	−6.833 −3	3.060 1	−5.882 −3
15	7.832 −1	−1.258 −2	6.168	−1.394 −2	1.673 1	−1.375 −2
20	6.755 −2	4.382 −2	1.816	−2.065 −2	8.866	−2.386 −2
30	1.169 −2	1.804 −1	1.035 −1	5.135 −2	2.528	−4.866 −2
40	4.216 −3	3.003 −1	1.802 −2	3.510 −3	6.586 −1	−6.051 −2
60	8.568 −4	6.363 −1	4.359 −3	6.843 −1	2.312 −2	8.042 −1
80	2.781 −4	9.554 −1	1.553 −3	9.721 −1	1.027 −2	9.981 −1
100	1.434 −4	9.307 −1	8.455 −4	9.137 −1	6.403 −3	8.570 −1
120	1.084 −4	5.872 −1	6.530 −4	5.697 −1	5.242 −3	5.173 −1
140	9.964 −5	2.546 −1	6.042 −4	2.469 −1	4.930 −3	2.238 −1
160	9.808 −5	6.041 −2	5.954 −4	5.912 −2	4.871 −3	5.379 −2
180	9.806 −5	0	5.949 −4	0	4.867 −3	0

Equations (4.68) and (4.62) and from Equation (4.69):

$$p_k = \frac{A-1}{A+2}; \quad A(m,\, k) = \left(\frac{k-1+m^2}{mk}\right)^4 \left(\frac{(k-2)(k-1+m^2)}{k(k-1-m^2)}\right)^{2k-4} \quad (4.69)$$

Table 4.2 gives exact values of $\phi_{rain}^{(k)}$, $\gamma_{rain}^{(k)}$ and p_k relating to particles of marine suspensions. Approximate formulae may be obtained for these quantities if it is assumed that $m = 1 + \alpha$ and $\alpha \ll 1$. From Equations (4.68) and (4.62) we easily find that:

$$\phi_{rain}^{(k)} = \pi/2 - \sqrt{2\alpha}/\sqrt{k^2 - 2k}; \quad \gamma_{rain}^{(k)} = 2\sqrt{2\alpha}\sqrt{k^2 - 2k} \quad (4.70)$$

In order to estimate the accuracy of Equation (4.70), we note that when $m = 1·02$ and $k = 3, 4$ and 5, we find $39°42'$, $64°48'$ and $88°45'$ for $\gamma_{rain}^{(k)}$ using Equation (4.70). These values are close to the values given in Table 4.2. Approximately, $\gamma_{rain}^{(k)} = 2\sqrt{2\alpha}(k - 1)$, ie as the order k increases, the angle at which the rainbow is observed is rotated by $2\sqrt{2\alpha}$. At small α, the quantity $A(n, k)$ and the degree of polarization of light in the rainbow p_k will be $A = 1 + 4\alpha$ and $p_k = (A - 1)/(A + 1) = 2\alpha$. According to this formula, the value of p_k for the three cases considered in Table 4.2 is equal to 4, 10 and 30%. These figures are close to those given in Table 4.2. Thus, for particles of marine suspensions with m tending to unity, the angles of incidence of rays giving a rainbow approach $\pi/2$ and the direction of the first rainbow tends to $\gamma = 0$. In contrast to the case $m = 1.3300$, considered in [90], the degree of polarization of light in rainbows is small and approximately the same in all orders.

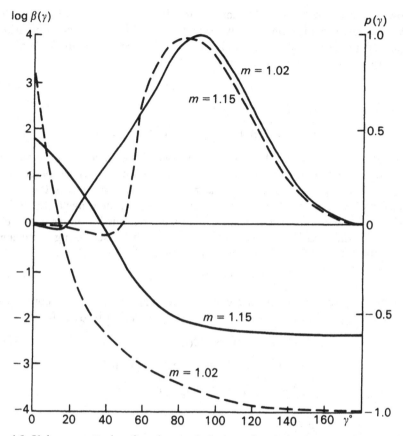

Fig. 4.6. Volume scattering function and degree of polarization of scattered light according to geometrical optics for typical particles of marine suspensions.

Table 4.2.

k	$\varphi_{rain}^{(k)}$			$\gamma_{rain}^{(k)}$			$\rho_k(\%)$		
	1.02	1.05	1.15	1.02	1.05	1.15	1.02	1.05	1.15
3	83°20′	79°21′	70°52′	39°17′	61°08′	100°47′	4.19	11.1	39.4
4	85 56	83 29	78 26	64 14	100 13	171 29	4.16	10.9	36.5
5	89 51	85 15	81 35	90 37	137 20	131 34	4.15	10.8	35.8

We now return to the 'external' beam diffracted by the contour of the particle. In [90, chapter 6, paragraph 2], it was shown that if in the exact summation of Equation (4.12) we let $\rho \to \infty$ and $\gamma \to 0$ so that $\rho\gamma = z$ is finite, then the intensity of the scattered light will be:

$$I_d(\gamma) = I_0 a^2 J_1^2(\rho\gamma)/\gamma^2 \qquad (4.71)$$

Thus, we obtain for the coefficient $b_d(\gamma)$:

$$b_d(\gamma) = a^2\rho^2 F(z)/4; \; F(z) = [2J_1(z)/z]^2, \; z = \rho\gamma \qquad (4.72)$$

where $J_1(z)$ is the first Bessel function. The graph of the function $F(z)$ is shown in Fig. 4.7. We see that as z increases, the function $F(z)$ rapidly decays, practically all its values being concentrated in the region $z < 6$. This means that light diffracted by the surface of the particle is concentrated in the range of scattering angles $\gamma < \gamma^0 = 6/\rho$. In particular, for the above-mentioned biological particle with $\rho = 115$, the angle $\gamma^\circ = 0.052 \text{ rad} = 3°$. The forward scattering coefficient $b_d(0)$ and the total scattering coefficient b_d will be:

$$b_d(0) = \pi^2 a^4/\lambda^2; \; b_d = \pi a^2 \qquad (4.73)$$

It is often necessary to know the energy flux $\Phi(z)$ scattered in a cone with divergence angle γ. In [90, p. 149], the following equation is given for it:

$$\Phi(z) = \pi a^2[1 - J_0^2(z) - J_1^2(z)] \qquad (4.74)$$

The total scattering coefficient of a large particle is given by:

$$b(\gamma) = a^2[\beta(\gamma) + \rho^2 F(z)]/4; \; b = 2\pi a^2 \qquad (4.75)$$

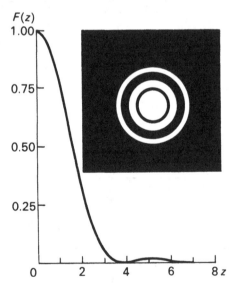

Fig. 4.7. Distribution of the intensity of diffraction by a sphere. A photograph is shown of the diffracted beam in the top right-hand inset.

We can add that as has been shown in [119], the area of application of the asymptotic Equation (4.72) may be significantly widened if a correction factor is introduced to it: $q(\rho, m) = K^2(\rho, m)/4$, where $K(\rho, m)$ is the attenuation cross-section of the particle. The corrected equation will take the form:

$$I_d(\gamma) = I_0[a^2/(4\gamma^2)]K^2(\rho, m)J_1^2(\rho\gamma) \qquad (4.76)$$

For $m = 1.33$, it is already satisfactory for $\rho \geqslant 5$, whereas the limit of applicability of Equation (4.72) is four to six times higher. It is very important that the correction q introduced in [119] does not depend on the angle γ, and it does not change the angular structure of the scattered beam.

4.4 'Soft' particles

4.4.1 The Rayleigh-Hans-Debye (RHD) approximation

In this section, we consider two other limiting cases, in which the problem of the scattering of light by a homogeneous sphere has a relatively simple solution. Both these cases related to particles with m close to 1. Such particles are called optically 'soft'. Although Equation (4.12) is suitable for any m, including $m \approx 1$, we shall, in both cases, construct solutions based on the direct physical picture of the phenomenon. This approach seems to be more elegant than the formal derivation from Equation (4.12).

The Rayleigh-Hans-Debye (RHD) approximation was suggested by J. Rayleigh. It was used by R. Hans, P. Debye and other authors. In order to distinguish it from the usual Rayleigh scattering (Section 4.3.1), it is called the Rayleigh-Hans approximation or the Rayleigh-Debye approximation. Milton Kerker considers the second name to be more correct ([154], p. 414). We call it the RHD approximation.

Rayleigh, Hans and Debye used the approach considered below in the first approximation. Subsequently, the method may be developed with the help of the integro-differential equations indicated in [90, chapter 8]. Recently, Acquista [131] used this equation to solve the problem in the second approximation. In quantum mechanics, this method is called Born's approximation.

At the beginning of this chapter, we presented an exact solution of the problem of diffraction by a sphere. The basic feature of this method is the use of curvilinear coordinates in which the boundary conditions of the problem can be written in the simplest manner. This approach, however, cannot be extended to irregularly shaped particles. Another drawback, as we have seen, is the cumbersome final equations, even for the case of small particles.

In this section, we present another method of analysing the problem of scattering of light by a particle, the method of integro-differential equations. This general method can be applied to soft particles with small and medium dimensions. This case is important for impurities in the ocean and in the atmosphere. It makes it possible to obtain a simple solution for spheres and for non-spherical particles. This method was developed in [90, chapter 8] for small optically soft particles. In Chapter 5, we use it to analyse light scattering by an ensemble of soft particles of any shape.

In [90], it was shown that in the steady state case, Maxwell's equations may be reduced to the following integro-differential equation for the electric field:

$$\mathbf{E}_{ef}(\mathbf{r}) = \mathbf{E}_0 \, e^{-ikr} + \text{rot rot} \int_V \alpha \mathbf{E}_{ef}(\mathbf{r}') \frac{e^{-ikR}}{R} \, d\mathbf{r}' - \frac{8\pi}{3} \alpha \mathbf{E}_{ef}(\mathbf{r}) \qquad (4.77)$$

Here, $\mathbf{E}_0 \, e^{-ikr}$ is the electric field of the incident wave; $\mathbf{E}_{ef}(\mathbf{r})$ is the effective electric field over the entire space including the scattering volume V; $R = |\mathbf{r} - \mathbf{r}'|$ is the distance between the observation point and the scattering volume point; α is the specific polarizability of the substance of the particles, $\alpha = (3/4\pi)|(m^2 - 1)/(m^2 + 2)|$.

If we introduce the operation rot rot under the integral sign, then according to the rule of differentiation of this type of integral (see [8, appendix V]), Equation (4.77) takes the form:

$$\mathbf{E}_{ef}(\mathbf{r}) = \mathbf{E}_0 \, e^{-ikr} + \int_V \text{rot rot } \alpha \mathbf{E}_{ef}(\mathbf{r}') \frac{e^{-ikR}}{R} \, d\mathbf{r}' \qquad (4.78)$$

For an optically soft particle, the parameter α is small. We also assume that the phase shift δ at the particle is small, ie the following inequality is satisfied:

$$|m - 1| \ll 1; \quad \delta = 2\rho|m - 1| < 1 \qquad (4.79)$$

We shall construct the solution inside and outside the particle in the form of a series in α. Then, in order to calculate the first approximation in the scattered field we must take the zeroth approximation in the internal field $\mathbf{E}_{ef} = \mathbf{E}_0 \, e^{-ikr}$. When the point of observation falls within the Fraunhofer diffraction zone, it is not difficult to show (see [90]) that the field E_{sc} scattered by particles with volume v, will be:

$$\mathbf{E}_p = \alpha[4\pi^2/\lambda^2 R]v \, e^{-ikr} \mathbf{E}_{0\perp} R(q) \qquad (4.79')$$

where $\mathbf{E}_{0\perp} = \mathbf{E}_0 - \mathbf{R}_0(\mathbf{R}_0, \mathbf{E}_0); |\mathbf{R}_0| = 1; R(q) = (3/q^3)(\sin q - q \cos q); q = (4\pi a/\lambda) \sin (\gamma/2)$.

Passing from the field to the intensity, we find that on scattering linearly polarized light, the intensity of the scattered light will be:

$$I = I_0 |a|^2 (16\pi^4/\lambda^4)v^2(1 - \sin^2 \gamma \cos^2 \phi)R^2(q) \qquad (4.80)$$

This equation differs from Equation (4.56) for Rayleigh scattering only by the factor $R^2(q)$. It is not difficult to explain the appearance of this factor. We are dealing with particles whose refractive indices are close to the refractive index of the medium and whose dimensions are not too large. In addition, each small volume element of the particle produces Rayleigh scattering. Since the radiation of different elements is coherent, then the waves scattered by them interfere with each other and partly extinguish each other due to the difference in position of the elements in space. As a result of this, the factor $R(\gamma)$, the so-called internal interference factor, appears in the equation for the field. Using Equation (4.80), we find that for $b(\gamma)$ and b:

$$b(\gamma) = \sigma(90°)(1 + \cos^2 \gamma)R^2(\gamma)/R^2(90°)$$

$$b(90°) = (8\pi^4 a^6/\lambda^4)|(m^2 - 1)/(m^2 + 2)|^2 R^2(90°) \qquad (4.81)$$

$$b = (9a^2/16\pi)|(m^2 - 1)/(m^2 + 2)|^2 F^{(1)}(z), \quad z = 4\rho = 8\pi a\lambda$$

$$F^{(1)}(z) = (2\pi^2/z^2)[z^4/4 + 5z^2 + (4z^2 - 16)(ciz - \ln z - c) -$$

$$- 2z \sin z + 14(\cos z - 1)]$$

$$ci\, z = -\int_z^\infty (\cos t/t)\, dt \qquad 4.82)$$

Tables of the functions $R(q)$ and $F^{(1)}(z)$ are given in [90, p. 239]. A comparison of the calculation data obtained from Equations (4.81) and (4.82), with the exact values taken from [123], shows that for $m = 1.15$, the discrepancy between the approximate and exact values does not exceed 15–20% over the entire range of scattering angles if $\rho \leqslant 2$; for $m = 1.02$, Equations (4.81) and (4.82) are satisfactory up till $\rho = 5$.

4.4.2. The physical optics approximation

Van de Hulst [87] calls this case anomalous diffraction. In it light perturbation is calculated as the interference of rays with phase differences superimposed on them. This is the usual approximation used to solve diffraction problems in physical optics.

We retain the conditions $|m - 1| \ll 1$ although we will assume that the phase shift δ may take any value. Consequently, ρ must be large. We are therefore dealing with large optically soft particles. It may be assumed for such particles that a ray passing through them does not change its direction, ie does not undergo refraction and reflection (the coefficient of reflection tends to zero as $m \to 1$). The perturbation produced by the particle only results in a change in phase of the wave. The ray passing through the particle at a point whose polar angle is equal to τ (Fig. 4.8) acquires the phase difference $\delta \sin \tau$, compared with the unperturbed field and the field acquires the factor $e^{-i\delta\sin\tau}$.

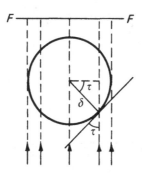

Fig. 4.8. Passage of a plane wave through a "soft" sphere.

In order to find the field in the Fraunhofer approximation, we must sum all parallel rays passing through the particle. This corresponds to the field at the focus of a converging lens. Thus, for a field scattered forward, we find that (see [13, p. 830]):

$$E(0) = i\rho^2 \int_0^{\pi/2} (1 - e^{-i\delta \sin \tau}) \sin \tau \cos \tau d\tau = i\rho^2 R(i\delta)$$

$$R(x) = \tfrac{1}{2} + e^{-x}/x + (e^{-x} - 1)/x^2$$

(4.83)

For any waves, the so-called optical theorem applies, which relates the field scattered directly forwards and the attenuation coefficient k (see [90, p. 139]). In our notation:

$$k = \pi a^2 (4/\rho^2) Im[E(0)]$$

(4.84)

Hence, we find for the attenuation cross-section $K(\rho, m)$ of a large soft particle:

$$K(\delta) = 2 - (4/\delta) \sin \delta + (4/\delta^2)(1 - \cos \delta)$$

(4.85)

The similarity principle applies: the attenuation cross-section does not depend separately on ρ and m but on the quantity δ (see [90, p 218]).

The graph of the function $K(\delta)$ is shown in Fig. 4.9. The extreme values of the function are given in Table 4.3. They are defined as the roots of the equation:

$$(2\delta^2)(\cos \delta - 1) + (2/\delta)\sin \delta - \cos \delta = 0$$

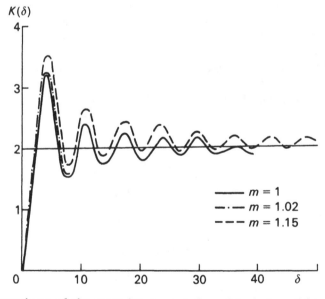

Fig. 4.9. Dependence of the scattering cross-section of typical particles of marine suspensions on the phase shift δ. The limiting curve for $m = 1$ is also shown.

Table 4.3. Extreme values of the attenuation curve $\delta*$ and the value of $K(\delta*)$ at the extreme points

Number of the extreme value	$\delta*$	$K(\delta*)$	Number of the extreme value	$\delta*$	$K(\delta*)$
1	4.086	3.173	5	17.16	2.246
2	7.634	1.542	6	20.33	1.814
3	10.79	2.404	7	23.52	2.178
4	14.00	1.734			

At large values of δ, they coincide approximately with the roots of the equation $\tan \delta = \delta/2$, ie they are close to the points $\delta = 2k\pi \pm \pi/2$. For extreme values 6 and 7 we find: 20·42 and 23·56 which are close to the values given in Table 4.3. For small and large values of δ, we have:

$$k = \begin{cases} (\delta^2/2)(1 - \delta^2/18 + \cdots), & \delta \ll 1 \\ 2[1 - (2/\delta)\sin \delta], & \delta \gg 1 \end{cases}$$

For large particles, $(\rho \to \infty)K \to 2$. This is a correct value. For small particles $K \sim \rho^2 \sim \lambda^{-2}$ which is incorrect since K must be $\sim \rho^4$ (Rayleigh dependence). This error is a consequence of the fact that Equation (4.85) was derived for large particles.

One more important result follows from Equation (4.83) relating to the structure of the field at small angles. We have already pointed out that the correction $q(\rho, m)$ to Equation (4.72) does not depend on the angle γ. From Equation (4.83), we find for the intensity of light scattered at an angle of $\gamma = 0$ that:

$$I(\rho, 0, m) = I_0(a^2\rho^2/4) \cdot 4|R(i\delta)|^2$$

Comparing with Equation (4.73) for $b_d(0)$, we find that for a large soft particle:

$$b_d(\gamma) = (a^2/4)\rho^2 F(z)P(\delta)$$

$$P(\delta) = 4|R(i\delta)|^2 \tag{4.86}$$

For the correction factor $P(\delta)$, we can write:

$$P(\delta) = 1 + (4/\delta^2)[1 + (1 + 2/\delta^2)(1 - \cos \delta - \delta \sin \delta)] \tag{4.87}$$

When $\delta \gg 1$ $P(\delta) \approx 1 - (4/\delta)\sin \delta$ so that Equation (4.86) turns into Equation (4.72). The values of the function $P(\delta)$ are given in Table 4.4.

Comparison of the calculated data with the exact scattering values for particles with refractive indices 1.02 and 1.15 presented in [123] shows that Equation (4.86) provides correct values of scattering in the region of small angles for particles of practically any size. The range of angles in which this equation holds true is the same as for Equation (4.72). An appreciable increase in errors begins on approaching the first zero of the function $J_1(z)$ which lies at $z = \rho\gamma = 3.83$.

In this approximation, a useful equation can also be obtained for the

Table 4.4 Correction factor $P(\delta)$ in the equation for diffraction by soft particles

δ	$P(\delta)$	δ	$P(\delta)$	δ	$P(\delta)$	δ	$P(\delta)$
0.1	4.44 -3	6.0	1.31	24.0	1.16	60.0	1.02
0.5	1.10 -1	7.0	7.12 -1	26.0	8.90 -1	70.0	9.57 -1
1.0	4.19 -1	8.0	6.26 -1	28.0	9.76 -1	80.0	1.05
1.5	8.74 -1	9.0	9.58 -1	30.0	1.14	90.0	9.61 -1
2.0	1.40	10.0	1.34	32.0	9.35 -1	100.0	1.02
2.5	1.90	12.0	1.21	34.0	9.48 -1	110.0	1.00
3.0	2.30	14.0	7.52 -1	36.0	1.12	120.0	9.81 -1
3.5	2.53	16.0	1.12	38.0	9.72 -1	130.0	1.03
4.0	2.57	18.0	1.18	40.0	9.32 -1	140.0	9.72 -1
4.5	2.42	20.0	8.32 -1	45.0	9.27 -1	150.0	1.02
5.0	2.11	22.0	1.03	50.0	1.02	160.0	9.95 -1

absorption cross-section. We will assume that in the case of large weakly refracting particles, each ray passing through the particle is attenuated by the factor $e^{-4\rho k\cos\psi}$,* ie the proportion of the ray's energy absorbed in the particle is equal to $1 - e^{-4\rho k\cos\psi}$ (see [90, p. 135]). Summing all rays passing through the particle, as was done for the calculation of $E(0)$, we find for K_a of large particles that:

$$K_a = 1 - \phi(v), \quad v = 4\rho k$$

$$\phi(v) = (2/v)[1/v - e^{-v}(1 + 1/v)]$$

$$(4.88)$$

The values of the function $K_a(v)$ are given in Table 5.1. Equation (4.88) correctly describes the picture for particles of any size. In fact, for the limiting cases we find that:

$$K_a = 2v/3 - v^2/4 = 8\rho k/3 - 4\rho^2 k^2, \quad v \ll 1 \qquad (4.89)$$

$$K_a = 1 - 2/v^2 = 1 - 1/(8\rho^2 k^2), \quad v \gg 1 \qquad (4.90)$$

Comparing Equation (4.89) when $v \ll 1$ and Equation (4.61), we can see that they coincide since in this case $24n/|m^2 + 2|^2$ is close to 8/3. This is very successful and enables Equation (4.88), derived for large particles, to be used for small particles also. At large v at the limit, the cross-section $K_a = 1$, ie all the light beam which intersects the contour of the particle is absorbed. In order to estimate the accuracy of the equation, we compared the exact values K_a given in [81] with the approximate values K_a^{app} calculated from Equation (4.88). The relative error $\Delta = (K_a^{app} - K_a)/K_a$ is given in Table 4.5.

We can see that in nearly all cases, $\Delta < 0$, ie $K_a > K_a^{app}$. The error is at a maximum for moderate ρ and decreases towards the edges. For $k = 0.1$ when $\rho \gtrsim 8 - 20$, the exact values of K_a are even larger than unity. Due to diffraction, the particle absorbs more radiation than falls on its contour.

The van de Hulst Equation (4.85) was obtained above by using the expression

---

*If ψ is the refraction angle, then the path length of the ray in the particle is $2a\cos\psi$ and the proportion of intensity absorbed in the particle will be $2a\cos\psi(4\pi/\lambda)k = 4\rho k\cos\psi$.

Table 4.5 Relative error $\Delta(\%)$ for the calculation of K_a from Equation (4.88)

ρ	$n = 1.02$			$n = 1.16$		
	k					
	10^{-3}	10^{-2}	10^{-1}	10^{-3}	10^{-2}	10^{-1}
0.5	0.2	0.2	−6.1	2.9	3.1	−3.1
1	−0.7	−0.2	−9.1	−4.3	−4.1	−12
2	−1.5	0.3	−10.1	−14	−11	−19
5	−2.3	−3.4	−8.8	−22	−22	−21
12	−1.8	−3.8	−6.7	−22	−23	−18
20	3.0	−3.8	−5.4	−23	−23	−14
40	−0.5	−3.7	−3.2	−24	−20	−7.3
60	−0.3	−3.7	−1.8	−23	−15	−4.0
100	0.2	−3.6	−0.3	−18	−9·3	−1.2

for the field $E(0)$ which was found in the physical optics approximation. Meanwhile, it would be important to perform a rigorous formal analysis of the exact expression [Equation (4.29)] of $K_s(\rho, m)$ for soft particles. This problem was successfully solved by A.Ya. Perel'man. By intensive analysis of the Mie series, he calculated a general expression of $K_s(\rho, m)$ for soft particles [166, 167]. The equation he obtained is somewhat more complicated than the simple expression of Equations (4.81) and (4.85). However, it turns into these equations exactly at small and large δ, respectively, and provides the best agreement with the Mie-based calculations for all intermediate values of δ.

4.5. Tabulation of the Mie formulae: tables and their analysis

The analytic formulae given above considerably simplify the analysis. In particular, they are useful for studying different limiting cases. However, it is necessary to use tables for exact calculations. For marine suspensions, such tables are given in [123] and [160].

The tables of [123] form the fifth volume of the series *Light scattering tables* published by Gidrometoizdat in Leningrad [81]. The first volume of this series contains the values of the angular functions $Q_l(\gamma)$ and $S_l(\gamma)$ for l from 1 to 400, for $\gamma = 0(0.1)5(1)90°$ and $\gamma = 135(0.2)140°$. They may be used for calculating scattering on calculators and also for checking programs on computers. The second and fourth volumes relate to atmospheric optics. They contain the results of the calculations of the fields and intensities for scattering of visible and infrared radiation by drops of water suspended in the air. There are three sections in the third volume. In the first and third sections, data are presented on the cross-sections K, K_s and K_d. In the first section for $n = 1.02(0.02)1.60$; $k = 0$;

10^{-4}; 5×10^{-4}; 10^{-3}; 5×10^{-3}; 10^{-2} and further 0.20(0.02)0.50 and ρ in the range 0.5–100. In the third section, values are given for $n < 1$ (including bubbles of air in water). In the second section, the coefficients k, k_s and k_d are given for a drop of water in air in the spectral range 0.5–40 μm.

The fifth volume [123] contains the optical characteristics of models for seawater. It consists of two sections. We shall look at the first section, which considers monodispersive models. It gives the cross-sections of scattering and radiation pressure for spherical particles with diffraction parameters $\rho = 0.3$–200 as well as the values of the elements of the scattering matrix for these particles for scattering angles γ^0: 0, 0.2, 0.5, 1, 2, 3, 5(2.5)175, 177, 178, 179, 179.5, 179.8, 180. Calculations are presented for two values of the relative refractive index $m = 1.02$ (biological suspensions) and $m = 1.15$ (terrigenous suspensions).

We make two observations concerning the use of the tables [123] (these observations equally apply to all volumes of the series); (1) the numbers in the tables are given in standard form. This means, for example, that the notation $3.641 - 1$ represents the number $3.641 \times 10^{-1} = 0.3641$; (2) all values in the tables are based on unit sol concentration $N = 1$. If we wish to obtain the scattering coefficient in m^{-1}, then if the common convenient units are used: microns for the radii of the particles and cm^{-3} for the concentration N, it is necessary to introduce a scale factor 10^{-10}. Of course, the same applies to the scattering coefficient in the given direction $b(\gamma)$ and to the elements of the scattering matrix which in this case are obtained in m^{-1} sr^{-1}.

Let us examine an actual example. Suppose it is necessary to determine the optical characteristics of the monodispersive system of terrigenous particles ($m = 1.15$) with radius $a = 0.2\,\mu$m and concentration $N = 10^{10}\,cm^{-3}$ for $\lambda_0 = 546$ nm. The diffraction parameter ρ for such spheres suspended in water will be:

$$\rho = 2\pi an/\lambda_0 = 6.283 \times 0.2 \times 1.334/0.546 = 3.07$$

Here, we have taken n of ocean water equal to 1.334. By linear interpolation between the values of K_s for $\rho = 3$ and $\rho = 3.5$ given in [123, p. 12], we find that $K_s = 0.384$. Thus from Equation (4.30), taking into account the scale factor, we find that $10^{10} \times 3.14 \times 0.04 \times 0.384 \times 10^{-10} = 4.82 \times 10^{-2}\,m^{-1}$.

We now examine some results of the calculations. Scattering cross-sections $K_s(\rho)$ are shown in Fig. 4.10 for typical particles of marine suspensions. For biological particles with $m = 1.02$, the first maximum is situated at $\rho^{(1)}_{max} = 102$, and K_s is equal to 3.236; the first minimum is situated at $\rho^{(1)}_{min} = 190$, and at it $K_s = 1.569$. For terrigenous particles with $m = 1.15$, the curve $K_s(\rho)$ contains a series of damped vibrations. The first main maximum is situated at $\rho^{(1)}_{max} = 13.5$, and K_s is equal to 3.62; $\rho^{(1)}_{min} = 26$, and at it $K_s = 1.689$. If the curves of K_s are plotted as a function of the phase shift $\delta = 2\rho(m - 1)$ (Fig. 4.9), then in accordance with the rule of similarity (see [90, p. 218]), the curves of K_s for $m = 1.02$ and $m = 1.15$ will be very similar. The phase shift corresponding to $\rho^{(1)}_{max}$ will be the same for both curves: $\delta^{(1)}_{max} = 4.09$. The similarity of the curves can be seen clearly if the curves are compared for $m = 1.15$ and $m = 1.02$ with the limiting curve for $m = 1$.

We make the following observations.

1. It is usually assumed that large particles scatter light neutrally. This is true generally speaking but the limiting radius a^* for which the particle may be considered large depends on m, as can be seen from Figs 4.9 and 4.10. Thus, for $\lambda_0 = 546$ nm, assuming that for δ, the limit lies at $\delta = 20$ (Fig. 4.9), we find that a^* is equal to 4.3 and 32.6 μm for $m = 1\cdot15$ and $m = 1.02$ respectively.

2. In regions where $dK_s/d\rho > 0$, the spectral transparency has a normal variation and short waves are scattered more strongly than long waves; in regions where $dK_s/d\rho < 0$, it has an anomalous variation—short waves are scattered less strongly than long waves. In poly-dispersive systems, the curves of $K_s(\rho)$ smooth out, although there are also anomalous areas in the smoothed-out curves. When observed through a similar system, the setting sun will appear blue rather than red. In fact, in 1951 when a cloud

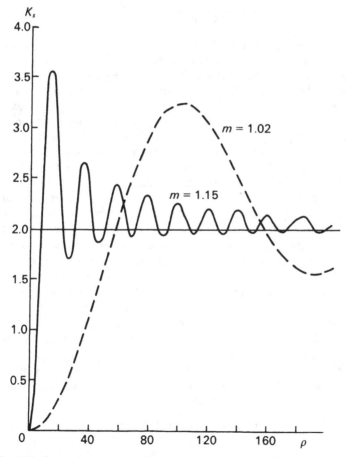

Fig. 4.10. Scattering cross-section for typical particles of marine suspensions.

of smoke from the Canadian forest fires passed over Europe, many observed a blue sun. The particles in the woodsmoke fall in the anomalous section of the curve of $K_s(\rho)$. The first anomalous region for particles with $m = 1.02$ in the visible part of the spectrum ($\lambda_0 = 546$ nm) comes at the range of particle radii 6.6–12.3 μm and for particles with $m = 1.15$ at the range of radii 0.88–1.7 μm. In this connection, we point out that the use of the equation $b(\lambda) = B\lambda^{-m}$, for $b(\lambda)$ where m varies from 4 to 0 (depending on the size of the particles) is a crude approximation which does not take into consideration the anomalous region.

3. For detailed calculations, a high-frequency small amplitude 'ripple' appears in the curves of Fig. 4.9. We have not shown this ripple in Figs 4.9 and 4.10.

The data given in tables [123] make it possible to determine the volume scattering coefficient $\beta(\gamma)$ for the particles of marine suspensions. According to the equation for $\beta(\gamma)$ given in Chapter 1, we have:

$$\beta(\gamma) = \frac{2}{\rho^2 K_s(\rho)}[i_1(\gamma) + i_2(\gamma)] \tag{4.91}$$

The values of $i_1(\gamma) + i_2(\gamma)$ and $K_s(\delta)$ are given directly in [123]. According to the data of [123], it is easy to calculate the table of $\beta(\gamma)$. In order to illustrate this, we give the graph of the scattering function in Figs 4.11 and 4.12.

When $\rho = 0.3$, the scattering function is almost Rayleigh in form (for $m = 1.15$, the asymmetry is somewhat larger than for $m = 1.02$). As ρ increases, the functions are stretched out in front (Mie effect), departing more and more from the symmetric Rayleigh form, and the scattering minimum shifts from $\gamma = 90°$ to $\gamma \approx 115°$; at small angles γ, the effect of the refractive index decreases. All these trends, shown in Fig. 4.11, are more clearly marked in Fig. 4.12 where larger particles are considered. In this case, there is another important factor: the functions start to oscillate. The curve of $\beta(\gamma)$ is no longer monotonic and a number of maxima and minima appear. In particular, the function $\beta(\gamma)$ acquires an irregular, interference character at large ρ values $\geqslant 5$–10. These irregularities of the angular structure of $\beta(\gamma)$ are well marked if for example we consider the ratio $\beta(\gamma_1)/\beta(\gamma_2)$ as a function of ρ; for example the variable $\psi(\rho) = \beta(0)/\beta(180°)$.

In the sea, for a number of reasons, the interference fringes are smoothed out and usually the observed functions are smooth. However in many cases, interference extrema are observed in real functions [55, 56].

An interesting property of the form of the function is the scattering cross-section in the direction $F(\gamma, \rho)$. This variable is proportional to the intensity of the light $I(\gamma, \rho)$ scattered by the particles at a scattering angle of γ. We introduced it in [93] for the dependence of the scattering intensity at a given angle γ on the size of the particle. The quantity F is related to $\beta(\gamma)$ and the scattering intensity $I(\gamma, \rho)$ by the equations:

$$I(\gamma, \rho) = I_0[\pi a^2/r^2]F(\gamma, \rho)$$

$$F(\gamma, \rho) = [K_s(\rho)/4\pi]\beta(\gamma, \rho) = \frac{b(\gamma)}{\pi a^2} \tag{4.92}$$

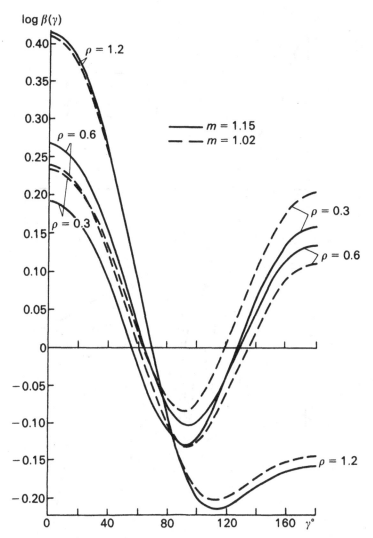

Fig. 4.11. Volume scattering functions for typical particles of marine suspensions (small particles). The curves for $\rho = 0.3$ are almost symmetric—the scattering is close to Rayleigh scattering.

In [93], the quantity $F(\rho)$ was calculated for the angle $\gamma = 90°$ for particles of different substances. In Fig. 4.13, a graph is shown of the cross-section $F(\rho)$ for $\gamma = 90°$ for particles in a marine suspension ($m = 1.15$). In order to understand the nature of this curve, we compare it with the curve $F(\rho)$ for $m = 1.33$ in [93]. We can see that the curve oscillates sharply and slowly approaches a geometrical optical asymptote. It is obvious that the curve for $m = 1.15$ has the same nature but the geometrical optical asymptote occurs at a significantly larger ρ.

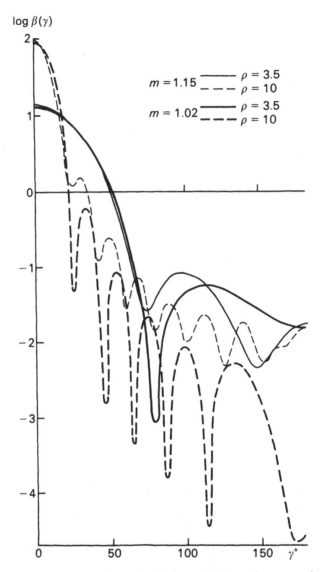

Fig. 4.12. Volume scattering functions of typical particles in marine suspensions (average particles). The refractive index m has an appreciable effect on the shape of the curves.

The sharp oscillations of $\psi(\rho)$ show how uncertain it is to use this function to define the degree of elongation of the scattering function. In this case, the integral characteristics are more suitable: the asymmetry coefficient and factor. Using the light pressure cross-section K_p data calculated in [123] we calculated the asymmetry factor $g(\rho)$ from Equation (4.36) for typical particles in marine suspensions. Some of the results of these calculations are shown in

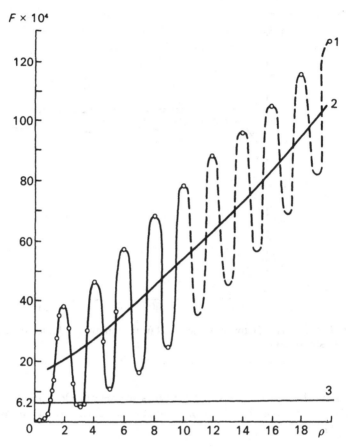

Fig. 4.13. Scattering cross-section $F(\rho)$ for a particle in a marine suspension ($m = 1\cdot15$). 1, exact values; 2, average curve; 3, geometric asymptote.

Fig. 4.14. The ripples of $g(\rho)$ for $m = 1\cdot15$ which are observed in the graph in Fig. 4.14 are the remains of interference lobes in the scattering function.

Based on the data of the tables of [123], we calculated the degree of polarization of scattered light p from the equation.

$$p = (i_1 - i_2)/(i_1 + i_2) \qquad (4.93)$$

It follows from an analysis of these calculations that the degree of polarization of scattered light $p(\rho)$ is gradually deformed from the smooth symmetric Rayleigh curve as ρ increases, with a maximum equal to unity at $\gamma = 90°$, producing a jagged curve with a large number of extreme values. Such a curve is shown in Fig. 4.15 where the curve of $p(\gamma)$ is plotted for $\rho = 200$ with coefficients $m = 1\cdot02$ and $m = 1\cdot15$, characteristic of particles of marine suspensions. Of course, in the sea the curve is smoothed out and its extreme values vanish for a number

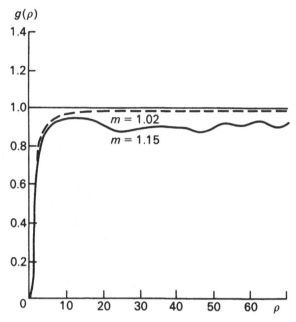

Fig. 4.14. Asymmetry factor $g(\rho)$ calculated from Equation (4.36) for typical particles in a marine suspension.

of reasons. In this case, the curve $p(\gamma)$ is in general close to that which was calculated from the equations of geometrical optics for a transparent sphere (Fig. 4.6).

4.6. Scattering of an ensemble of particles

4.6.1. Coherent and non-coherent scattering

The first question which arises as we move from one particle to the investigation of the scattering of light by an ensemble of particles is how to take into account the interaction between the particles. For ocean water, the suspended particles are usually considered as independent scatterers. This means that the intensity of light I observed at a certain point is the sum of the intensities produced by every turbidity element in the illuminated volume I_j:

$$I = \sum_j I_j \tag{4.94}$$

Equation (4.94) can be explained by the fact that the particles suspended in seawater are randomly distributed in the volume under observation. Thus, there are no systematic relationships between the phases of the waves scattered by the

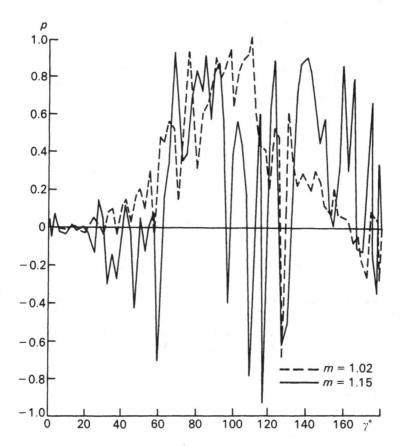

Fig. 4.15. Degree of polarization of scattered light (monodispersive suspension, $\rho = 200$).

individual particles and interference effects which are created on the addition of coherent vibrations disappear.

However, this explanation is not quite exact. First, the random location of particles in space does not guarantee the cancellation of the interference components. It is important that the phases of the waves arriving from two neighbouring particles are considerably different from each other. In this case, the average distance l between the particles must be significantly larger than the wavelength λ. Second, it is not difficult to see that for any location of the particles, the degree of coherence of the scattering depends on the scattering angle.

Let us look at Fig. 4.16. Let a plane wave propagated along the z-axis be received by two receivers D and D'. We will measure the phase from the plane z in which the wave meets the first particle. The phases of waves arriving at the receivers from two particles A and B are determined by the time which light takes in propagating from the plane z to D or D'. It can be seen from Fig. 4.16

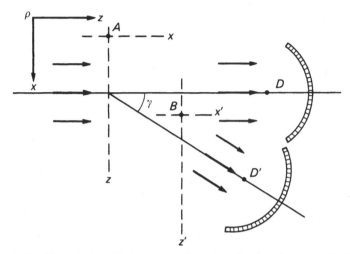

Fig. 4.16. The relationship between coherent and non-coherent scattering.

that if the time necessary for re-emission of the absorbed photon by both particles is the same, then the time of arrival of the waves scattered from A and B at the receiver D will also be the same and the path traversed by the wave scattered from A in the direction of D' will be longer by Δl than the path from the particle B:

$$\Delta l = (z' - z)(\cos \gamma - 1) + (x' - x)\sin \gamma$$

Thus, the phase of the wave arriving from A at D' will be behind the phase of the wave from B by $\delta = 2\pi\Delta l/\lambda$:

$$\delta = (2\pi/\lambda)[(z' - z)(\cos \gamma - 1) + (x' - x)\sin \gamma] = \delta_z + \delta_x \qquad (4.95)$$

Here, δ_z and δ_x are the phase shifts connected to the differences in the coordinates of the particles A and B with respect to the z- and x-axes, respectively. It can be seen from Equation (4.95) that when $\gamma \to 0$, the phase shift also tends to 0 and $\delta_z \sim \gamma^2$, whereas $\delta_x \sim \gamma$, ie δ_z disappears earlier than δ_x. In the strictly forward direction when $\gamma = 0$, the quantity $\delta = 0$ for any pair of particles; ie in this direction, the scattering is always coherent.

Let us imagine now that our volume is approximately uniformly filled with particles.

The average phase shift between waves scatterd by two particles inside our turbid volume depends on the radius of the volume R and on the scattering angle γ. We write it in the form $\delta = (2\pi R/\lambda)\phi(\gamma)$. The expression for $\phi(\gamma)$ may be obtained from Equation (4.95); note, due to the random distribution of the particles with respect to both the z and x coordinates and the randomness of the differences $(z' - z)$ and $(x' - x)$, characterizing the position of the particles inside the volume, it can be assumed that $\overline{|z' - z|} \approx \overline{|x' - x|} \approx R; \delta^2 = \delta_z^2 + \delta_x^2$. Thus, we find $\phi^2(\gamma) = (\cos \gamma - 1)^2 + \sin^2 \gamma = 4 \sin^2 (\gamma/2)$. We now obtain for δ:

$$\delta = (4\pi R/\lambda)\sin(\gamma/2) \qquad (4.96)$$

In Equation (4.96), the effect of each of the variables R and γ can be seen on the phase shift $\bar{\delta}$. For scattering at small angles γ, it is easy to determine the critical angle γ_{crit} when $\bar{\delta} = 2\pi$. We find from Equation (4.96) that $\gamma_{crit} = \lambda/R$. This means that at angles $\gamma \ll \gamma_{crit}$, the scattering of light will be coherent and its total intensity will be $\sim N^2$ where N is the number of scattering particles in the illuminated volume. For angles $\gamma \gg \gamma_{crit}$ the scattering is incoherent and the total intensity will be $\sim N$. Thus, the ratio of the intensity scattered at an angle $\gamma = 0$ to the intensity scattered at a large angle makes it possible to determine the number of scattered particles N. In hydrooptics the exact value of the angle at which one form of scattering turns into another form may be obtained from a rigorous solution of the problem of coherent scattering of light by an ensemble of particles. An analytical solution of this problem presents great difficulty. Numerical experiments using the Monte-Carlo method could be useful here. However, a correct estimate of this quantity may be obtained using the arguments given in [90, p. 264]. Here, it is shown that the intensity of light scattered at any scattering angle γ by a system of N identical particles situated inside a sphere of radius R is determined by the equation:

$$I(\gamma) = NI_0[1 + Nf^2(q)] \tag{4.97}$$

where:

$$f(q) = 3(\sin q - q\cos q)/q^3, \quad q = (4\pi R/\lambda)\sin(\gamma/2)$$

The degree of coherence of the scattering is determined by the value of the parameter η:

$$\eta = Nf^2(q) \tag{4.98}$$

The critical angle γ_{crit} can obviously be found from the condition $\eta = 1$. At large q, assuming that:

$$f^2(q) \approx 9/(2q^4)$$

we find that:

$$\eta = k\lambda^4/[Rl^3 \sin^4 (\gamma/2)], \quad k = 3/(128\pi^3) = 7.6 \times 10^{-4} \tag{4.99}$$

The scattering will be coherent when $\gamma < \gamma_{crit}$ and non-coherent when $\gamma > \gamma_{crit}$, where:

$$\sin(\gamma_{crit}/2) = \alpha(\lambda/l)\sqrt[4]{l/R}, \quad \alpha = \sqrt[4]{k} = 0.166 \tag{4.100}$$

For a given λ, the critical angle depends on the two lengths l and R. The quantity $2R$ corresponds to the linear dimensions of the area where the scattering is measured. We can assume that they are equal to ~ 1 cm. For the average distance l, it is related to the concentration n by the equation $l = n^{-1/3}$. The maximum value of n given in [109] is $\sim 10^7$ cm^{-3}. Thus for $\lambda = 0.5 \mu$m, we find that $l = 46 \times 10^{-4}$ cm, $\sin (\gamma_{crit}/2) = 5.4 \times 10^{-4}$ and $\gamma_{crit} = 36'$.

Thus, at scattering angles $\gamma \gg 36'$, the phases of the scattered secondary waves emitted by particles are randomly distributed and the particles of marine suspensions can be considered as independent scatterers.

In this case:

$$\bar{b}(\gamma) = \int_{r_1}^{r_2} b(\gamma, r)n(r) \, dr; \; \bar{b} = \int_{r_1}^{r_2} b(r)n(r) \, dr \qquad (4.101)$$

where $b(\gamma, r)$ and $b(r)$ refer to a particle with radius r; $n(r) \, dr$ is the number of particles with a radius between r and $(r + dr)$; r_1 and r_2 are the limiting radii of marine suspended particles.

Equations (4.101) are the starting points for calculating the optical characteristics of real systems.

The function $n(r)$ is usally written in the form $n(r) = Nf(r)$ where N is the total number of particles in $1 \, cm^3$ and $f(r)$ is the distribution function of particles over size. It is also called the microstructure of the system or the particle spectrum and its graph is called the distribution function.

4.6.2 Distribution functions

The theory of distribution functions is a subject of mathematical statistics. A large number of different distribution functions are studied here: uniform and normal distributions and various distributions due to the latter (logarithmic normal, ZOLD, etc), the gamma and generalized gamma distribution, the exponential and beta distribution. In the optics of the sea, the normal, gamma and exponential distributions are most common.

The two-parameter distribution of the form:

$$f(x) = \begin{cases} 0, & x < 0 \\ \dfrac{\beta^{\mu+1}}{\Gamma(\mu+1)} \, x^\mu \, e^{-\beta x}, & x \geq 0; \mu > -1, \beta > 0 \end{cases} \qquad (4.102)$$

In statistics this is usually called a gamma distribution (parameters μ and β). It belongs to the group of distribution functions studied by Karl Pearson (1894). The gamma distribution belongs to the so-called type III according to Pearson. The Pearson curves are studied in detail in statistics. They form a family depending on four parameters: this family is classified into 12 types. The normal distribution and some other distributions also belong to it. In a number of cases, it is better to use the more general equation in calculations on the optics of dispersive systems:

$$f(x) = \begin{cases} 0 & x < 0 \\ \dfrac{\gamma \beta^{(\mu+1)/\gamma}}{\Gamma[(\mu+1)/\gamma]} \, x^\mu \, e^{-\beta x^\gamma}, & x \geq 0 \end{cases} \qquad (4.103)$$

When $\gamma \neq 1$, we call distribution (4.103) the generalized gamma distribution. It already contains three parameters (μ, β and γ). The use of these three parameters gives the family considerable generality and makes it possible to describe practically any single-peak distribution curve. By summing expressions of the type in Equation (4.103), any multi-peak curve can be described with the necessary accuracy. Apart from its great flexibility, Equation (4.103) possesses another very important property: simplicity. With its help, the distribution moments and other characteristics of the suspension can be calculated simply from the equation:

$$\int_0^\infty x^v \, e^{-\beta x^\gamma} \, dx = \frac{\Gamma[(v + 1)/\gamma]}{\gamma \beta^{(v+1)/\gamma}} \tag{4.104}$$

The equations for the gamma distribution look especially simple. We find for the mode and first moments that:

$$x_m = \frac{\mu}{\beta}; \quad \bar{x} = \frac{\Gamma(\mu + 2)}{\Gamma(\mu + 1)} \frac{1}{\beta}; \quad \overline{x^2} = \frac{\Gamma(\mu + 3)}{\Gamma(\mu + 1)} \frac{1}{\beta^2}$$

$$\overline{x^s} = \frac{\Gamma(\mu + s + 1)}{\Gamma(\mu + 1)} \frac{1}{\beta^s} \tag{4.105}$$

The distributions given in Equations (4.102) and (4.103) have been widely used in the optics of turbid media. They are simple and give a good description of a wide range of dispersive systems. An essential feature is that they are non-symmetric. To the left of the mode, they behave as a parabola of the form x^μ and on the right they decay as $\exp(-\beta x^\gamma)$ which usually produces a long slowly falling 'tail' (Fig. 4.17). Such a shape, corresponds better to the actual data than the symmetric Gaussian distribution. The relative width of the distribution $\Delta\varepsilon$ only depends on the parameter μ. The simple equation applies:

$$\Delta\varepsilon = 2 \cdot 48/\sqrt{\mu} \tag{4.106}$$

In McCartney's book [54] and also in some other studies, it is stated that the application of the distributions given in Equations (4.102) and (4.103) for the calculation of the optical characteristics of polydispersive systems was first suggested by D. Deirmendjien in 1963 [171]. This is not so. In 1930, I Rocard

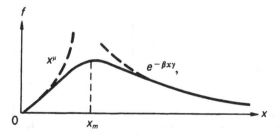

Fig. 4.17. Generalized gamma distribution.

[173] had already used the particular distribution of the form in Equation (4.102) ($\gamma = 1$, $\mu = 2$) for the calculation of atmospheric functions. Since 1951, we have been systematically using the total distributions given in Equations (4.102) and (4.103) for calculating the optical characteristics of polydispersive systems. The first of our publicatons appeared in 1953 [73] and a more detailed article appeared in 1955 [92].

Both parameters of the distribution given in Equation (4.102), β and μ, can be simply expressed in terms of x_m and $\Delta\varepsilon$, characteristics with a simple geometrical meaning:

$$\mu = 6\cdot15/(\Delta\varepsilon)^2; \quad \beta = [6\cdot15/(\Delta\varepsilon)^2](1/x_m) \tag{4.107}$$

As μ increases, the width of the distribution $\Delta\varepsilon$ tends to zero and the distribution becomes a δ function with a maximum at $\varepsilon = 1$. It describes a monodispersive system. The skew coefficient γ_1 and excess γ_2 of the gamma distribution are equal to:

$$\gamma_1 = 2/\sqrt{\mu + 1}; \quad \gamma_2 = 2/(\mu + 1) \tag{4.107'}$$

The integral function of the gamma distribution is expressed in terms of the incomplete gamma function. Its properties and detailed table have been published by E. E. Slutsky.

The exponential distribution takes the form:

$$f(x) = \begin{cases} 0, x < x_{min} \quad \text{and} \quad x > x_{max}, \ R = x_{max}/x_{min} \\ \dfrac{(v - 1)x_{min}^{v-1}}{1 - R^{1-v}} x^{-v} = C_v x^{-v}, \ x_{min} \leqslant x \leqslant x_{max} \end{cases} \tag{4.108}$$

The exponential distribution is defined with finite limits between x_{min} and x_{max}. This is a three-parameter family of curves (parameters: v, x_{min}, x_{max}). Its graphs are shown in Fig. 4.18. When the experimental data are being processed, it is convenient to use logarithmic scales along the axes. In this case, the distribution graph is a straight line. The angular coefficient of the straight line is determined by the parameter v. The value $R = x_{max}/x_{min}$ is usually large. Therefore, if $v > 1$, in the region (x_{min}, x_{max}) the function $f(x)$ can be approximately represented by the equation:

$$f(x) = (v - 1)x_{min}^{v-1} x^{-v} \tag{4.109}$$

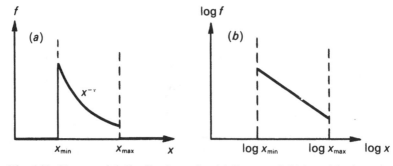

Fig. 4.18. Exponential distribution using (a) linear and (b) logarithmic scales.

When $v = 4$, Equation (4.109) is often called Junge's distribution and when $v \neq 4$, a Junge-type distribution. Junge suggested that Equation (4.109) could be used to describe the particle size distributions of atomspheric aerosols over continents.

We note the following equation for the distribution moments:

$$\overline{x^{n-1}} = A_v[x_{min}^{n-v}/(n-v)](R^{n-v}-1)$$
$$A_v = (v-1)x_{min}^{v-1}/(1-R^{1-v})$$

(4.110)

Here, we assume that $n \neq v$. If $n = v$, then:

$$\overline{x^{n-1}} = A_v \log R$$

(4.111)

We find for the mode and first moments that:

$$x_m = x_{min}; \quad \bar{x} = A_v x_{min}^{2-v}(R^{2-v}-1)/(2-v)$$
$$\overline{x^2} = A_v x_{min}^{3-v}(R^{3-v}-1)/(3-v)$$

(4.112)

In studying marine suspensions Sheldon and Parsons drew attention to the fact that in the case of Junge's distribution ($v = 4$), we find for the elementary volume $dv = (\frac{4}{3})\pi a^3 f(a)da$ that:

$$dv = (\frac{4}{3})\pi C_v \, da/a$$

(4.112′)

Consequently:

$$dv/dlgd = \text{constant}$$

(4.113)

Thus, if we construct the distribution of $dv/dlgd$ versus $d(d = 2r)$, then the experimental data must lie on a horizontal line. Data processed in this way are shown in Fig. 4.19 according to [28] for particles of ocean suspended material based on observations in the western part of the Mediterranean Sea.

The distributions given in Equations (4.108) and (4.109) have a disturbing property; they are discontinuous at the points: $x = x_{min}$ and $x = x_{max}$

However, smoothing out the experimental data shows that they are described significantly better by the analytical beta distribution (x and y in micrometres), especially in the region of small particles [71]:

$$f(x) = [\gamma^n/B(l, n)][x^{l-1}/(x+\gamma)^{l+n}](l > 0, n > 0, \gamma > 0)$$

(4.114)

The calculation of the density and moments of this distribution results in the calculation of the beta function and is not difficult. The density of Equation (4.114) is continuous, whereas the density of Equation (4.108) has a sharp discontinuity at $x = x_{min}$ equal to a full oscillation of this function (when $R < \infty$ there is another discontinuity at the point $x = x_{max}$). The density of Equation (4.114) has greater generality than the density of Equation (4.108) since Equation (4.114) has more shape parameters.

In some cases, it is convenient to write the beta distribution in another form:

$$f(x) = \frac{1}{\mu_1 - \mu_2} \frac{\Gamma(\gamma + \eta)}{\Gamma(\gamma)\Gamma(\eta)} \left(\frac{x - \mu_0}{\mu_1 - \mu_0}\right)^{\gamma - 1} \left(1 - \frac{x - \mu_0}{\mu_1 - \mu_0}\right)^{\eta - 1}$$

(4.115)

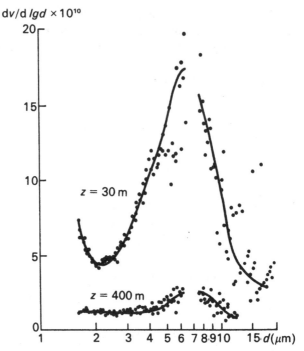

Fig. 4.19. Distribution of $dv/d\lg d$ versus d for marine suspended particles. At a depth of $z = 30\,\text{m}$, the distribution appreciably differs from an exponential distribution. At $z = 400\,\text{m}$, it is close to an exponential distribution with $v = 4$.

In this notation, μ_0 and μ_1 are the limits of the distribution (they correspond to x_{min} and x_{max}). If we set $y = (x - \mu_0)/(\mu_1 - \mu_0)$, then Equation (4.115) becomes simpler:

$$f(y) = \frac{\Gamma(\gamma + \eta)}{\Gamma(\gamma)\Gamma(\eta)}\, y^{\gamma - 1}(1 - y)^{\eta - 1},\ 0 \leqslant y \leqslant 1 \qquad (4.116)$$

The graphs of the distribution density given in Equation (4.116) for several γ and η are given in Fig. 4.20. By making the substitution $y = x/(1 + x)$, Equation (4.116) turns into Equation (4.114).

For the distribution given in Equation (4.115), the mode x_m, the average value x and other characteristics will be:

$$x_m = \frac{\gamma - 1}{A - 2}\mu + \mu_0; \quad \frac{\sigma}{\bar{x}} = \left(\frac{B}{A + 1}\right)^{1/2}\frac{\mu}{B\mu_1 + \mu_0}$$

$$\bar{x} = \frac{B\mu_1 + \mu_0}{B + 1}; \quad \gamma_1 = \frac{\mu_3}{\sigma^3} = \frac{4(1 - B)^2(A + 1)}{B(A + 2)^2}$$

$$\sigma^2 = \frac{B}{(A + 1)(B + 1)^2}\mu^2 \qquad (4.117)$$

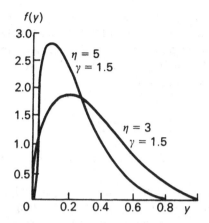

Fig. 4.20. The beta distribution given in Equation (4.116) for $\gamma = 1.5$ and $\eta = 3$ and $\eta = 5$.

where we have introduced the notation:

$$\mu = \mu_1 - \mu_0; \; A = \gamma + \eta; \; B = \gamma/\eta \qquad (4.117')$$

The integral function of the beta distribution is expressed in terms of the incomplete beta function. Its table can be found in [165].

We note in conclusion that log-normal distribution often used is very similar to the gamma distribution [51]. The latter is considerably simpler to use.

4.6.3. Limiting cases for small and large particles, small angle scattering, soft particles

Within the framework of the equivalent sphere model, the optical properties of polydispersive systems may be calculated from Equation (4.101). The results of the calculations are given in the tables in [123, 160]. They depend on the characteristics of single scattering and on the form of the distribution curves. A general conclusion arising from such calculations is that on averaging over size, the spectral or angular dependences become smoother and the interference oscillations disappear.

In the limiting cases considered above, the results become especially transparent. Thus, for Equations (4.56) and (4.61), the polydispersive characteristics are described by the sixth and third distribution moments, respectively and by the second moment for the limiting case of large particles (in the geometrically optical part of the angles).

Of special interest are cases when the general expressions given in Equation (4.28) represent analytic equations. These are cases of scattering of large particles at small angles, the scattering of soft particles. Here, the distribution of intensity for polydispersive systems has been specially investigated in a number of studies. We shall look at them briefly.

The diffraction part of the scattering function in the approximation of

Equation (4.71) was analytically investigated in [94] for gamma distributions. It was shown that the calculation of the light distribution scattered by a polydispersive system at small angles makes it necessary to calculate the definite integrals containing the square of the Bessel function of the first order. We are dealing with integrals of the type:

$$\int_0^\infty z^k \, e^{-\eta z} J_1^2(z) dz = \phi_k(\eta) \tag{4.118}$$

Equation (4.118) may be regarded as the definition of a certain class of special functions $\phi_k(\eta)$.

The study of the properties of the functions $\phi_k(\eta)$ and methods for calculating them was begun in [94]. Here, the case of integral k was considered. Subsequently, cases were considered for any k, integral and non-integral and Equation (4.118) was generalized for the case when the exponent coefficient z differs from unity, ie the case when the generalized gamma distribution given in Equation (4.103) is used instead of the gamma distribution.

It follows from Equation (4.118) that the function $\phi_k(\eta)$ satisfies the differential equation:

$$\phi_k(\eta) = -\partial\phi_{k-1}(\eta)/\partial\eta \tag{4.119}$$

Since all $\phi_k(\eta)$ are positive, it follows from Equation (4.119) that $\phi_k(\eta)$ are monotonically decaying functions of η.

The equations for integral indices were obtained in [94]. We write them here using the notation:

$$x = 2/\sqrt{4 + \eta^2}$$

$K(x)$ and $E(x)$ are full elliptical integrals of x.

The equations given below enable the functions $\phi_k(x)$ to be calculated using the tables for the functions $K(x)$ and $E(x)$:

$$\phi_0(x) = \frac{1}{\pi x}[(2 - x^2)K - 2E]$$

$$\phi_1(x) = \frac{1}{2\pi(1 + x^2)^{1/2}}[(2 - x^2)E - 2(1 - x^2)K]$$

$$\phi_2(x) = \frac{x^3}{4\pi(1 - x^2)}[(2x^2 - 1)E - (1 - x^2)K] \tag{4.120}$$

$$\phi_3(x) = \frac{x^4}{8\pi(1 - x^2)^{3/2}}[(-8x^4 + 13x^2 - 3)E + (1 - x^2)(-4x^2 + 3)K]$$

It is not difficult to show that the following approximate formulae can be derived from the equations in [94] for $\phi_k(\eta)$ for small and large η:

$$\phi_k(\eta) = (k-1)!/(\pi\eta^k) \ (\eta \ll 1)$$

$$\phi_k(\eta) = (k+2)!/(4\eta^{k+3}) \ (\eta \gg 1) \tag{4.121}$$

Thus, $\phi_k(\eta)$ has a pole of order k at $\eta = 0$; at large η as η increases, $\phi_k(\eta)$ decays as $\eta^{-(k+3)}$. The table of $\phi_k(\eta)$ for $k = 0(1)7$ is given in [94].

A complete analysis of the question of the structure of the light field diffracted by a gamma system at small angles is given in [118]. Tables are given there, calculated using quadratures with the maximum algebraic degree of accuracy. These tables provide convenient interpolation for all variables of the problem.

Analytical equations for the coefficient $b(\lambda)$ and the function $\beta(\gamma)$ were obtained in a number of studies for optically soft particles in the RHD approximation. The departure point was the study of I. Rocard [173] who found equations for the function $\beta(\gamma)$ for gamma structures with parameter $\mu = 2$. However, a small error was made in his work which was corrected in [120]. For any μ, equations for $\beta(\gamma)$ are derived in the studies referred to in [121], where a table of $\beta(\gamma)$ is given for $\mu = 0(2)10$. For exponential distributions given in Equation (4.108), the equations and tables of $\beta(\gamma)$ and $b(\lambda)$ for integral ν are given in [121] and [71].

In the approximation of physical optics for optically soft particles, analytical equations have been suggested for the polydispersive attenuation coefficient. For gamma distributions, equations are given in [120–122] and especially useful equations in [115] and for exponential distributions in [71]. In [115], the optical characteristics of quasi monodispersive systems were also calculated. With the help of the family of gamma distributions, a sequence of functions was constructed with increasing μ whose limit was the Dirac δ function (it corresponds to $\mu = \infty$). Expressions for the coefficient $b(\rho)$ and the function $\beta(\gamma)$ (in the region of small angles) were obtained for similar systems in the form of series in μ^{-1}. The first term in this series describes the monodispersive approximation.

4.6.4. The general case: tables and their analysis

We now look at numerical calculations of the optical properties of polydispersive systems using the exact Mie formulae. The results of such calculations for models of ocean suspensions are given in [123] and [160]. We have already discussed the first section of tables [123]. We now look at the second section in which polydispersive models are considered. This gives three-figure polydispersive scattering coefficients and light pressure coefficients as well as the elements of the scattering matrix and the degree of polarization of the scattered light for the same angles γ as for monodispersive particles. Two types of distribution are considered: an exponential distribution of the form in Equation (4.108) and a normal distribution of the form $f(a) = [1/(\sigma\sqrt{2\pi})]\cdot\exp\{-[(a-\bar{a})/2\sigma]^2\}$. The parameters of the models for which calculations have been made are given in

Table 4.6. Characteristics of polydispersive models

Exponential distribution						Normal distribution					
No	v	ρ_{max}	No	v	ρ_{max}	No	$\bar{\rho}$	ρ_σ	No	$\bar{\rho}$	ρ_σ
1	3	20	9	5	20	1	3.6	1.2	9	72	36
2	3	50	10	5	50	2	5.3	1.8	10	96	24
3	3	100	11	5	100	3	7.2	2.4	11	96	48
4	3	200	12	5	200	4	25	12	12	96	96
5	4	20	13	6	20	5	36	18	13	144	72
6	4	50	14	6	50	6	48	12	14	144	144
7	4	100	15	6	100	7	48	24	15	192	96
8	4	200	16	6	200	8	72	18			

Table 4.6. For all exponential models it is assumed that: $\rho_{min} = 0.3$ and six values of the refractive index $m = 1.02$, $1.05(0.05)$, 1.25. The values of the parameters of the normal distribution $\bar{\rho} = 2\pi\bar{a}/\lambda$, $\rho_\sigma = 2\pi\sigma/\lambda$ were based on the experimental data on the spectra of biological suspended particles. The calculations were carried out for two values of m: 1.02 and 1.05.

Due to the inaccuracy in the normalization of the truncated distribution, the absolute values indicated for 12 and 14 models of normal distribution must be increased by 19%. Of course, the introduction of a constant correction has no effect on the form of the scattering function and on the other relative variables. Strictly speaking, the question of whether it is necessary to introduce a correction depends on how it is agreed to understand truncated distributions (see Proc. containing [112]). In the tables presented by A. Morel [160], data are given on the intensities I_1, I_2, $I = I_1 + I_2)/2$ and the degree of polarization of scattered light p for three types of distributions: (1) the Junge distribution ($\rho_{min} = 0.2$, $\rho_{max} = 200$); $m = 1.02$, 1.05, 1.075, 1.10, 1.15; (2) logarithmic normal distribution ($\rho_{min} = 0.2$; $\rho_{max} = 200$); $m = 1.02$, 1.05; (3) Junge's distribution with variable limits ($\rho_{min} = 0.2$, $\rho_{max} = 50$, 100, 200; $\rho_{max} = 200$, $\rho_{min} = 1.2$, 2.5, 5, 10).

The results are shown in the form of tables and in the form of graphs of the scattering functions and graphs of the function:

$$F(\rho, \gamma) = \int_{\rho_{min}}^{\rho} f(x)\{i_1(\gamma, x) + i_2(\gamma, x)]/2\} \, dx$$

for eight scattering angles $\gamma = 0$, 2, 10, 20, 40, 90, 140, 180°. In the tables of A. Morel, the parameter ρ is designated by α and the scattering angle γ by θ.

In accordance with the definitions given in Chapter 1, the scattering coefficients of polydispersive models b coincide with the variables α_r calculated in [123] and the polydispersive functions $\beta(\gamma)$ are expressed in terms of the variables α_r, I_1 and I_2 tabulated in [123] by the equation:

$$\beta(\gamma) = 2\pi(I_1 + I_{\bar{2}})/\alpha_r$$

In the tables given in [123], values are given for α_r which are equal to the scattering coefficient b in dimensionless numbers. In order to obtain b in μm^{-1}, it is necessary to multiply them by $N\lambda^2$ also in μm^{-1}. If we substitute N in cm^{-3}, λ in μm then in order to obtain b in m^{-1}, it is necessary to introduce the scale factor 10^{-6}. For example, for $m = 1.02$, $v = 3$, $\rho_{max} = 200$ for α_r we find 6.24×10^{-2} in [123, p. 126]. This means that for $\lambda_0 = 546\,nm$ and $N = 10^8\,cm^{-3}$, we have:

$$b = 10^8(0.546/1.344)^2 \times 6.24 \times 10^{-2} \times 10^{-6} = 1.04\ m^{-1}$$

Based on the data of tables [123], we plotted the graphs for the polydispersive functions for $m = 1.02$ and $m = 1.15$. They are shown in Figs 4.21 to 4.23. For

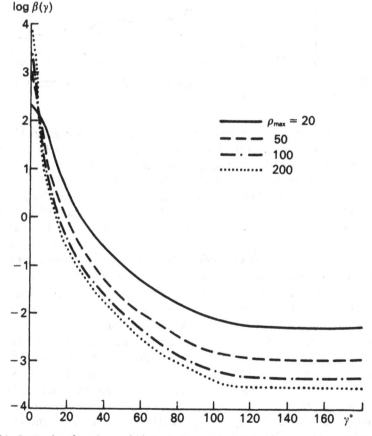

Fig. 4.21. Scattering functions of biological particles with exponential distribution for $v = 3$ for different ρ_{max}, $m = 1.02$.

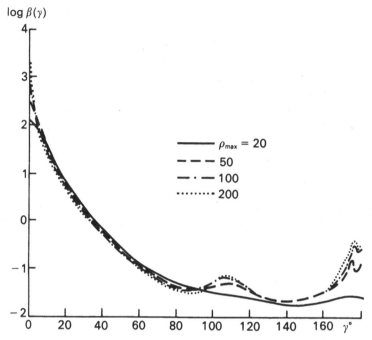

Fig. 4.22. Scattering functions for terrigenous particles with exponential distribution, for $v = 3$, for different ρ_{max}, $m = 1.15$.

$m = 1.02$, the curves corresponding to different ρ_{max} are appreciably different and as ρ_{max} increases, they are strongly elongated. For $m = 1.15$, the difference between curves with different ρ_{max} is smaller. Here, the ends of the functions are considerably different. A remarkable feature of the curves given in Fig. 4.22 is the peak at $\gamma = 106°$. This peak, which is unnoticeable at $\rho_{max} = 20$, is already well marked at $\rho_{max} = 50$ and particularly at $\rho_{max} = 200$. The peak in the function corresponds to a rainbow. When $m = 1.15$, according to the equations of geometric optics, ie for $\rho = \infty$, the first rainbow must lie at $\gamma = 101°$. In our model of a large number of small particles and in accordance with the wave theory of Airy for these particles, the maximum in intensity of the scattered light is shifted in the direction of large angles γ. As $(m - 1)$ decreases, the energy arriving at the rainbow drops sharply. Therefore, the rainbows are unnoticeable on the curve for $m = 1.02$. In addition, as v increases, the proportion of large particles decreases. According to this, the rainbows on the functions for $m = 1.15$ are also unnoticeable for $v = 4, 5, 6$. Comparing the curves in Figs. 4.21 to 4.23 with the curves in Figs 4.11 and 4.12, we can see that they are considerably different. The interference maxima and minima have disappeared, and the curves with the exception of the peak at the rainbows, have become monotonic.

Figure 4.24 shows the degree of polarization of scattered light according to [123] for three polydispersive models. Averaging over the distribution eliminates the jagged nature of the curves and they become smooth. For biological particles

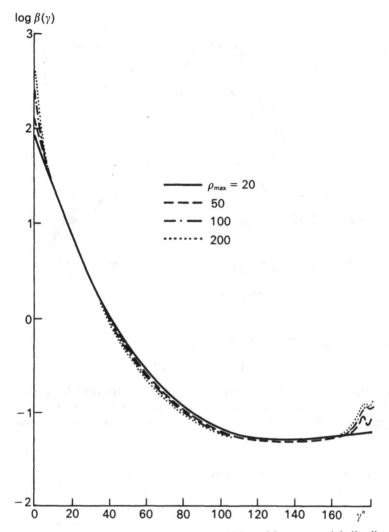

Fig. 4.23. Scattering functions for terrigenous particles with exponential distribution, for $v = 4$, for different ρ_{max}, $m = 1·15$.

with $m = 1.02$, the curve of $p(\gamma)$ hardly differs from the Rayleigh curve for all considered values $v = 3–5$. For terrigenous particles with $m = 1.15$, it is also close to the Rayleigh curve at $v = 5$. On reducing v, ie on increasing the fraction of large particles, the degree of polarization at the maximum falls sharply reaching only 45 % and in the region of small angles ($\gamma < 40°$) the polarization is negative. The peak in the neighbourhood of 170° and the sharp negative minimum at $\gamma = 178°$ are especially interesting.

In conclusion, we look at the asymmetry factor g for the scattering function. Its values determined from [123] are given in Table 4.7. The calculation

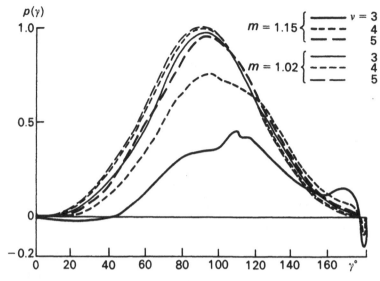

Fig. 4.24. Degree of polarization of scattered light.

was carried out using the equation:

$$g = 1 - K_p/K_s$$

We can see that g appreciably decreases as v increases (this happens as the result of the reduced contribution of large particles) and decreases more slowly as m increases. This is associated with the reduced amount of light scattered forward.

Table 4.7. Asymmetry factor g of polydispersive functions

m	v			
	3	4	5	6
1.02	9.97 −1	9.83 −1	8.43 −1	4.64 −1
1.05	9.85 −1	9.60 −1	8.03 −1	4.58 −1
1.10	9.54 −1	9.19 −1	7.55 −1	4.50 −1
1.15	9.22 −1	8.78 −1	7.16 −1	4.37 −1
1.20	8.90 −1	8.38 −1	6.81 −1	4.27 −1
1.25	8.60 −1	8.00 −1	6.52 −1	4.18 −1

Chapter 5

Optical properties of real ocean water

5.1. Absorption by suspended particles

5.1.1. Relationship between the spectrum of the suspension and the spectrum of the particulate material

We described the optical properties of the individual components of ocean water in Chapter 3. However, we omitted an important component, the absorption by suspended particles. We did this because a sequential analysis of this phenomenon can only be carried out on the basis of the theory developed in Chapter 4. In order to use the equations given there for calculations, it is necessary to have data on the quantity $k(\lambda)$, the imaginary part of the complex refractive index of the substance of the particles. For phytoplankton cells, $k(\lambda)$ will be appreciably different from zero at the wavelengths of the absorption by pigments. Phytoplankton particles contain a large number of different coloured substances—pigments. These are chlorophyll a, b and c, carotenoids, phaeophytins (products of oxidation of chlorophylls with the central magnesium atom in the chlorophyll molecule replaced by two atoms of hydrogen), chlorophillides and phaeophorbides, etc. The pigment system in phytoplanktons depends on their species, age and a large number of biotic and abiotic factors which continuously vary under natural conditions. Within the framework of the homogeneous sphere model, it is only important for us to know how all these changes influence the constants n and k of the substance of the cell and its dimensions. These data can be used to calculate the absorption spectra of different species of phytoplankton.

The important question now arises of how the spectra of the substance of the cells are related to the spectrum of the suspended cells. We are concerned with how the dimensions of the particles affect the absorption spectra of dispersed systems (suspended or precipitated on filters). In practice in oceanography, the difference between the absorption spectra of the substance and of the dispersive system is either simply ignored or a certain correction factor is introduced to allow for it. We note that a similar problem was examined in the optics of fogs and clouds where the absorption spectrum was estimated using the method of 'the precipitated substance layer'. In this method, the absorption spectrum was simply calculated from the number of molecules penetrated by the light beam. In [90, 95], it was pointed out that this approach is unsatisfactory and may lead to considerable errors. In [95], a specific absorption coefficient a' related to unit mass was introduced, and it was shown that in dispersive systems, one of the basic laws of physical optics, Beer's law, is violated. The absorption depends on the quantity of the absorbing substance traversed and also on how this substance is dispersed.

Following [95], we compare the absorption in a plane layer and in an equivalent column of suspended matter. Let dM be the mass of substance in a column of cross-section 1 cm^2 and length dl. On passing through the layer of the substance, the amount of absorbed radiation $d\Phi_a$ will be:

$$d\Phi_a = adl = adM/\sigma \tag{5.1}$$

Here, a is the absorption coefficient and σ is the density of the substance. On passing through a layer of particles:

$$d\Phi'_a = k_a \, dN \tag{5.2}$$

where dN is the number of particles in layer dl'; k_a is the coefficient of absorption of one particle.

We will first assume that our system consists of monodispersed spheres of radius r. In this case, the mass of substance in the column dl' will be $dM' = (4/3)\pi r^3 \sigma \, dN$. Thus, we find that:

$$d\Phi'_a = k_a \cdot 3dM'/(4\pi r^3 \sigma) \tag{5.3}$$

Setting $dM = dM'$ and comparing Equations (5.1) and (5.3), we find that the specific absorption coefficients a' in m^2/kg will be:

$$\begin{aligned} a'_1 &= a/\sigma \\ a'_2 &= 3k_a/(4\pi r^3 \sigma) \end{aligned} \tag{5.4}$$

The prime is used to distinguish the Beer coefficient a' from the normal coefficient a related to unit length.

Substituting $k_r = \pi r^2 K_a$, we find that:

$$\begin{aligned} a'_2 &= [3\pi/(2\sigma\lambda)](K_a/\rho) = \eta(v)a'_1 \\ \eta(v) &= (\tfrac{3}{2})[K_a(v)/v]; \quad v = 4\rho k \end{aligned} \tag{5.5}$$

where the coefficient $\eta(v)$ characterizes the tranformation of the absorption spectrum from the layer to the dispersive system. Equation (5.5) gives a general solution of the problem in the monodispersive case.

In the limiting case of small particles [see Equation (4.61)]:

$$K_a = K = 24nk\rho/|m^2 + 2|^2; \quad a'_2 = \eta_s a'_1; \eta_s = 9n/|m^2 + 2|^2 \tag{5.6}$$

Here, both specific coefficients are simply proportional to each other. If, furthermore, it is assumed that for phytoplankton particles, $n = 1.02$ or 1.05 and k even at the absorption maximum does not exceed 0.01, then η is equal to 0.993 or 0.982, ie for small particles, a'_2 and a'_1 practically coincide. The dispersive system consisting of small particles is similar to a molecular solution.

In the limiting case of large optically 'soft' particles, we use Equation (4.88) for K_a. For $\eta_l(v)$, we then find that:

$$\eta_a(v) = \frac{3}{2v} \left\{ 1 - \frac{2}{v} \left[\frac{1}{v} - e^{-v} \left(1 + \frac{1}{v} \right) \right] \right\} \tag{5.7}$$

Equation (5.7) like Equation (4.88) actually gives a good description of the picture for large and for small particles. In the limiting cases, we find that:

$$\eta_l(v) = 1 - \frac{3}{8} v + \frac{v^2}{10} - \ldots v \ll 1$$

$$\eta_l(v) = \frac{3}{2v} \left(1 - \frac{2}{v^2} \right) = \frac{3}{2v} - \frac{3}{v^3} + \ldots v \gg 1 \tag{5.8}$$

The values of $K_a(v)$ from Equation (4.88) and $\eta_a(r)$ from Equation (5.7) are given in Table 5.1. We note that according to the asymptotic Equation (5.8), we have at the limits of Table 5.1:

When

$$v = 0{\cdot}1, \quad K_a = 0{\cdot}06423, \quad \eta = 0{\cdot}9635$$

and when

$$v = 10, \quad K_a = 0{\cdot}9800, \quad \eta = 0{\cdot}1470$$

Thus, Table 5.1 in conjunction with Equations (5.8) and (4.89) make it possible to find values of $K_a(v)$ and $\eta(v)$ with high accuracy for any value of v.

Graphs of the functions $K_a(v)$ and $\eta_l(v)$ are given in Fig. 5.1. We can see that for large values of v, the quantity $\eta_l(v)$ is appreciably less than unity. We remember that Equations (4.88) and (5.8) give an approximate value for $K_a(v)$ and $\eta_l(v)$. In order to obtain exact values, it is necessary to turn to Mie's formulae. In order to illustrate the relationship between a_1' and a_2', we compare the two spectra for a layer of water drops in air in Fig. 5.2. We see that the spectra are considerably different. The main conclusion from Fig. 5.2 is that the absorption spectra of the dispersive system not only depends on the absorption spectrum of the substance of the particles but also on the size of the particles. In addition, since $\eta(v) \leqslant 1$, then $a_2' \leqslant a_1'$ at all times.

For a polydispersive system of spheres with a distribution curve $f(r)$ it is obvious that where τ is the volume of the sphere:

$$\overline{a_2'} = \overline{k}_a / (\overline{\tau}\sigma) \tag{5.9}$$

where:

$$\overline{k}_a = \int_{r_1}^{r_2} k_a \cdot f(r) \cdot dr$$

$$\overline{\tau} = \int_{r_1}^{r_2} \tau \cdot f(r) \cdot dr. \tag{5.10}$$

Table 5.1. Absorption cross-section and transformation coefficient of the absorption spectrum

v	$K_a(v) \times 10$	$\eta_l(v) \times 10$	v	$K_a(v) \times 10$	$\eta_l(v) \times 10$	v	$K_a(v) \times 10$	$\eta_l(v) \times 10$
0	0	10.000	1.7	6.493	5.729	4.6	9.108	2.970
0.1	0.642	9.635	1.8	6.684	5.570	4.8	9.173	2.867
0.2	1.238	9.288	1.9	6.863	5.418	5.0	9.232	2.770
0.3	1.792	8.960	2.0	7.030	5.272	5.2	9.286	2.679
0.4	2.306	8.648	2.2	7.333	5.000	5.4	9.334	2.593
0.5	2.784	8.351	2.4	7.599	4.479	5.6	9.378	2.520
0.6	3.228	8.069	2.6	7.783	4.519	5.8	9.418	2.436
0.7	3.641	7.801	2.8	8.038	4.306	6.0	9.454	2.364
0.8	4.025	7.546	3.0	8.220	4.110	6.5	9.532	2.200
0.9	4.382	7.304	3.2	8.381	3.929	7.0	9.595	2.056
1.0	4.715	7.073	3.4	8.524	3.761	7.5	9.646	1.929
1.1	5.025	6.853	3.6	8.651	3.604	8.0	9.688	1.817
1.2	5.314	6.643	3.8	8.764	3.459	8.5	9.724	1.716
1.3	5.584	6.443	4.0	8.864	3.324	9.0	9.753	1.626
1.4	5.835	6.252	4.2	8.955	3.198	9.5	9.779	1.544
1.5	6.070	6.070	4.4	9.035	3.080	10.0	9.800	1.470
1.6	6.288	5.896						

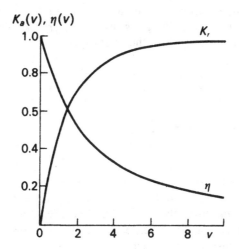

Fig. 5.1. Graphs of the functions $K_r(v)$ and $\eta(v)$.

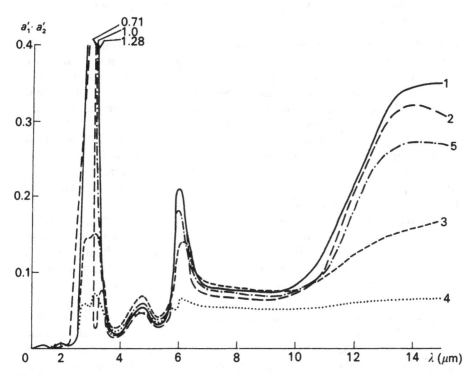

Fig. 5.2. Comparison of the specific absorption coefficient (in mg/m²) of the substance and an equivalent layer of particles. 1, water; 2, 3 and 4, monodispersive drops of water with radius 0·2, 6, 14 μm; 5, polydispersive system according to Junge. $v = 4$, $r_{min} = 0.08$ μm, $r_{max} = 10$ μm.

In order to obtain an equation analogous to Equation (5.5), we transform from the variable r to v in the integrals of (5.10) $[r = \lambda v/(8\pi k)]$, we denote $v^3 f[\lambda v/(8\pi k)] = \psi(v)$ and introduce the quantity:

$$\mu = \int_{v_1}^{v_2} \eta(v)\psi(v) \, dv / \int_{v_1}^{v_2} \psi(v) \, dv \tag{5.11}$$

Thus for a polydisperse system, we find that:

$$\overline{a_2'} = \mu a_1' = \mu \cdot 4\pi k/(\lambda\sigma), \tag{5.12}$$

In order to illustrate the relationship between a_2' and a_1', we have also plotted $\overline{a_2'}$ and a_1' in Fig. 5.2 from the data of [81] for a polydisperse system of drops of water in air. In this case, the difference is smaller than for monodisperse systems but it is still appreciable, especially in the absorption bands. This is understandable because the coefficient k is high in the absorption bands, v increases accordingly, and as v increases, the function $\eta(v)$ decreases (Fig. 5.1).

5.1.2. Absorption by phytoplankton cells, terrigenous particles and detritus

We apply the theory developed above to suspensions of plankton cells in ocean water. The variety of cell sizes has to be taken into consideration in analysing the absorption spectra of phytoplankton. Unfortunately, application of the theory developed above to the calculation of absorption spectra of different species of phytoplankton is made difficult by the lack of data on the coefficient $k(\lambda)$ for pigments in different cells. Nevertheless, simple estimates show that this effect is considerable. According to different data, the value of $k(435)$ is approximately 0·004 at the absorption maximum. This means, that for cells of radius 12 μm, $v = 1$ and for larger particles, v is even greater. Thus, the coefficient $\eta(v)$ is appreciably less than unity, ie the spectra of pigments and plankton suspensions will be appreciably different.

Data are given in Fig. 5.3 for direct measurements of the optical density (with respect to absorption) of some species of live plankton based on the data of [179]. The values of a are given in relative units along the ordinate axis. The absolute values of $a(\lambda)$ will be proportional to the concentrations of the corresponding species. It can be seen from Fig. 5.3 that the absorption by plankton has local maxima in contrast to continuous absorption by yellow substance. We can see a large maximum at $\lambda = 435$ nm (the chlorophyll blue band of 420–460 nm), a relatively transparent region between 530 and 650 nm and a second smaller maximum at 675 nm (the red chlorophyll band of 660–710 nm). The presence of two absorption maxima is a characteristic feature of plant pigments. We can add that the red maximum owes its origin to chlorophyll and its derivatives, whereas in the blue region of the spectrum, all pigments and their transformation product are absorbers.

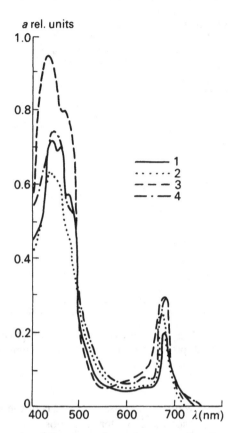

Fig. 5.3. Absorption spectra of the suspensions of some species of marine phytoplankton. 1, diatoms (*Ciclotella* sp.); 2, Dinoflagellata (*Amplidium* sp.); 3, green flagellates (*Chlamydomonas*); 4, natural populations in the waters of Woods Hole.

It is striking that the spectra of different species of plankton shown in Fig. 5.3 are qualitatively similar to each other and in general terms, coincide with the absorption spectrum of chlorophyll a, the main pigment in plankton cells. This is connected with the fact that the absorption spectra of different pigments are close to one another. Furthermore, the total concentration of pigments contained in the upper euphotic layer is closely correlated with the content of chlorophyll a. This can be seen from Fig. 5.4 in [179]. In the same study, C. Yentsch gave values for the specific optical density $E(\lambda) = \log (I_0/I)$ over a path of 1 m. The quantity E is related to the specific absorption by pigments of the phytoplankton suspension a_p^{sp}, ie the absorption relative to unit concentration C of chlorophyll a by the equation $a_p^{sp} = 2.30E$. It is given in Table 5.2.

The absolute value of the absorption coefficient a will be:

$$a_p = C \cdot a_p^{sp} \tag{5.13}$$

According to the value of the concentration C, ocean waters are divided into

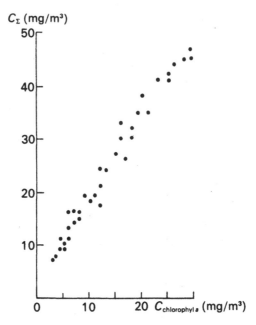

Fig. 5.4. Relationship between the concentration of chlorophyll a, $C_{chlorophyll\ a}$ and the total concentration of all pigments C_Σ in natural populations of phytoplankton.

three types. Oligotrophic waters are those where C is of the order of $0.1\ mg/m^3$ in the top 100 m layer, mesotrophic waters are those where $C = 0.3\ mg/m^3$ in the top 100 m layer and eutrophic waters are those where C is of order of $0.5\ mg/m^3$ and above in the top 100 m layer. At the blue absorption maximum ($\lambda = 440\ nm$), a will be on average 0.01, 0.03 and $0.05\ m^{-1}$ for the above three types, respectively and at the red maximum ($\lambda = 670\ nm$), a will be 0.004, 0.012 and $0.02\ m^{-1}$. Equation (5.13) and the concept of a_p^{sp} associated with it are based on Beer's law. It is clear from the above that this law is only true when the characteristics of the distribution of the cells over size $[f(r), r_1$ and $r_2]$ and the optical constants of the substance do not vary. If this condition is not observed, then a will not be simply proportional to the concentration C. In

Table 5.2. Specific absorption coefficient of phytoplankton pigments a_p^{sp} (m^2/mg)

$\lambda(nm)$	400	410	420	430	440	450	460	470	480
$a_p^{sp} \times 10^4$	506	598	736	805	966	874	736	598	575
$\lambda(nm)$	490	500	510	520	540	560	580	625	630
$a_p^{sp} \times 10^4$	437	253	172	115	71	60	69	69	87
$\lambda(nm)$	645	650	655	660	665	670	675	700	
$a_p^{sp} \times 10^4$	69	80	92	138	230	414	246	41	

order to check the situation for real ocean water, a_p^{sp} was calculated by two different methods in [48] for the spectral region 430–435 nm: (1) theoretically from the data on the spectrum of suspended particles and the values of k determined by the equations given in Chapter 4; (2) directly as the difference between the total a and the sum of the absorption coefficients of pure ocean water a_w and yellow substance a_y. For all samples, the concentration of chlorophyll a was measured simultaneously. The results are given in Table 5.3.

Table 5.3 gives the average values for the samples and the variation $\delta_{a_p^{sp}}$. We can see that the averages are close to the data given in Table 5.2 but the variation in a_p^{sp} is large, amounting to almost 100%. Possibly some of the variation in a_p^{sp} should be ascribed to measurement or calculation errors. However, most of it undoubtedly indicates the considerable variability of the cells and the composition of pigments. The quantity $\delta_{a_p^{sp}}$ determines the accuracy with which it is possible to use the data given in Table 5.2 to determine a_p.

Experimental data on the absorption by suspended particles are either obtained from suspension samples or as the difference between the absorption of filtered and unfiltered samples. When making measurements on the samples, a comparison is made between the clean filter and the filter with particles. The attenuation is actually determined but this is done so that all of the scattered light together with the direct beam enter the receiver. An integrating sphere or an opal glass plate is used since the volume scattering function of the suspension is strongly elongated and in practice it is adequate to collect the light scattered in the front hemisphere. In addition, for such measurements. Yentsch suggested the "anti-reflection" filter method [180]. He coated a filter with a specimen of cedar oil and assumed that this caused the scattering to disappear since the refraction coefficients of oil and the suspension are similar. This is the same idea which is used in Christiansen filters. However, the explanation of Yentsch does not hold. In fact, the scattering does not disappear: (1) it is impossible for n_{abs} to coincide exactly; (2) the particles and the medium differ in terms of k. However, the closeness of n_{rel} to unity leads to a strong elongation of the scattering function and in the scheme of [180], the fraction of scattered light entering the receiver is increased sharply. Therefore, the data in [180] with the coated filter turned out to be similar to the data obtained with the opal plate.

Some of the results of measuring the absorption by suspended particles taken

Table 5.3. Variation in the specific absorption coefficient of phytoplankton pigments

	$a_p^{sp} \times 10^4 (m^2/mg)$	$\delta_{a_p^{sp}}(\%)$	Number of samples
Theoretical calculation from the suspension spectrum	800–1180	80	45
Difference $a - (a_w + a_y)$	940	90	69

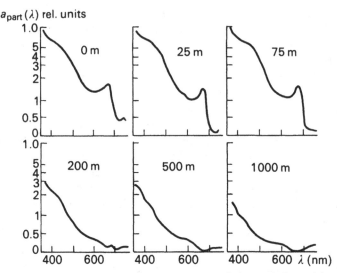

Fig. 5.5. Examples of the absorption spectrum of suspended particles $a_{part}(\lambda)$. The samples were taken at specified depths in December 1960 in the north-western part of the Atlantic Ocean.

from [180] are shown in Fig. 5.5. We can see that although in the upper layers pigment absorption bands can be seen clearly, below 100 m a steady increase in absorption in the short-wave region is observed. It is very similar to the absorption by yellow substance. Oscillations in the gradient of $a(\lambda)$ curves indicate the effects of different attenuating components. Figure 5.5 refers to the entire suspension. Ch. Yentsch notes that in living cells at depths > 5–10 m there is only a blue absorption band left, since only blue light penetrates there (Fig. 1.5).

A detailed analysis of the absorption spectra of suspensions was carried out by B.V. Konovalov [40]. He studied more than 1600 spectrophotograms for samples taken within the euphotic layer in the Pacific, Atlantic and Indian Oceans as well as the Sea of Japan, the Barents Sea and the Black Sea. Some of the results obtained by him are shown in Fig. 5.6. It must be emphasized that both in the complete spectra of suspensions and in the spectrum of extracts of phytoplankton pigments a large number of absorption bands can be seen. This indicates the large variety of absorbing substances in ocean water. Their absorption bands are spread practically over the entire visible spectrum. The author points out that among these substances it is possible to find up to 50 carotenoids, a number of chlorophylls and more than 10 products of their transformation. However, apart from chlorophylls and the products of their transformation, only fucoxanthin, peridinin, α- and β-carotenes, diatoxanthin and α-zeacarotin are found in appreciable concentrations. When the phytoplankton pigments are destroyed (this is done by treating the suspension with hydrogen peroxide and subjecting it to ultraviolet radiation), only the absorption due to detritus and mineral particles remained. The modified spectrum indicated a

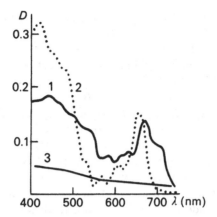

Fig. 5.6. Absorption spectra (optical density) for (1) a marine suspension, (2) a mixture of pigments extracted from the suspension and (3) suspensions in which the pigments have been destroyed.

steady increase in absorption at the short-wave end. Naturally, the possibility of 'seeing' bands of specific pigments depends on the amount of absorption due to mineral particles and detritus. In terms of the increase in this amount and the gradual disappearance of spectral details, the author divides all the absorption spectra which he obtained into four types. In the fourth type, there is only one absorption band left at $\lambda = 675$ nm and a transmisson band at $\lambda = 652$ nm, with the amount of absorption due to detritus and mineral particles reaching 80%. Below $\lambda = 652$ nm, towards shorter wavelengths, the absorption steadily increases.

5.2. Absorption of light by ocean water

5.2.1. Overview

We now turn to an analysis of the overall effect—the absorption coefficient of ocean water, ie an analysis of the function $a(\lambda, z)$. Although the specific shape of the absorption spectrum $a(\lambda)$ depends on the mutual contribution of different components, three general characteristics can be used to give some basic guidelines to reduce the chaos of the experimental spectra. First, since, as a whole, the spectral variation of the scattering coefficient $b(\lambda)$ is small, all $a(\lambda)$ curves are similar to the $c(\lambda)$ curves shown in Fig. 1.3. They have a minimum in the blue-green region. An increase in a with increasing λ is due to pure water and an increase in a with decreasing λ is due to dissolved substances and particles. The position of the minimum is associated with the absolute absorption values. In pure ocean water, the minimum absorption is found in the vicinity of 510 nm and in more transparent waters the minimum is shifted to 490 nm or even to 470 nm, ie it coincides with the position of the absorption minimum of pure water. As the turbidity increases, the minimum is shifted to larger λ. Thus, for the turbid waters of the Baltic Sea, it lies in the yellow-green or even yellow section of the spectrum (up to 570 nm). The transformation of the spectrum can be seen from Fig. 5.7 from [179] where it was shown that the absorption minimum is shifted for different concentrations of chlorophyll a.

This first main feature in the structure of the spectrum $a(\lambda)$ determines the spectral composition of daylight at different depths in the ocean. The scattering by itself is weakly wavelength selective; it only increases the free path of photons in water. As a result, daylight in the sea becomes blue with increasing depth (Fig. 1.5). Curves similar to those shown in Fig. 1.5 have been obtained by many authors. We should not be surprised that they do not contain absorption bands of dissolved or suspended substances, although it is well known that in the atmosphere, the detailed spectrum of solar radiation is very jagged. In ocean water, as in any condensed system, there are only blurred broad absorption bands due to the strong interaction between the molecules. In addition, in standard hydrooptical measurements which are carried out at large wavelength intervals, all spectral details disappear. When instruments with high spectral resolution are used and what is more important with continuous recording of the spectrum (eg submerged grating monochromators), it can be seen that in waters rich in plankton, the spectrum of downwelling, irradiance has a two-peak distribution separated by a trough corresponding to the absorption band of chlorophyll and associated pigments in the region of $\lambda = 430$ nm. By way of

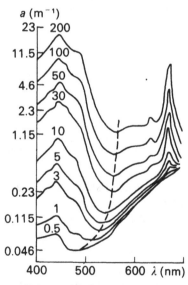

Fig. 5.7. Total absorption coefficient $a(\lambda)$ of a mixture of pure water and pigments. The numbers on the curves indicate the concentration of chlorophyll a in mg/m^3. The dotted line indicates the position of the minimum.

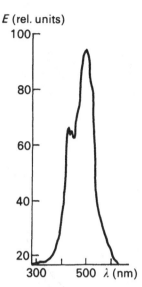

Fig. 5.8. Example of the spectrum of underwater irradiance obtained by means of an underwater monochromator (a trough can be seen in the spectrum in the region of $\lambda = 430$ nm).

example, Fig. 5.8 based on [26] shows the recording of spectral irradiance in the inland waters of the Java Sea distinguished by their increased turbidity due to high biological productivity. A similar trough is observed in the spectrum of upwelling irradiance. The depth of the trough at $\lambda = 430$ nm was found to be approximately proportional to the concentration of chlorophyll with a correlation coefficient of $r = 0.88$ [26].

The second common characteristic relating to $a(\lambda)$ can be seen from the curves in Fig. 5.9 and from the data given in Table 5.4 (based on [47, 48]). In the red region, the various curves of $a(\lambda)$ are coincident. A great variation in the absorption spectra is observed at short wavelengths. In the waters of the Baltic Sea, for example, absorption in this part of the spectrum is almost two orders of magnitude higher than the absorption in ocean waters. In the red region, the absorption is mainly due to pure water. In relatively unproductive ocean waters, the boundary where all curves of $a(\lambda)$ coincide, lies at approximately $\lambda = 570$ nm. For $\lambda \leqslant 570$ nm, the $a(\lambda)$ spectra are different and indicate the composition of the individual absorption components; for $\lambda > 570$ nm, the $a(\lambda)$ curves practically coincide. In the red region, the absorption by pure water is so high that in ocean waters as a rule it is impossible to see the red absorption maximum of chlorophyll at 670 nm which usually does not exceed $0.02\,\mathrm{m}^{-1}$. Its value is small in comparison to the absorption by pure water (which is $0.34\,\mathrm{m}^{-1}$ at 670 nm). In contrast to this, in lakes and rivers where there is a large amount of chlorophyll, the red maximum can be detected. In Fig. 5.10, we give an example of the recording of the absorption spectrum of a surface sample of water from the

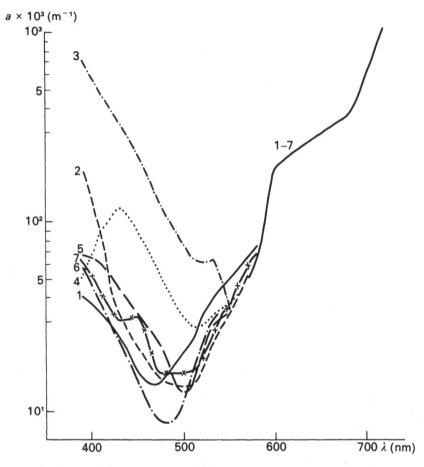

Fig. 5.9. Spectral absorption coefficient of light for different waters (m^{-1}). 1, Sargasso Sea; 2, Caribbean Sea; 3, The Gotland Trench in the Baltic Sea; 4, Pacific Ocean, area of the Galapagos Islands at a depth of 20 m; 5, Pacific Ocean area of the Galapagos Islands at a depth of 200 m; 6, The Tonga Trench; 7, The Trade Wind current in the Indian Ocean.

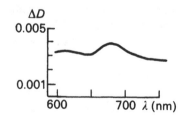

Fig. 5.10. Absorption spectrum of the waters of the Baikal (a clear maximum can be seen in the region of the red band of chlorophyll).

Table 5.4. Values of the spectral absorption coefficients $a(\lambda) \times 10\,(m^{-1})$ in different oceanic regions

Region	Level (m)	Wavelength (nm)									
		390	410	430	450	470	490	510	530	550	570
Atlantic Ocean											
North Trade Wind current	0	0.32	0.34	0.21	0.18	0.14	0.12	0.18	0.30	0.34	0.55
Sargasso Sea	0	0.41	0.34	0.25	0.16	0.14	0.18	0.23	0.39	0.50	0.67
Gulf Stream	10	0.41	0.44	0.44	0.41	0.30	0.30	0.25	0.39	0.46	0.60
Caribbean	0	2.0	0.9	0.34	0.23	0.16	0.14	0.14	0.23	0.32	0.55
Pacific Ocean											
Northern subtropical convergence	85	0.85	0.71	0.48	0.37	0.23	0.14	0.14	0.29	0.39	0.63
The Gulf of Panama	10	1.2	1.4	1.3	1.1	0.80	0.60	0.44	0.48	0.51	0.62
The Tonga Trench	10000	0.58	0.37	0.23	0.12	0.09	0.09	0.16	0.25	0.32	0.55
Waters of the Continental Shelf near Peru	10	3.8	3.0	2.5	1.8	1.4	1.0	0.8	0.73	0.74	0.70
	500	1.6	1.3	0.82	0.63	0.41	0.27	0.26	0.3	0.43	0.49
Indian Ocean											
Trade Wind current	0	0.55	0.37	0.29	0.31	0.17	0.15	0.18	0.28	0.41	0.50
Southern tropical convergence	0	0.53	0.46	0.43	0.38	0.28	0.24	0.27	0.36	0.46	0.52
	500	0.25	0.16	0.07	0.03	0.03	0.04	0.10	0.22	0.35	0.47
Central part of the Bay of Bengal	10	0.53	0.37	0.36	0.28	0.21	0.18	0.16	0.25	0.38	0.50
Close to the Ganges Delta	10	0.33	0.6	2.0	1.6	1.3	0.98	0.68	0.56	0.54	0.58
Baltic Sea											
Gotland Trench	0	7.1	4.8	3.4	2.3	1.5	0.90	0.64	0.64	0.32	0.55
The Gulf of Irbensk	0	18.0	12.0	—	6.9	4.8	3.4	2.3	2.3	1.7	1.5
The Gulf of Riga	0	27.0	19.0	12.0	9.0	8.3	6.2	4.4	3.9	3.0	2.5

Baikal Lake, 7 km from the Ust-Anga Bay (concentration of chlorophyll 0·6 mg/m³) [2].

The form of the $a(\lambda)$ spectrum in the short-wave region ($\lambda < 570$ nm) which is of particular interest, depends on the ratio of the two main absorption components: dissolved organic matter and phytoplankton pigments. The pigments have characteristic bands and if there is a large amount of phytoplankton as is observed in the upper eutrophic layer, then a characteristic maximum can be seen on the $a(\lambda)$ curves. In Fig. 5.11, we show $a(\lambda, z)$ data for different depths in the Gulf Stream obtained during the fifth voyage of the scientific research ship 'Dmitry Mendeleyev'. At depths of 50 and 100 m, a blue maximum can be clearly seen. It is also observed at other depths but it less marked and is sometimes just noticeable. Thus, we can formulate a third common principle concerning the form of the $a(\lambda)$ function in the short-wave part of the spectrum. The absorption spectra may be divided into two types: the first in which the absorption steadily increases towards shorter wavelengths and the second in which there is a small local maximum at 410–430 nm. A predominance of dissolved substances leads to the first type of spectrum and 'living' suspensions to the second type. Typical results are shown in Figs 5.12 and 5.13 in which absorption exceeding that of distilled water is shown. Figure 5.12 refers to the waters of the Baltic Sea rich in yellow substance and dissolved organic matter. Although some variation in the absorption spectrum with depth is observed here, all the spectra belong to the first type. Figure 5.13 shows absorption spectra in the region of transparent waters to the north-west of the island of Rarotonga.

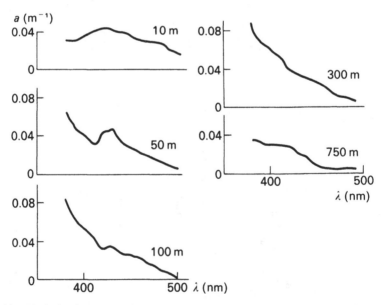

Fig. 5.11. Variation in the absorption spectrum with depth (Gulf Stream, February 1971, fifth voyage of the scientific research ship 'Dmitry Mendeleyev').

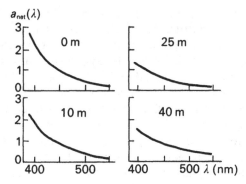

Fig. 5.12. Values of $a_{net}(\lambda)$, ie the spectra of absorption which exceeds that of pure water for samples taken from different depths of the Baltic Sea (Gulf of Riga).

The sharp maximum at 430 nm at the 50 m depth decreases appreciably at the 200 m depth but on the other hand the absorption increases at 380–390 nm. At the depths of 500 and 2000 m, these changes in the spectrum are found to be even stronger. The picture shown in Fig 5.13 is typical for ocean waters. It can be seen clearly that the absorption by suspensions descreases with depth. The absorption by dissolved substances on the other hand is greater at greater depths than at the surface. At the 200 m depth, it is approximately two times greater than at 50 m; at 2000 m it decreases somewhat but still exceeds the absorption by dissolved substances at 50 m. The variation in the absorption spectra with depth is related to the variation in the composition of the water. In the surface

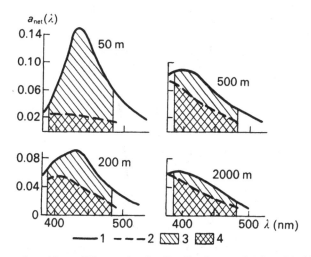

Fig. 5.13. Values of $a_{net}(\lambda)$ at different depths (Pacific Ocean, April 1971, fifth voyage of the scientific research ship 'Dmitry Mendeleyev'). 1, absorption by unfiltered samples; 2, absorption by filtrates; 3, absorption by suspended matter; 4, absorption by dissolved substances.

layer, cells of living phytoplankton play an essential role. They decompose at greater depths so the contribution of dissolved organic matter increases. Of course, the division of absorption spectra into two types should be regarded as a simplified model. In actual conditions, curves are possible which are intermediate between the two types (Fig. 5.11). We can add that the second type (curves with a maximum) are found only in the upper 200 m layer but even here it is less widespread than the first type. In the near ultraviolet region, the absorption sharply increases at shorter wavelengths due to yellow substance and possible due to other components and also as a consequence of inorganic salts. Therefore, when considering the ultraviolet region, spectra of the second type must have the form shown in Fig. 5.11 at the 50 m depth.

The variation in absorption with depth is shown in Fig. 5.14. As a rule, the absorption decreases with depth and this decrease is greater for 430 nm than for 390 nm. However, in some cases, a local increase in absorption is observed in intermediate layers and in particular in the vicinity of the bottom (Fig. 5.14).

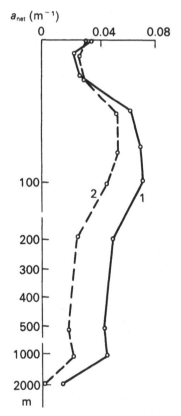

Fig. 5.14. Examples of variation in absorption which exceeds that of pure water, a_{net} with depth (Indian Ocean, Equator, 76° longitude east). 1, $a_{net}(390)$; 2, $a_{net}(430)$.

5.2.2. Variation in the absorption coefficient

In [48], the variation $a_{net}(\lambda)$ was examined. Data from the Indian and Pacific Oceans were considered but only values of $a_{net}(\lambda)$ at $\lambda \leqslant 490$ nm were analysed; at larger λ the coefficient $a_{net}(\lambda)$ is small and the accuracy of its measurement is inadequate. Furthermore, as λ increases, the absorption by pure water increases so that the variation in the absorption coefficient of seawater $a(\lambda)$ decreases. For surface samples from the Indian Ocean, for example, the variation coefficients were found to be 43, 19, 10 and 6 % for 510, 530, 550 and 570 nm, respectively. The results of the calculations are given in Table 5.5.

We note that the absorption in surface waters is higher than in deep waters and that all average spectra (even surface spectra) belong to the first type. This is due to the greater occurrence of spectra of this type. The spectra of deep waters were found to be steeper. In surface waters, when λ changes from 390 to 490 nm, a_{net} decreases by a factor of 3·4–3·8 and in deep waters considerably more (a factor of 5·8–5·5). What is interesting is that it is not only the shapes of the spectra which are similar in the two oceans but also the absolute values of the absorption coefficients of the deep waters. Not too much importance should be attached to the difference in $\overline{a_{net}}$ of surface waters. Simply, the majority of samples taken from the Pacific Ocean belong to the turbid waters in the neighbourhood of the Peruvian coast and to the east of the equatorial region. Let us examine the data on the magnitude of the variation δ_a. First of all, it should be pointed out that the variations are highly significant. For the surface samples, the variations are 130–170 %. For deep samples from the Pacific Ocean (where the sampling is more representative), they are appreciably less: 50–80 %. This seems understandable but the difference in δ_a in the deep waters of both oceans is more likely to be caused by the small number of

Table 5.5. Average values of the absorption coefficients $\overline{a_{net}}(\lambda)$ (m^{-1}) and average variation coefficients for these quantities $\delta a(\%)$ per sampling

λ(nm)	Indian Ocean				Pacific Ocean			
	$\leqslant 100$ m		> 100 m		$\leqslant 100$ m		> 100 m	
	$\overline{a_{net}}$	δ_a	$\overline{a_{net}}$	δ_a	$\overline{a_{net}}$	δ_a	$\overline{a_{net}}$	δ_a
390	0.068	156	0.064	133	0.095	132	0.072	57
410	0.056	146	0.054	124	0.077	136	0.058	53
430	0.046	143	0.039	130	0.063	149	0.042	50
450	0.035	140	0.026	142	0.047	151	0.028	54
470	0.028	132	0.017	182	0.036	161	0.018	61
490	0.020	145	0.011	182	0.025	172	0.011	82
No. of samples	69		24		77		79	

samples and the particular local measuring conditions in the Indian Ocean. The spectral variation of the mean square deviations is similar to the variation of the mean values—they increase as λ decreases. The opposite change in the variation coefficients is associated with an increase in the contribution of measurement errors of a_{net}. The mean square deviations in both oceans is higher in surface waters than in deep waters. Of course, the data given in Table 5.5 are not universal. This is the result of treating a specific set of samples but since there are a large number of these samples, they can be used for various estimates.

Since absorption is essentially determined by a small number of varying components, the question naturally arises as to the possibility of an approximate determination of $a(\lambda)$ for any wavelength from its value for some fixed wavelengths. In spectra of the first type, which constitute the majority, it is likely that one fixed wavelength is sufficient. In this connection, the correlation coefficients $r[a(\lambda_i), a(\lambda_j)]$ were calculated for all four samples specified in Table 5.5. The coefficients were greater than 0.96 over the entire spectral range investigated. The only exceptions were the values $r[a(490), a(\lambda)]$ for deep waters of the Pacific Ocean (0.75–0.95) but his could be the effect of experimental errors which are more pronounced for these samples, since natural variations in the absorption are smaller here.

Taking into account the high values of the correlation coefficients, the coefficients $A(\lambda)$ and $B(\lambda)$ in the regression equation:

$$a(\lambda) = B(\lambda) \cdot a(\lambda_0) + A(\lambda) \qquad (5.14)$$

were calculated in [44] for the combined data of Table 5.5. A wavelength of 430 nm was selected as λ_0. It was found that the regression errors do not exceed $0.02 \, m^{-1}$.

This procedure for the calculation of the $a(\lambda)$ spectrum suffers from the drawback that it is essentially based on the value of $a(430)$. Random errors in this quantity may play an unjustifiably large role. Therefore, another procedure was considered in [48]. The eigen values and vectors of the covariant matrices were determined for the data from the same array. We remember that the expansion of a random function in eigen vectors of its covariant matrix ensures the most accurate representation (in the sense of the mean square approximation) by a small number of terms of the series. It can be proved that the eigen vectors are not correlated and that the total dispersion is equal to the sum of the eigen values of the matrix. The value of the method is explained by the fact that usually two to three eigen values are sufficient to obtain the dispersion. Thus, in our case it was found that the first eigen value practically corresponded to the overall dispersion (98.7% for surface waters and 87.5% for deep waters). Therefore, in expanding the $a(\lambda)$ spectrum it was sufficient to limit the expansion to only one eigen vector $\psi_1(\lambda)$:

$$a(\lambda) = \overline{a(\lambda)} + c_1 \psi_1(\chi) \qquad (5.15)$$

where c_1 is the corresponding expansion coefficient. The values for $\overline{a(\lambda)}$ and $\psi_1(\lambda)$ for surface and deep waters are given in Table 5.6.

Table 5.6. Average values of the absorption coefficient $\overline{a(\lambda)}\,(\mathrm{m}^{-1})$ of ocean water and the first eigen vectors $\psi_1(\lambda)\,(\mathrm{m}^{-1})$

λ(nm)	Surface waters $\leqslant 100\,\mathrm{m}$ (135 samples)		Deep waters $> 100\,\mathrm{m}$ (103 samples)	
	$\overline{a(\lambda)}$	$\psi_1(\lambda)$	$\overline{a(\lambda)}$	$\psi_1(\lambda)$
390	0.085	0.595	0.078	0.668
410	0.071	0.496	0.063	0.524
430	0.057	0.433	0.045	0.379
450	0.044	0.326	0.030	0.270
470	0.033	0.261	0.020	0.209
490	0.026	0.194	0.015	0.141

The average theoretical error over the spectrum of Equation (5.15) is only $0.003\,\mathrm{m}^{-1}$ for surface waters and $0.002\,\mathrm{m}^{-1}$ for deep waters. The coefficient c_1 may be determined from the measured value of $a(\lambda_0)$ and in this method any λ from the investigated spectral range may be taken as λ_0. Of course, it is better to select λ_0 in the short-wave part of the range as the experimental error is smaller here.

A verification of the possibility of recalculating $a(\lambda)$ using this method for 69 surface samples from the waters of the Indian Ocean (based on the data of the tenth voyage of the scientific research ship 'Dmitry Mendeleyev') showed that in the majority of cases, the errors did not lie outside the limits of the theoretical estimate. The maximum deviations on using a wavelength of 390 nm as λ_0 did not exceed $0.01\,\mathrm{m}^{-1}$ in 43 cases and in only six cases exceeded $0.02\,\mathrm{m}^{-1}$; five of these referred to the stations No. 783 and 784 situated in the vicinity of the Ganges Delta. The coordinates of these stations are given in L.M. Brekhovskikh [9]. Apart from these unique waters, low accuracy was obtained for the second type of spectra. This is natural since Table 5.6 is based on the first type of spectra. If $a(390)$ is used, then Equation (5.15) gives a reduction in the region of the maximum for these spectra. Fortunately, the second type of spectra is quite rarely encountered in the ocean, the values of the maxima are small so that correspondingly the errors resulting from Equation (5.15) will not be very large.

5.2.3. Relative role of the different components

The absorption of light by ocean water is governed by a complex set of many components. Attempts to explain the relative roles of different components have been made by measuring the absorption by filtered and unfiltered samples of ocean water [44] and suspended matter collected on filters with different pore sizes [40, 48, 180] with intact or destroyed pigments.

According to the 14 samples obtained during the fifth voyage of the scientific research ship 'Dmitry Mendeleyev', the percentage of suspended matter which contributes to the total absorption in the upper 200 m layer is approximately 40% at 390 nm and 70% at 430 nm. As a rule, the absorption spectra for suspended materials do not have a clearly marked maximum at 430 nm as in Fig. 5.11 at the 50 m depth; they only have a well-marked red absorption maximum for pigments while the blue maximum is usually smooth. We have already noted the hypothesis of C. Yentsch that this arises from the absorption by yellow substance adsorbed on small suspended particles. This hypothesis was also confirmed by new data obtained at the Institute of Oceanology of the USSR Academy of Sciences [40, 48]. According to these data, the absorption by suspended particles with degraded pigments is usually 0·3–0·8 of the total absorption by the precipitate on the filter.

We represent the absorption coefficient of seawater in the form of a sum:

$$a(\lambda) = a_y^{sp}(\lambda) \cdot C_y + a_p^{sp}(\lambda) C_{chl\,a} + a_w(\lambda) \qquad (5.16)$$

where $a_y^{sp}(\lambda)$ and $a_p^{sp}(\lambda)$ are the specific absorption coefficients for yellow substance and phytoplankton pigments; C_y and $C_{chl\,a}$ are their concentrations; $a_w(\lambda)$ is the absorption coefficient of pure water.

For the sake of simplicity, we will assume that the spectral functions $a_y^{sp}(\lambda)$ and $a_p^{sp}(\lambda)$ are invariant for different waters; for $a_y^{sp}(\lambda)$ we assume the function $e^{-0\cdot015\lambda}$ and for $a_p^{sp}(\lambda)$ we take the values from Table 5.2 for the lack of anything better. Since the concentration of yellow substance is determined from its absorption (Chapter 3), we write the first term in Equation (5.16) in the form $a_y(390)e^{-0\cdot015(\lambda-390)}$, where $a_y(390)$ is the absorption coefficient of yellow substance at 390 nm. Its value can be found from the difference between the measured value of $a_{net}(390)$ and the value of $a_p^{sp}(390)$. $C_{chl\,a}$. Equation (5.16) represents a two-parameter model: its input parameters are the absorption coefficient $a(390)$ at 390 nm and the concentration of chlorophyll a, $C_{chl\,a}$. The model of absorption described by Equation (5.16) is very crude. It is based on Beer's law and as we saw at the beginning of this chapter it does not apply to the second or even to the first term in Equation (5.16). Nevertheless, in [48], it was used to calculate the average absorption spectrum for the surface waters of the Indian Ocean. The results were compared with the data given in Table 5.5. The deviation between the calculated and measured values does not exceed 0·004 m^{-1} (Fig. 5.15). For individual samples, the agreement is of course worse, but here, the contribution of the measurement errors of the input paameters in Equation (5.15) increases: $a(\lambda)$ and the chlorophyll concentration $C_{chl\,a}$. However, the deviations are not very large for individual samples: in 69 surface samples from the Indian Ocean, only in 12 cases do they exceed 0.01 m^{-1} and not once were they greater than 0.02 m^{-1}.

Using Equation (5.15), it is possible to assess the role of various factors in the overall absorption of light by seawater. The results of such calculations on the Indian Ocean (data from the tenth voyage of the scientific research ship 'Dmitry Mendeleyev') are given in Table 5.7. In the blue region, the predominant

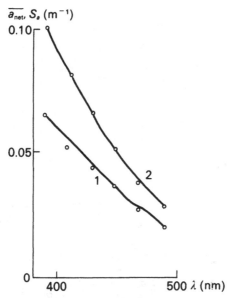

Fig. 5.15. Calculation using Equation (5.16) of the average spectrum $a_{net}(\lambda)$ (1) and the mean square deviations $S_a(\lambda)$ (2) for the surface waters of the Indian Ocean. The curves are the measured values and the circles are the calculated values.

factor in the absorption is yellow substance even in the surface waters where the contribution by pigments does not exceed 35% and the contribution by water 20%. For individual samples, the contribution by pigments is a maximum at 440–450 nm and varies from 3 to 82%. In the overall spectrum, the blue absorption maximum for pigments can only be seen when the contribution by pigments is 70% and above. However, such situations are rarely found. In fact, as observations show, there is a marked correlation between $C_{chl a}$ and C_y. For the same 69 samples from the Indian Ocean, the correlation coefficient between $a_y(390)$ and $C_{chl a}$ is 0.577 which with a confidence level of more than 0.999 indicates the existence of a significant relationship between these variables. This means that if, for example, the contribution by pigments begins to increase then the contribution by yellow substance also increases and the blue maximum does not appear [48]. Therefore, the majority of absorption spectra belong to the first type even in the presence of significant concentrations of chlorophyll.

It is also possible to calculate the dispersion $D[a(\lambda_i)]$ from Equation (5.15). Obviously, it is expressed in terms of the dispersion of C_y and $C_{chl a}$ and the covariation between these variables (since they are closely related). Such calculations are given in [48] and they reveal that the main part of the dispersion is associated with yellow substance (from 50 to 90%), whereas the variation due to pigments does not exceed 9% even at the absorption maximum at $\lambda = 450$ nm. Figure 5.15 also gives a comparison of the mean square deviations.

Table 5.7. Percentage contribution by various components in the spectral absorption of light by seawater

Components	Spectral region (nm)						
	400–430	430–490	490–510	510–540	540–600	600–670	670–680
Surface waters Oligotrophic							
Pigments	15–30	30–35	10–30	2–10	1–2	1	1
Yellow substance	60–70	50–60	30–50	10–30	1–10	0.5	0.5
Pure water	15	15–20	20–60	60–90	90–98	99–99	98–99
Mesotrophic							
Pigments	15–25	25–35	10–30	3–10	1–3	1	1–2
Yellow substance	65–80	55–65	45–55	20–45	2–20	2	1
Pure water	9	5–15	15–40	40–80	80–95	96–97	97–98
Eutrophic							
Pigments	10–20	20–25	10–25	4–10	1–4	1	2–4
Yellow substance	80–90	70–80	65–70	40–65	4–40	4	1
Pure water	3	3–7	7–30	30–60	60–95	95	96
Deep waters							
Yellow substance	90	70–90	40–75	10–45	1–15	1	0
Pure water	10	10–30	25–60	55–90	85–99	100	100

Equation (5.15) reproduces them approximately with the same accuracy as the average values $\overline{a_{net}}(\lambda)$.

5.3. Light scattered by ocean water

5.3.1. Scattering by non-spherical and inhomogeneous particles

We now turn to an analysis of the scattering properties of real ocean water. As we saw in Chapter 2, suspended particles are heterogeneous and non-spherical. There is no general method for the exact calculation of the scattering of light by such complex particles. The model of equivalent spheres which we considered in Chapter 4 only allows the picture of scattering to be described to a first approximation. We look at some more detailed models.

A simple method for calculating the light scattered by an ensemble of particles in the RHD approximation was suggested in [111]. We give a brief description. The volume under consideration contains a large number of particles distributed randomly. This difficulty which usually complicates the calculations since integration is required over the ensemble of scatterers may be overcome by averaging. By this method, the scattering by a system of complex particles may be reduced to the scattering of one particle of random structure. The statistical properties of this structure reflect the integral characteristics of the investigated ensemble. The ensemble approach, developed below, appreciably simplifies the calculation in a number of cases. For a large number of particles, which are not even completely randomly oriented, the ensemble acquires properties of symmetry which the individual particles do not have. In particular, the situation is simplified if the particles are oriented completely randomly, as in the case of white noise.

Since our aim is the analysis of light scattering by irregularly shaped particles, we have to discard the method of dividing the variables in curvilinear coordinates right from the beginning and use the integrodifferential equation method described in Chapter 4 [Equations (4.77) and (4.78)]. In the Fraunhofer approximation, we find for the amplitude of the scattered field of a system of N particles:

$$\mathbf{E}_p = \varepsilon_1 \sum_{m=1}^{N} e^{-i q \mathbf{r}_m} u_m(\mathbf{q}),$$

$$\mathbf{q} = \mathbf{k} - k\mathbf{R}_0, \quad \varepsilon_1 = \alpha \mathbf{E}_0 \perp k^2 / R \tag{5.17}$$

where $\mathbf{r}_m$ is the radius vector from the centre of the ensemble to the centre of the mth particle, $\mathbf{R}$ is the radius vector to the point of observation ($\mathbf{R}_0 = \mathbf{R}/R$), $\mathbf{r}''_m$ is the radius vector from the centre of the mth particle to the volume element. The quantities α and $E_0 \perp$ are the same as in Chapter 4; $u_m(\mathbf{q})$ is the shape factor of the particle:

$$u_m(q) = \int_{v_m} e^{-i q \mathbf{r}''_m} d\mathbf{r}''_m \tag{5.18}$$

Due to the random displacements of the centres of the particles, and random rotations with respect to the centres O_m (Fig 5.16a) [the value of $u_m(q)$ varies with this], there is a random variation in the electric field intensity.

The electric field of the scattered radiation $\langle E(r) \rangle$ in the wave zone is determined by averaging the expression:

$$\langle E_s(r) \rangle = \varepsilon_1 \left\langle \sum_{m=1}^{N} e^{-iq\mathbf{r}_m} u_m(q) \right\rangle \tag{5.19}$$

We assume that the coordinates of the polarizability centre of the particle and its shape factor are independent random variables. Then, as has been shown in [111], the estimate for the average field [Equation (5.19)] will be:

$$\langle E_s(\mathbf{r}) \rangle = N\varepsilon_1 f_1(\mathbf{q}) \frac{1}{N} \sum_{m=1}^{N} u_m(\mathbf{q}) \tag{5.20}$$

In this case, $f_1(q)$ is a factor characterizing the interference between the particles (Chapter 4). For a spherical volume, it is analogous to the quantity $R(q)$ in Section 4.4.1. It is a characteristic function of the probability $\omega(r)$ of finding the particle centre inside the illuminated volume V:

$$f_1(q) = \langle e^{-iqr} \rangle = \int_V e^{-iqr} \omega(\mathbf{r}) \, d\mathbf{r} \tag{5.21}$$

When $q = 0$, the function $f_1(q) = 1$. As q increases, the function $f_1(q)$ rapidly decreases and in the region $\gamma \gtrsim \lambda/L$ becomes close to zero [L is the size of the region, see Equation (4.96)].

It is possible to simplify the calculations of $u_m(q)$ appreciably if it is noted that the three-dimensional integral may be transformed to a one-dimensional integral; since the value of $\mathbf{q}\mathbf{r}_m''$ in the phase factor is constant in the plane perpendicular to q (Fig. 5.16a):

$$u_m(q) = \int_{-\infty}^{\infty} B_m(q, b) e^{-iqb} \, db \tag{5.22}$$

where $B_m(q, b)$ is the cross-sectional area of the mth particle perpendicular to q at a distance of b from the centre of the particle. Denoting $\overline{B(\mathbf{q}, b)}$ as the average of the sampled values of the cross-sectional areas of the particles:

$$\overline{B(\mathbf{q}, b)} = \frac{1}{N} \sum_{m=1}^{N} B_m(\mathbf{q}, b) \tag{5.23}$$

and n as the concentration of particles in the volume $V(n = N/V)$, we obtain the final expression for the estimate of the average field:

$$\langle E(r) \rangle = nV\varepsilon_1 f_1(q) \int_{-\infty}^{\infty} \overline{B(\mathbf{q}, b)} \, e^{-iqb} \, db \tag{5.24}$$

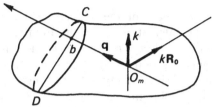

Fig. 5.16a. Arrangement of the vectors in the ensemble scattering problem. The point O_m is the centre of the mth particle; R_0 is the direction vector to the point of observation; CD is the cross-section of the particle perpendicular to the vector q at a distance of b from O_m.

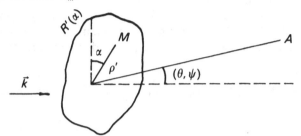

Fig. 5.16b. Scattering geometry by a "random" screen.

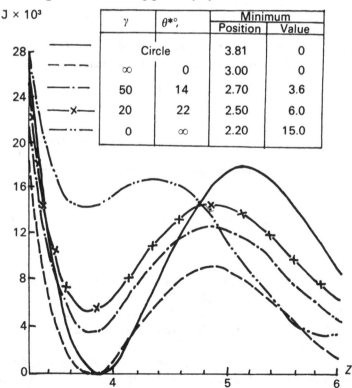

	γ	$\theta^{*\circ}$,	Minimum	
			Position	Value
	Circle		3.81	0
	∞	0	3.00	0
	50	14	2.70	3.6
	20	22	2.50	6.0
	0	∞	2.20	15.0

Fig. 5.16c. Intensity distribution for the small angle light scattering by an ensemble of random particles. Statistical characteristics of the ensembles are shown in the insert.

The calculation of the estimate of the average field from Equation (5.24) significantly simplifies the calculations since the Fourier transformation only applies to the "averaged" particle and not to all N particles in V. In addition, this particle has a more regular (smooth) shape than the individual particles which may have irregular shapes. Using Equation (5.24) together with Equation (5.23) also makes it possible to simplify the calculation of the estimate of the average field. Here, instead of the three-dimensional Fourier transformation, the one-dimensional transformation of the cross-sectional areas perpendicular to q are estimated. The number of cross-sections determines the accuracy of the estimation of the average field.

The intensity of the scattered light $I(\gamma)$ for scattering angles γ where interference can be neglected [where the average field, Equation (5.24), is equal to zero] is equal to the sum of the scattered intensities at individual particles:

$$ I = nV|\varepsilon_1^2| \iint_{-\infty}^{\infty} \overline{B(\mathbf{q}, b_1)B(\mathbf{q}, b_2)} \, e^{-iq(b_1 - b_2)} \, db_1 \, db_2 \qquad (5.25) $$

where:

$$ \frac{1}{N} \sum_{m=1}^{N} B_m(\mathbf{q}, b_1)B_m(\mathbf{q}, b_2) = \overline{B(\mathbf{q}, b_1)B(\mathbf{q}, b_2)} \qquad (5.26) $$

is the correlation function of the cross-sectional areas of the particles. We introduce the centred values of the cross-sectional areas:

$$ B^0(\mathbf{q}, b) = B(\mathbf{q}, b) - \overline{B(\mathbf{q}, b)} $$

It is obvious that

$$ \overline{B(\mathbf{q}, b_1)B(\mathbf{q}, b_2)} = \overline{B(\mathbf{q}, B_1)} \; \overline{B(\mathbf{q}, b_2)} + \overline{B^0(\mathbf{q}, b_1)B^0(\mathbf{q}, b_2)} $$

Here, the double integral in Equation (5.25) may be rewritten in the form (see [111])

$$ \iint_{-\infty}^{+\infty} = \left| \int_{-\infty}^{+\infty} B(\mathbf{q}, b)e^{-iqb} \, db \right|^2 + \iint_{-\infty}^{+\infty} \overline{B^0(\mathbf{q}, b_q)B^0(\mathbf{q}, b_2)} e^{-iq(b_1 - b_2)} \, db_1, \, db_2 $$

$$ (5.25') $$

The first term in Equation (5.25') describes the scattering at an 'average' particle. The second term contains the correlation function of the centred values of the cross-sectional areas. In the case of 'white noise' it is equal to zero and the problem reduces to scattering at the 'average' particle. In the general case, we will speak of a 'random' particle.

According to Equation (5.25), the calculation of the average volume scattering function of a sample of soft particles of any shape reduces to an estimate of the average scattering intensity at one random particle. The statistical characteristics of the sizes and shapes of such a particle are defined by the correlation functions of its cross-sections $\overline{B(q, b_1)B(q, b_2)}$ which are determined from Equation (5.26) averaged over the particles of the investigated volume. When defining individual forms of the function $B_m(q, b)$, it is convenient to imagine the particles at a single centre taking into account their orientations. In the calculation procedure ascribed to Equation (5.26), an analytical expression for the shape of the particles is not used. The cross-sectional area of the particles may be estimated for example on the basis of micro-photography. In this case, the laborious procedure of calculating the Fourier transformation is only carried out on one rather smooth function describing the averaged particle and not over N particles with complex shapes. Furthermore, by using the estimate of the intensity from Equation (5.25) and taking into account the definitions for Equation (5.26) the calculaton is simplified as a result of passing from a three-dimensional Fourier transform to a one-dimensional Fourier transform.

It follows from an estimate of the accuracy of the method carried out in [111] that the error depends on the accuracy with which the cross-sections of the particle are measured and on the number of cross-sections. Examples show that the requirements of these characteristics are moderate. The method can be generalized to particles with complex structure. Of course, in this case it is assumed that the correlation functions of the polarizability of different elements of suspended particles are known.

In [109, paragraph 6.7], it is shown that a good model for the volume scattering function for actual marine suspensions is the Kirchoff approximation for small angles [Equation (4.72) or (4.86)] and the RHD approximation for all other angles. In Kirchoff's approximation, the calculation of the small-angle volume scattering function of a population of N large particles of any shape reduces to the determination of the average scattering intensity at a screen with random shape. This problem was examined in [125a]. We look at it briefly. Omitting the factor $(i\lambda)^{-1}$ which is unimportant in the following, we have for the field from the individual jth particle in the Fraunhofer zone:

$$f_j = \int_0^{2\pi} d\alpha \int_0^{R_j'(\alpha)} \rho' \, d\rho' \, e^{ik\rho' \sin\theta \cos(\psi - \alpha)} \quad \text{where } 0 \leqslant \rho' \leqslant R_j'(\alpha)$$

Here, k, θ and ψ are the wave number of the incident radiation, the angle of scattering and the direction azimuth at the point of observation A. $R_j'(\alpha)$ is the equation of the particle contour in polar coordinates and $M(\rho, \alpha)$ is a point running over the screen (see Fig. 5.16b). We introduce the notation:

$$a = \frac{1}{2\pi N} \sum_{j=1}^{N} \int_0^{2\pi} R_j'(\alpha) \, d\alpha, \quad z = ka \sin\theta, \quad R_j(\alpha) = \frac{1}{a} R_j'(\alpha)$$

$$F_j(\alpha) = [1 - izR_j(\alpha) \cos(\psi - \alpha)] \exp[izR_j(\alpha) \cos(\psi - \alpha)] - 1,$$

$$\Phi(\alpha_1, \alpha_2, z) = \frac{1}{N} \sum_{j=1}^{N} F_j(\alpha_1)F_j^*(\alpha_2) = \langle F(\alpha_1)F^*(\alpha_2) \rangle.$$

Now for the field f_j and for the average intensity of the light $J(z)$ of the scattered ensemble of particles, we find

$$f_j = \frac{a^2}{z^2} \int_0^{2\pi} \frac{F_j(\alpha) \, d\alpha}{\cos^2(\psi - \alpha)}$$

$$J = \sum_{j=1}^N J_j = \sum_{j=1}^N |f_j|^2 = N \frac{a^4}{z^4} \iint_0^{2\pi} \frac{\Phi(\alpha_1, \alpha_2, z) \, d\alpha_1 \, d\alpha_2}{\cos^2(\psi - \alpha_1) \cos^2(\psi - \alpha_2)}. \quad (5.26')$$

The latter expresson significantly simplifies the calculation of the small-angle function of scattering of light by the ensemble. Here, it is not necessary to calculate the diffraction at individual particles, the ensemble is represented by the averaged values of the geometrical parameters of the particles. The algorithm holds true for any N (even if N is small). If N is large, there is statistical stability and the problem reduces to the determination of the average intensity of light scattered by a 'random' particle. Then the functions $R_j'(\alpha)$ will be individual realisations of the contour $R'(\alpha)$ of our 'random' particle.

If the ensemble consists of chaotically oriented particles of irregular shape (no predominant orientation) and if the light is unpolarised then the scattering function will have cylindrical symmetry. On average, the 'random' particle coincides with the circle of radius a. In the symmetrical problem, in view of the independence of the statistical characteristics of the contour on the azimuth ψ, we set $\psi = 0$; we will assume that the one-dimensional moments $\langle R''(\alpha) \rangle$ do not depend on the angle α and the two-dimensional moments $\langle R''(\alpha_1)R'''(\alpha_2) \rangle$ only depend on the difference $(\alpha_2 - \alpha_1)$. Expanding the exponents in the equation for $F_j(\alpha)$ in a power series we find after the averaging operation:

$$J(z) = a^4 \sum_{k,l=0}^\infty \frac{i^{3l+k}z^{k+l}}{(k+2)(l+2)k! \, l!} \iint_0^{2\pi} \cos^k\alpha_1 \cos^l\alpha_2 < R_1^{k+2}R_2^{l+2} > d\alpha_1 d\alpha_2.$$
$$(5.26'')$$

Here, $R_1 = R(\alpha_1)$ and $R_2 = R(\alpha_2)$, $N = 1$ and the expression $\langle R_1^{k+2}R_2^{l+2} \rangle$ denotes the two-dimensional initial moments of the contour of the 'random' particle. We separate out the terms in Equation (5.26'') corresponding to diffraction at an average circle. For this, we pass from the initial moments to the central moments in the double sum. Limiting ourselves to two-dimensional moments of the second order, we write:

$$J(z) = 4\pi^2 a^4 \left\{ \frac{J_1^2(z)}{z^2} + \sigma_0^2 \frac{J_1(z)}{z} \left[J_0(z) - zJ_1(z) \right] \right.$$

$$\left. + \frac{1}{4\pi^2} \sum_{n,m=0}^\infty \frac{i^{3m+n}z^{m+n}}{n! \, m!} \iint_0^{2\pi} \cos^n\alpha_1 \cos^m\alpha_2 \mu_{11}(\alpha_2 - \alpha_1) \, d\alpha_1 \, d\alpha_2 \right\}$$

$$(5.26''')$$

Here, $\sigma_0 = \sqrt{\mu_{02}} = \sqrt{\mu_{20}}$ is the mean square deviation of the contour of the 'random' particle from a circle. The first term in (5.26''') describes the diffraction at an average screen, the second indicates the contribution of non-correlated deviations of the contour boundaries of the particle and the third containing the double sum determines the contribution to the scattering of the effect of correlation of the deviations of the contour boundaries at different angles.

Calculations using Equation (5.26''') show that in comparison with the 'average' disc, random deviations increase the total scattering. Furthermore, they change the form of the scattering function, elongating it in the neighbourhood of the main maximum. This advanced calculation procedure makes it possible to estimate the accuracy of the method of equivalent spheres in the calculation of the small-angle scattering function for particles with irregular shape.

We introduce the correlation function $\mu_{11}(y)(y = \alpha_2 - \alpha_1)$ in the form $\mu_{11}(y) = \sigma_0^2 \exp(-\gamma y^2)$. For small correlation radii (γ large), using the Laplace method for the calculation, we find:

$$J(z) = 4\pi^2 a^4 \left\{ \frac{J_1^2(z)}{z^2} + \sigma_0^2 \frac{J_1(z)}{z} [J_0(z) - zJ_1(z)] + \frac{\sigma_0^2}{2} \frac{1}{\sqrt{\pi\gamma}} \right\} \quad (5.26'''')$$

Thus, when taking the correlation into account for any distribution of the deviations of the contour boundaries from a circle, the first term which leads to smoothing out the diffraction zeros has the order σ_0^2.

Setting $\exp(-3) \simeq 0$, we find that the area of angles θ^* in which the deviations are correlated is related to the parameter γ by the simple relationship: $\theta^* = \sqrt{3/\gamma}$ Calculations using Equation (5.26'''') are illustrated in Fig. 5.16c. Here, it is assumed that $\sigma_0^2 = 0.15$ and 4 values of γ have been taken indicated in the inset. For the sake of comparison, a curve is also given for a circle in Fig. 5.16c. The ordinate axes give the values of the function $\tilde{J}(z) \times 10^3$. For a circle, the function $\tilde{J}(z)$ coincides with $F(z)$ shown in Fig. 4.7. The position of the first minimum and the values of $\tilde{J}(z) \times 10^3$ at this point are also given in the table in the inset.

In the case of limitingly large correlation radii ($\gamma = 0$, $\theta^* = $ infinity), we are actually dealing with a polydisperse system of spherical particles. In fact, it is shown in [125a] that in this case, Equation (5.26''') reduces to Equation (4.101).

We note another simple model which is useful for analysing the scattering by an ensemble of irregular particles. This is a sphere with an anti-reflection coating. An ensemble of 'soft' isotropically arranged rods, star-shaped particles, fluffy particles, etc belong to this model. Calculations of the scattering of light at a coated sphere show that in such a sphere, the oscillations of the different optical characteristics are appreciably reduced. The scattering indicatrix $x(\gamma)$ for angles $\gamma \geqslant 60°$ is practically isotropic, there is no deep well in the area $\gamma = 135°$, there is no aureole effect in the source direction. The degree of polarization for large coated particles ($\rho \geqslant 30$) is practically equal to zero for all scattering angles $\gamma \leqslant 150°$ [117a]. All this is in agreement with the actual data for marine suspensions.

There have been few experimental studies of scattering by irregularly shaped particles. Important research was carried out by E. A. Fedorova. The thesis was defended in June 1952 but only published in full in 1957 [85]. She studied the volume scattering function $\beta(\gamma)$ for the scattering of light by fragments of glass. The average linear dimensions of the particles were 44–50, 105–125, 177–210 and 210–250 μm and the refraction indices $n = 1\cdot5$–$1\cdot9$. The particles fell in air inside a cylindrical chamber which had photographic film over its inner surface. The photographic film recorded the scattered light from a large number of particles of a given sort which made it possible to obtain the average values $\overline{\beta}(\gamma)$. The measurements were carried out for the green mercury line with $\lambda = 0\cdot546$ μm, in natural light and in polarized light. The scattering functions of irregular fragments were found to be significantly different from the scattering functions of spheres. Figure 5.17 shows some of her measurement results. We can see that for fragments in the range of angles γ 50–180°, $\overline{\beta}(\lambda)$ is practically constant. For spheres, it decreases it appreciably as γ increases (Fig. 4.6). In [85], data are given of the distribution of the scattered flux inside a certain solid angle $\Delta\omega$, the quantity $\beta(\gamma) \cdot \Delta\omega$ as well as the scattering asymmetry coefficient G. For glass fragments, the angular distribution is significantly smoother. For three types of glass fragment, with $n = 1.515$, 1.717 and 1.930, the value of G was found to be 6.7, 5.2 and 4.0 and for spheres 32, 16 and 12, respectively.

R. Hodkinson studied light scattered by suspensions of quartz, diamond, anthracite and coal particles in water [171]. The data on quartz are especially significant for ocean optics for which $m = 1\cdot16$ in water. An interesting result was obtained for the attenuation coefficient for a particle. It is shown in Fig. 5.18. For irregular particles, the sharp resonance maximum characteristic for monodispersive spheres disappears. A scarcely noticeable maximum is observed similar to that which is obtained for broad polydispersive distributions. This is an obvious result of the effect of averaging the phase shift $\delta = 2\rho(m - 1)$, arising when light passes through particles oriented in different ways. The scattering functions were measured for a collection of particles with average ρ in the range

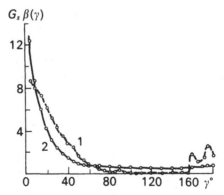

Fig. 5.17. Scattering functions for glass spheres (1) and irregular glass fragments (2) of the same type ($m = 1\cdot5147$).

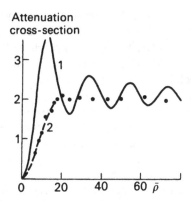

Fig. 5.18. Attenuation cross-section for transparent particles. 1, calculation for spheres with $m = 1\cdot16$ according to the exact formulae; 2, measurements for different quasi monodispersive fractions of suspensions of quartz particles in water.

12–74. All curves were found to be smooth as distinct from the multi-lobe curves for monodispersive spheres. On comparison with the calculations for the spheres it was found that the calculated curves agree with the experimental curves approximately up to $\gamma = 30°$. For large angles, $\beta(\gamma)$ for irregular particles was found to be significantly larger than for spheres in [171] as well as in [85].

Turning now to heterogeneous particles, we note that the exact solution of the problem of the diffraction of electromagnetic waves at bilayer and multi-layer spheres was published at practically the same time in three studies [91, 130, 146]. It is used widely for studying the scattering of light by cell models and in many other problems. In the monograph by M. Kerker [154], all cases are noted when for radially symmetric particles, ie particles for which $m = m(r)$ when the solution of the equation for the radial function may be expressed in terms of known tabulated functions.

Diffraction at particles with complex shape was studied experimentally by F. S. Harris, Jr [147]. Using a 35 m chamber without lenses and long exposures (up to 1000 hours and more), he obtained unique photographs of the field in the Fraunhofer zone from different models. Some of his models are very similar to actual plankton particles.

In conclusion, we add that a large number of calculations of diffraction by particles with complex shape and structure is being carried out by radiophysicists in conjunction with antennae calculations. Here, not just specific antenna are examined but also different statistical ensembles. A general approach to this problem and the main physical results are given in the monograph by Ya. S. Shifrin [126].

5.3.2. General properties of the scattering functions

It is possible to note three general properties of marine scattering functions: (1) a sharp elongation in the directon of the incident beam; (2) a blurred minimum in the range 100–130°; (3) a small increase in the reverse direction. If an asymmetry coefficient G equal to one for molecular scattering and for very small particles is used, then for seawater, its value may reach 60–80 (in the atmosphere, the values of G do not exceed 35 [90]). This is related to the presence in seawater of large biological particles whose complex refractive indices m are close to unity. Molecular scattering may make a significant contribution to the scattering at large angles (more than 80% at $\gamma = 135°$) but, as the scattering angle decreases, its contribution rapidly decreases (Chapter 3). At small angles, the scattering is completely determined by the suspended particles and the fall off in intensity over the entire range of scattering angles may exceed five orders of magnitude. The quantity $b(\gamma)$ changes most strongly here. Using a semilogarithmic scale, this change is approximately 0.61 log/degree in the neighbourhood of 1° and 0.21 log/degree at 5°. As the angle increases, the rate of change of $b(\gamma)$ decreases—0·01 log/degree at 80°; the scattering minimum is reached at 105–120°.

In the studies [55, 56], V.I. Man'kovsky analysed more than 200 values of $b(\gamma)$ measured in the northern part of the Atlantic Ocean, Mediterranean Sea and Black Sea at depths down to 200 m. The measurements were made with an instrument *in situ* at angles of γ from 1·5 to 152·5° with an interval of 5°. In the range of angles 0–1.5°, the values were extrapolated along a parabola of the form: log $b(\gamma) = A + B\gamma + C\gamma^2$ with respect to three angles: $\gamma = 1.5, 7.5$ and 12.5°; for the angles 152.5–180° it was assumed that $b(\gamma) = $ constant. He gives the following extreme values for $b(\gamma)$ from the data which he investigated (Table 5.8).

The integral characteristics of the functions $b_1(\gamma)$ and $b_2(\gamma)$ are given in Table 5.9. The total scattering coefficients for them are: $b_1 = 0.404\,\text{m}^{-1}$, $b_2 = 0.023\,\text{m}^{-1}$. This is also typical in the general case: the greater the value of b,

Table 5.8. Functions $b(\gamma)$ $(\text{m}^{-1} \cdot \text{sr}^{-1})$ with maximum $[b_1(\gamma)]$ and minimum $[b_2(\gamma)]$ asymmetry

$\gamma°$	$b_1(\gamma) \times 10^4$	$b_2(\gamma) \times 10^4$	$\gamma°$	$b_1(\gamma) \times 10^4$	$b_2(\gamma) \times 10^4$
0	3.17 5	1.04 4	87.5	3.8	1.2
1.5	1.07 5	3.53 3	92.5	3.8	1.3
4.5	1.60 4	6.65 2	97.5	3.4	1.4
7.5	3.94 3	1.98 2	102.5	3.2	1.3
12.5	1.14 3	7.17 1	107.5	3.6	1.5
17.5	4.25 2	3.61 1	112.5	3.5	1.6
27.5	1.01 2	1.39 1	122.5	3.6	1.7
37.5	4.51 1	8.0	132.5	4.5	2.1
47.5	2.23 1	4.3	142.5	5.6	2.7
57.5	1.43 1	3.0	147.5	7.2	2.6
67.5	1.01 1	1.9	152.5	7.8	2.7
77.5	6.0	1.3	180	7.8	2.7

Table 5.9. Limiting values of the integral characteristics of the scattering functions

Characteristic	According to the data of [56]	According to the data of [48]
Asymmetry coefficient G	7.9–59.5	9.7–84.9
Average scattering angle $\bar{\gamma}^0$	7.1–28.9	7.7–30.1
Average cosine of the scattering angle $\overline{\cos \gamma}$	0.743–0.957	0.756–0.957
Root mean square scattering angle $\sqrt{\overline{\gamma^2}}^{\,0}$	—	18.8–46.5

the more strongly the functions are elongated. Most of the scattered light is concentrated in the vicinity of the direction of the incident beam. The scattering angle γ^*, corresponding to the solid angle in which half the scattered light of the beam is concentrated, usually lies between 2 and 10°; according to [55], its minimum value is 1·7° and maximum value 14°. In [42, 55–57] are tables of the ratios of the light flux scattered in the range of angles from 0 to γ to the total scattered light flux for different waters. Table 5.9 gives the limiting values for some integral characteristics of the scattering functions of seawater according to [56] and [48].

It can be seen from Table 5.9 that the limiting values of the integral characteristics agree with each other for different data despite the fact that the data of [56] were obtained *in situ*, whereas the data in [48] were obtained from samples using the procedure of [47]. This is described in Section 6.4. The minimum scattering angle there was only 20′. The only exception is produced by the maximum values of the asymmetry coefficient G which differ by a factor of almost 1.5. In [48], it is pointed out that the values of G higher than 80 were obtained often in the surface layers where they may be explained by the contribution of large biological particles. In deeper waters, where the amount of such particles is small, the maximum value of G is lower than in surface waters. At 100 m it was 69.6 and below 500 m it was 58.8.

It is shown in Table 5.10 how the values of $b(\gamma)$ and the integral characteristics vary with depth for two stations in the Indian Ocean: No. 740 (area of the southern tropical convergence) and No. 765 (area of the subequatorial divergence); for the coordinates of the stations see [9].

It can be seen from the table that the stations are appreciably different. First of all this is manifested in the absolute values of scattering where it is not only the surface levels but also the deeper levels which are different (except for the values of scattering at large angles). Stations No. 740 and 765 also differ in the variation of the integral characteristics with depth: at station No. 765, the values of G, $\cos \gamma$ and $\bar{\gamma}$ vary slightly and at station No. 740 their variation with depth is strongly marked. These changes are mainly associated with a reduction in forward scattering, whereas the values of the back scattering coefficient b_b for surface and deep samples are close to each other (and also to the values at the

PHYSICAL OPTICS OF OCEAN WATER

Table 5.10. Variation with depth of $b(\gamma)$ ($\mathrm{m}^{-1} \cdot \mathrm{sr}^{-1}$) and integral scattering characteristics

	Station No. 740				Station No. 765			
γ^0	0 m	50 m	200 m	1000 m	0 m	50 m	200 m	1000 m
0·5	1.6 1	1.2 1	5.0	5.0	6.9 1	3.9 1	3.4 1	3.2 1
1	9.3	6.5	2.5	2.6	2.2 1	1.5 1	8.5	7.4
2	3.3	2.6	7.4 −1	8.0 −1	9.2	6.3	2.5	2.3
4	9.5 −1	6.6 −1	2.1 −1	1.8 −1	2.8	2.1	7.3 −1	6.8 −1
6	4.1 −1	3.0 −1	9.8 −2	7.6 −2	1.2	8.4 −1	3.2 −1	2.4 −1
10	1.2 −1	1.3 −1	3.8 −2	2.1 −2	3.3 −1	2.0 −1	6.4 −2	3.7 −2
15	4.9 −2	5.4 −2	1.6 −2	8.8 −3	1.3 −1	6.9 −2	2.5 −2	1.4 −2
45	2.5 −3	2.5 −3	1.4 −3	9.5 −4	8.0 −3	3.5 −3	1.8 −3	1.2 −3
90	4.4 −4	6.0 −4	2.9 −4	3.3 −4	1.7 −4	5.3 −4	3.7 −4	3.5 −4
135	5.5 −4	6.3 −4	4.0 −4	4.2 −4	1.5 −3	6.3 −4	3.2 −4	3.7 −4
$b \, \mathrm{u}^{-1}$	1.09 −1	9.4 −2	3.3 −2	2.7 −2	3.13 −1	1.94 −1	8.7 −2	7.2 −2
$\delta \, \mathrm{u}^{-1}$	1.06 −1	9.0 −2	3.1 −2	2.5 −2	3.04 −1	1.9 −1	8.5 −2	7.0 −2
$\beta \, \mathrm{u}^{-1}$	2.8 −3	2.5 −3	2.1 −3	2.3 −3	8.7 −3	3.4 −3	2.0 −3	2.0 −3
G	3.78 1	2.45 1	1.52 1	1.07 1	3.51 1	5.52 1	4.34 1	3.14 1
$\cos \gamma$	9.21 −1	8.91 −1	8.39 −1	8.07 −1	9.14 −1	9.43 −1	9.33 −1	9.24 −1
$\bar{\gamma}^0$	1.27 1	1.6 1	2.06 1	2.19 1	1.33 1	1.03 1	1.06 1	1.05 1

200 and 1000 m depths at station No. 765). For the ocean as a whole, a reduction in the elongation of the volume scattering function with depth is characteristic. This can be seen, for example, from the average values of the integral characteristics calculated in [48] for the Indian Ocean (data from the tenth voyage of the scientific research ship 'Dmitry Mendeleyev')—<100 m (80 samples) and ⩾100 m (66 samples) (Table 5.11).

The majority of the volume scattering functions of seawater have a single minimum at 100–130° but for some of them, local extreme values are observed in the vicinity angles at 30, 60, 80, 100 and 135° [55]. The situation here is similar to that which is depicted in Fig. 2.4. The function under the integral in the formula for the polydispersive scattering function has a sharp maximum for particles of a specific size. Thus, it is found that in scattering at a given

Table 5.11. Average values of the integral characteristics for surface and deep samples

Level (m)	$b(\mathrm{m}^{-1})$	$b_f(\mathrm{m}^{-1})$	$b_b(\mathrm{m}^{-1})$	G	$\overline{\cos \gamma}$	$\bar{\gamma}^0$	$\sqrt{\bar{\gamma}^2}\,^0$
<100	0.146	0.142	0.0038	41.7	0.912	13.6	26.6
⩾100	0.073	0.070	0.0030	26.2	0.881	15.8	31.3

angle γ, a decisive contribution is made by particles of a specific size. Such 'monochromatization' leads to the appearance of local maxima similar to those which are observed in the scattering at monodispersed particles (Figs 4.11 and 4.12). This can be seen from the studies given in [123, 160]. In the graphs given in [160], it is possible to see that for refractive indices 1.05–1.15 and $v = 3.1$–3.5 appreciable local extreme values are observed for the scattering functions. The position of the local maximum is shifted regularly depending on the values of the complex refractive index m: 66° at 1.05, 76° at 1.075, 92° at 1.10 and 106° at 1.15. This means that for such scattering functions it is possible to estimate m from the position of the extreme value.

5.3.3. Statistical characteristics and models of the scattering functions

Measurements of seawater scattering functions have been carried out for a number of years at the Institute of Oceanology of the USSR Academy of Sciences with the help of the spectrohydronephelometer SGN developed by V.B. Veinberg and collaborators [39]. This instrument was designed for measurements in the range of scattering angles γ from 0.5 to 145°. However, whilst working with it, it was found that its measurements were unreliable in the most important range of small scattering angles $\gamma \leqslant 3°$. A great achievement was the application of a principle developed earlier for the determination of scattering functions of cloud drops [95] at small angles. This method is based on the measurement of the distribution of irradiance in the focal plane of the receiving lens (Section 6.4). The instrument described in [23, 99, 100] was designed on the basis of this principle. An advanced version of this instrument ('Strela') was used for measurements of water samples. The same principle was also used for measurements *in situ* ('Poseidon' instrument). Both instruments and the operational procedures are described in [39, 47]. In conjunction with SGN, these instruments made it possible to measure the scattering function in the angular range of $20'$–145°. As a result, a mass of data was obtained for $b(\gamma)$. The first statistical analysis of these data was carried out in the study of O.V. Kopelevich and V.I. Burenkov [43] and then the complete data were investigated in [48]. In this study, a sample of 41 scattering functions was used for the calculation of the correlation and covariant matrices, the eigen values and the eigen vectors. An analysis of the correlation matrix showed that in agreement with previously obtained results [43], there is a high correlation for the scattering at angles 45–135° and a poor correlation between the scattering at small angles (0.5–4°) and scattering at angles of 15–135°. This means that small-angle and large-angle scattering are governed by different factors independent of each other. It is obvious that these factors are the large and fine suspension fractions which we discussed in Chapter 2. Each of these fractions is responsible for scattering in the different ranges of angles: the large fraction is responsible for small-angle scattering and the fine fraction for large-angle scattering. The reason for the division into the 'zones of influence' of the two fractions is that the number of large particles is relatively small.

The large particles have a considerably elongated forward scattering function and they dominate in the scattering at small angles. However, beginning at the angle of 45° their contribution to the scattering is less than 3% [11]. In [48], it was pointed out that several hundred functions were measured with the help of the procedure of [47, 99] for different regions and depths. These data confirmed the ideas formulated above concerning the variation of $b(\gamma)$ in ocean waters. Figure 5.19 shows the correlation matrices r [log $b(\gamma_i)$, log $b(\gamma_j)$] calculated for surface and deep waters. On the whole, these matrices are similar to each other and to the correlation matrices calculated previously for other regions [43]. This provides evidence for the idea that the principles discussed above are general in nature.

The system of eigen vectors for the covariant matrix for $b(\gamma)$ makes it possible to approximate any form of this function in an optimum manner. It was found that the use of only two to three initial eigen vectors makes it possible to approximate the measured scattering functions with satisfactory accuracy. The system of eigen vectors for the sample of 41 scattering functions mentioned above turns out to be 'universal', ie it is possible to use it for approximating the scattering function of any region of the world's oceans. The required expansion has the form:

$$\log b(\gamma_i) = \overline{\log b(\gamma_i)} + \sum_{k=1}^{m} c_k \psi_k(\gamma_i) \tag{5.27}$$

The values of $\overline{\log b(\gamma_i)}$, $\psi_1(\gamma_i)$, $\psi_2(\psi_i)$ and $\psi_3(\gamma_i)$ are given in Table 5.12.

The expansion coefficients c_k for the given form of log $b(\gamma)$ in terms of $\psi_k(\gamma)$ can be found from the equation:

$$c_k = \sum_i \psi_k(\gamma_i)[\log b(\gamma_i) - \overline{\log b(\gamma_i)}] \tag{5.28}$$

Of course, it is also possible to calculate the coefficients c_k directly from Equation (5.27), having measurements for $b(\gamma)$ for two to three scattering angles. The accuracy of the approximation is not so good in this case but on the other hand it is possible to calculate the full function from two to three points. Of course, in this case it is necessary to select those angles γ_i for which the values of $b(\gamma_i)$

Table 5.12. Average value and eigen vectors of the covariant matrix for the scattering function

$\gamma°$	$\log b(\gamma)$	$\psi_1(\gamma)$	$\psi_2(\gamma)$	$\psi_3(\gamma)$	$\gamma°$	$\log b(\gamma)$	$\psi_1(\gamma)$	$\psi_2(\gamma)$	$\psi_3(\gamma)$
0.5	1.307	0.316	−0.326	0.548	10	−0.960	0.369	0.067	−0.345
1	0.912	0.331	−0.262	0.316	15	−1.357	0.347	0.209	−0.406
2	0.509	0.351	−0.219	0.086	45	−2.550	0.253	0.447	0.079
4	−0.044	0.388	−0.181	−0.116	90	−3.225	0.177	0.503	0.251
6	−0.474	0.382	−0.076	−0.289	135	−3.222	0.136	0.487	0.377

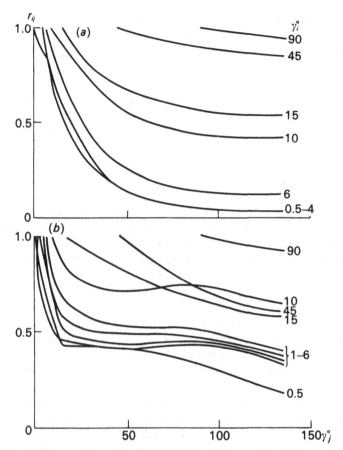

Fig. 5.19. Correlation matrices r [log $b(\gamma_i)$, log $b(\gamma_j)$] for waters of the Indian Ocean (a) < 100 m, 77 samples; (b) ≥ 100 m, 68 samples.

are not correlated. Otherwise, the system of equations for calculating c_k are ill-conditioned. Analysis showed that the optimum angles are 6° for calculating 1 vector, 1° and 45° for 2 vectors and 1°, 6° and 45° for 3 vectors. The errors involved (averaged over the angle) do not exceed 28, 19 and 16%, respectively. Examples of calculating various functions are given in [48].

The total scattering coefficient b may be determined either by integrating the restored function $b(\gamma)$ or with the help of the regression equations between log b and the values of log $b(\gamma_i)$:

$$\log b = \sum_{i=1}^{m} B_i \log b(\gamma_i) + A \qquad (5.29)$$

The values of the coefficients and the root mean square error s_m of the regression given in Equation (5.29) are given in Table 5.13.

The relative error in determining the value of b is only 9% when the angles

Table 5.13. Regression coefficients and error of nephelometric methods for determining b

m	Scattering angle (°)	B_1	B_2	B_3	A	s_m
1	6	0.86	—	—	−0.56	0.069
2	1, 45	0.658	0.449	—	−0.47	0.067
3	1, 6, 45	0.335	0.427	0.271	−0.40	0.041

1, 6 and 45° are used; it increases to 15% when two angles are used and up to 16% when one angle is used. The last estimate is the minimum value of the error when the nephelometric method is used for determining b; the angle 6° is optimum for this method [42] (according to [57] $\gamma = 4.5°$ is optimum).

On the basis of the statistical analysis carried out above, it is possible to construct physical models for light scattered by seawater. The simplest of these is the model of A. Morel [159]. It has the form:

$$b(\gamma) = b_w(\gamma) + b_h^*(\gamma)C \tag{5.30}$$

This is a one-parameter model. In it, $b_w(\gamma)$ is the scattering coefficient for pure water and $b_h^*(\gamma)$ is the scattering coefficient for hydrosol particles (relative to unit concentration of the suspension C). It is assumed that it is constant for all waters and variations in the form of the (total) scattering function depend only on C, the concentration of the suspension. This model is analogous to that which is used for atmospheric scattering functions. The calculations by Morel of the scattering function give reasonable agreement with actual scattering functions by marine suspensions. However, Morel's model does not make it possible to explain some characteristics peculiar to the scattering functions of seawater. Thus, for example, according to this model, the elongation of the scattering functions must increase as the magnitude of the scattering increases but sometimes the opposite effect occurs. According to Morel's model, the point of intersection of the scattering functions must lie at 26–27°, whereas it is located in the neighbourhood of 4–6° [42, 57].

Of course, the two-parameter model [48] gives better agreement with the real situation. In it, it is assumed that marine suspensions consist of two constant fractions: a large biological fraction and a fine mineral fraction.

We represent the scattering coefficient in the given direction $b(\gamma)$ in the form:

$$b(\gamma) = b_F(\gamma) \cdot V_F + b_L(\gamma) \cdot V_L + b_w(\gamma) \tag{5.31}$$

where V_F and V_L are the volume concentrations of the fine and large suspensions; $b_F(\gamma)$ and $b_L(\gamma)$ are the scattering coefficients for these fractions for unit concentration.

We assume that the functions $b_F(\gamma)$ and $b_L(\gamma)$ are invariant. We assume exponential distributions for the fine and large fractions with the following values for the parameters: $v = 4, 3$; $\rho_{min} = 0.3, 20$; $\rho_{max} = 20, 200$; $m = 1.15$,

1·02. Investigations show that such values for the parameters agree approximately with the values actually obtained at sea [11, 109]. The scattering coefficients $b_F(\gamma)$ and $b_L(\gamma)$ can be calculated for these fractions using the tables in [123].

The model given by Equation (5.31) agrees with the observation data. For a suitable selection of V_F and V_L, it even describes the extreme value functions given in Table 5.8 as can be seen from Fig. 5.20. This model also describes the variation of the scattering characteristics. In order to investigate this variation,

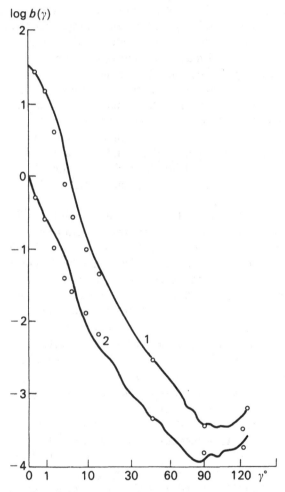

Fig. 5.20. Approximation of the extreme value scattering functions with the help of the model given by Equation (5.31). The curves are the measured values and the points are the calculated values: 1, with maximum asymmetry ($V_F \simeq 0.027$; $V_L \simeq 0.237\,\text{cm}^3/\text{m}^3$); 2, with minimum asymmetry ($V_F = V_L = 0.004\,\text{cm}^3/\text{m}^3$).

a sample of N functions of the form of Equation (5.31) was taken in [45] with random independent coefficients V_F and V_L. Fifty model functions were calculated from Equation (5.31) and the coefficients V_F and V_L were taken from tables of random numbers. The range of values for these coefficients was taken as $0.01–0.20 \, cm^3/m^3$ for fine suspensions and $0.01–0.40 \, cm^3/m^3$ for large suspensions which approximately corresponds to the values observed for ocean waters. For 41 samples, the average scattering coefficient calculated from the model agreed with the actual measured values. The same was true for the dispersion $D[b(\gamma_i)]$, the correlation matrix $r[b(\gamma_i), b(\gamma_j)]$ and the point of intersection of the normalized volume scattering functions $\beta(\gamma) = b(\gamma)/b$ [48].

As we saw above, the scattering functions of irregular particles (such as the particles of marine suspensions) differ appreciably from the scattering functions of spherical particles. Therefore, it may seem strange that the model given by Equation (5.30) and in particular the model given by Equation (5.31) give a good description of actual marine scattering functions with the values of the specific coefficients $b(\gamma)$ for spherical ensembles.

The reason for this is that in comparing the theoretical data with the observed data, we do not have independent data for marine suspensions but have selected plausible values. Furthermore, it is necessary to take into account that the logarithmic scale used in Fig. 5.20 conceals the deviation. Small particles are responsible for the scattering at medium and large angles. It can be seen from Fig. 5.17 that for angles $\gamma > 60°$ the scattering function of spherical particles is appreciably smaller than that for irregular particles. This difference may be compensated by increasing V_F.

We shall look at the spectral variation of the scattering functions. As was pointed out in [110], the elongation of the total function decreases as λ decreases. This occurs because of the increase in the contribution of molecular scattering and the scattering by small particles which enhances the curve at large angles. An analogous 'anomalous Mie effect' is observed in the atmosphere. Due to the division of the zones of influence of the fine and large suspension fractions which we described above, it is possible to establish a spectral variation $b(\gamma, \lambda)$ in different ranges of angles. We show in Chapter 6 that the scattering of the fine suspension fractions (in angles $> 15–30°$) is determined by the RHD equation. This means that when the Rayleigh correction is taken into account, the scattering does not relate to ρ and γ independently but to $\rho \sin(\gamma/2)$. ie there is similarity with respect to $1/\lambda$ and $\sin(\gamma/2)$. For small angles, if the correction factor $P(\delta) = 1$ is used, the scattering coefficient $b(\gamma) \sim \rho^2 F(\rho\gamma)$ [see Equation (4.72)]. In this case, there will also be similarity for the quantity $b(\gamma)/\rho^2$ but in terms of $1/\lambda$ and γ. As λ increases, the function $\rho^2 F(\rho\gamma)$ passes through a maximum. Setting $F(\rho\gamma) = 1 - \rho^2\gamma^2/4$, we can easily find that the maximum with respect to λ lies at the point $\rho\gamma = \gamma 2\pi r/\lambda_m = 2$. Thus, $\lambda_m \sim \gamma$. Actually, the observation data on the spectral scattering coefficients in the equatorial part of the Atlantic Ocean (at given γ), given in [10], show that for small angles, $b(\gamma, \lambda)$ as a function of λ has a maximum at a certain λ_m which increases with γ. At medium and large angles, $b(\gamma, \lambda)$ steadily decreases with λ.

5.3.4. Variation in the scattering coefficient

The variation in the scattering coefficient of light by ocean water is almost entirely related to the variation in the quantitative and qualitative composition of the suspension. Quantitatively, the scattering varies over a broad range, with values of the scattering coefficient b ranging from several hundred to $1 \, m^{-1}$ ($\lambda_0 = 546 \, nm$). The minimum value $b = 0.022 \, m^{-1}$ was measured 500 m deep in the Pacific Ocean in the transparent water region to the north-west of the island of Rarotonga. Values of b close to this ($0.026–0.028 \, m^{-1}$) were measured at depths of more than 1000 m in the zone of the southern subtropical convergence in the Indian Ocean. Data close to this value are given in [155] for the Sargasso Sea: $0.021 \, m^{-1}$ ($\lambda_0 = 633 \, nm$, 400 m depth) and in [55]: $0.023 \, m^{-1}$ ($\lambda_0 = 520 \, nm$). The values of b of the order of $0.02 \, m^{-1}$ for the green section of the spectrum are the lowest values which can be found in the world's oceans.

The values for the scattering coefficient given above relate to deep waters. Above 50–100 m, the values of b are appreciably larger, the lowest lying in the range $0.05–0.06 \, m^{-1}$ (in the region of transparent waters to the north-west of the island of Rarotonga at a depth of 10 m, $b = 0.066 \, m^{-1}$, at 50 m depth, $b = 0.054 \, m^{-1}$; in [29] values of $0.05–0.06 \, m^{-1}$ are given for the 0–50 m depths in the Tyrrenian Sea). Table 5.14 gives the average values of b for various depths calculated from the meaurements at 34 deep stations in the Pacific and Indian Oceans.

Table 5.14. Variation in the scattering coefficient with depth

Layer (m)	0–50	50–100	100–200	200–1000
$b \, (m^{-1})$	0.166	0.131	0.090	0.063

Intermediate waters at depths of the order of 500–1500 m and clearly identified by the salinity cannot be differentiated usually in terms of light scattering properties [16]. At ocean depths of the order of a few kilometres, waters below 200 m to the bottom are as a rule homogeneous in terms of their optical properties and are characterized by low scattering values. In the vicinity of the bottom, 'natural nephelloid layers' are observed in many cases. Their existence is associated with the special features of the relief of the bottom [18]. Typical values of b for surface waters of the open ocean are $0.10–0.15 \, m^{-1}$ and for deep waters $0.05–0.10 \, m^{-1}$. The highest values of b were measured in the vicinity of the coast of Peru: 2.7 and $3.3 \, m^{-1}$ [48].

We look at the spectral variation of b. The measured data show that the coefficient b weakly depends on the wavelength, slowly increasing as λ decreases. For a surface sample taken from the eastern part of the Pacific Ocean, for example, the following values given in Table 5.15 were obtained for $b(\lambda)$ [48].

Table 5.15. Spectral variation of coefficient b

λ (nm)	473	520	546	568	601	649
b (m^{-1})	0.16	0.14	0.12	0.12	0.11	0.10

According to Morel [161], the spectral dependence of the total scattering coefficient on average may be approximated by the relationship $b(\lambda) \approx \lambda^{-1}$. According to the data given in the literature, the exponent in this formula varies from -0.9 to -1.8.

In order to assess the effect of the different factors on the spectral dependence of the scattering, we represent $b(\lambda)$ in the form of a sum of three components analogous to the model for $b(\gamma)$ given by Equation (5.31). Out of the three factors, the most selective is molecular scattering. As we have seen in Chapter 3, $b_w(\lambda) \sim \lambda^{-n} (n \approx 4.2 - 4.3)$, but in terms of absolute value, its contribution to $b(\lambda)$ is small: when λ changes from 700 to 400 nm, the increase in b due to molecular scattering is only $\sim 0.005 \, \text{m}^{-1}$. The variation in the value of $b_L(\lambda)$ and $b_F(\lambda)$ over the spectrum may be estimated using the physical model of scattering. Optimum models are obtained for $v = 4.5$, $m = 1.15$ (fine suspended particles) and $v = 3$, $m = 1.03$ (large suspended particles). For these models, we find dependences of the type:

$$b_L(\lambda) \sim \lambda^{-0.3}; \quad b_F(\lambda) \sim \lambda^{-1.7} \tag{5.32}$$

The main factor governing the variation in the scattering coefficient with wavelength is the fine fraction: when it is predominant, the exponent at λ^{-1} reaches its maximum values. If the large fraction dominates, the selectivity of the scattering is considerably reduced.

The space-time variation of the field of the scattering coefficient has been studied by a number of authors [25, 28, 29, 36, 60a, 62, 63, 161, 164]. We are referring here to the variation in the field of b. Significantly more investigations have been carried out on the variation of c which essentially is also determined by the variation in b. This is discussed in Section 5.5. G.S. Karabashev and V.V. Yakubovich produced a nephelometer at an angle of $45°$ which they used to study the field $b(z, t)$ in the main oceans as well as in the Baltic Sea.

The main characteristics in the vertical variation of $b(z)$ in the active layer of the ocean is the decrease of the coefficient b with depth. This is simply related to the fact that the main sources of suspensions are found as a rule in the vicinity of the surface. In [129], different types of vertical profiles are identified. These profiles are shown in Fig. 5.21. They were obtained from measurements in the Indian Ocean and relate to the route of the tenth voyage of the scientific research ship 'Dmitry Mendeleyev'. Nevertheless, it is worth pointing out that the percentage occurrence of each type was 48, 34, 14 and 4%, respectively. The first steadily decreasing profile is typical of waters with a deep pycnocline and a low content of biogenic suspensions and the second is characteristic of waters

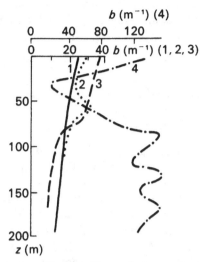

Fig. 5.21. Types of vertical profiles $b(z)$. The scale for profile 4 is shown separately.

with a typical turbidity layer in the area of the thermocline (usually coinciding with the fluorescence maximum of chlorophyll which indicates its biological nature). The third is characterized by a sharp fall below the thermocline and subsequently an almost constant value (deeper than 150–200 m). The fourth type has a multi-layer structure observed in the estuarine areas of rivers in the Baltic Sea.

In order to illustrate the spatial variation of b, Fig. 5.22 shows the observed data of $b(45°)$ along the meridian 100° west in March 1971 (section through the

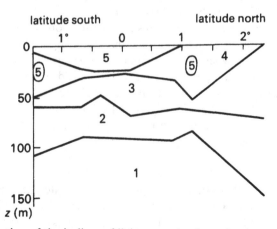

Fig. 5.22. Distribution of the isolines of light scattering intensity at an angle of 45° in relative units. (1) < 20; (2) 20–40; (3) 40–60; (4) 60–80; (50) > 80.

equator) [36, 62]. We can see that the distribution of turbidity is not symmetrical, the centre is shifted to the south by 20'; two turbid streams can be seen at depths of 25 m, and below 100 m there is a region of transparent waters.

5.4. Polarized light scattering

5.4.1. Degree of polarization

The scattering of polarized light by colloidal systems has been studied for a very long time. However, the first observations of polarized underwater light in the ocean only were made in 1954 by the American biologist Waterman. They were associated with the study of the mechanism of the orientation of invertebrates in the sea with respect to polarized light. In the period between 1959 and 1964, detailed measurements of the degree of polarization $p(\gamma)$ were carried out for seawater by A.A. Ivanoff. They were carried out on samples as well as *in situ* using a polarimeter with remote control. Investigations in model media were carried out by V.A. Timofeeva [164]. The total scattering matrix of ocean water has only been measured on samples. The first determinations were made in 1966 by G. Beardsley [133]. Detailed investigations in different seas and on model suspensions were carried out by G.V. Rosenberg, Y.S. Lyubovtseva, E.A. Kadyshevich, S.E. Kondrashov and others [33, 76].

First we shall examine data on the simplest polarization characteristic, the degree of polarization $p(\gamma)$. Figure 5.23 shows a summary of the data obtained by A.A. Ivanoff during measurements in 1961 in the western part of the Mediterranean Sea [164]. It can be seen from Fig. 5.23 that as the turbidity of the water increases, the degree of polarization $p(90°)$ decreases. The highest value $p(90°)$ approaches the value of 0·835 corresponding to optically pure water. In this case, the 90° relative scattering coefficient $b(90)/b_w(90)$ tends to unity. The lowest values of $p(90°) = 0.4$ were observed in turbid coastal waters where $b(90)/b_w(90)$ was found to be approximately 14.

In order to interpret the obtained results we must look at the physical picture of the phenomenon. We remember that the scattering function of a limitingly small isotropic sphere when it is irradiated by linearly polarized light is described by Equation (4.56). This equation corresponds to the distribution of the intensity of the scattered light which arises from the dipole oscillating along the electric vector of the incident wave. The degree of polarization $p(\gamma)$ for all angles γ will be positive and symmetric with respect to the maximum at $\gamma = 90°$. It can be written in the form:

$$p(\gamma) = p_{max} \sin^2 \gamma/(1 + p_{max} \cos^2 \gamma) \qquad (5.33)$$

For an isotropic Rayleigh sphere, $p_{max} = 1$. In pure water, due to the anisotropy of the molecules, the $p(\gamma)$ curve is slightly deformed, dropping at the edges by 9% and at the centre by 17·5% (Table 3.10) but its symmetry with respect to $\gamma = 90°$ is preserved. Equation (5.33) is also preserved but p_{max} is reduced to 0·835.

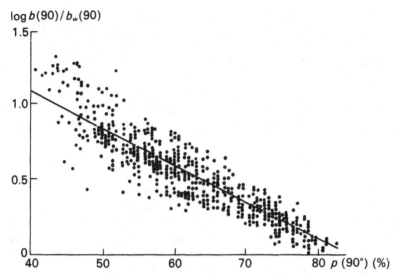

Fig. 5.23. Relationship between the relative scattering coefficient $b(\gamma)/b_w(\gamma)$ [the ratio of the actual $b(\gamma)$ to $b_w(\gamma)$ of optically pure water] for $\gamma = 90°$ and degree of polarization $p(90°)$.

The degree of polarization of scattering at a limitingly large transparent sphere is shown in Fig. 4.6. The essential feature here is the negative polarization when $\gamma < 45°$. The polarization curve for a large particle calculated from the Mie equations is shown in Fig. 4.15. It is very jagged but on average is similar to the curve given in Fig. 4.6. Assuming that the molecular and hydrosol scattering in ocean water are independent, it can be shown that the total function $p(\gamma)$ will be:

$$p(\gamma) = p_m(\gamma)\eta + p_h(\gamma)\,(1 - \eta), \quad \eta = b_m/(b_m + b_h) \qquad (5.34)$$

where η is the fraction of molecular scattering as whole; $p_m(\gamma)$ and $p_h(\gamma)$ refer to molecular and hydrosol scattering. The maximum $p(90°)$ corresponding to $\eta = 1$ agrees with the data of Fig. 5.23. The minimum value 0·4 is obtained when $\eta = 1/(1 + b/b_w) = 0·07$ in which case from Equation (5.34) we find the degree of polarization of a hydrosol suspension $p_h(90°) = 0·36$. The value $p(90°)$ in Fig. 4.6 is approximately unity and in Fig. 4.15 it is approximately 0·7, ie both are a long way from the experimental value of 0·36. The data for a polydispersive system of spheres are shown in Fig. 4.14. Of significance in this case is the large variation in $p(90°)$ as a function of the distribution parameters. In one of the six cases illustrated in Fig. 4.14 ($v = 3$, $m = 1·15$), the value of $p(90°)$ is 0·35, ie it almost exactly agrees with the experimental value. However, the sharp fluctuations observed in $p(90°)$ on changing the distribution parameters indicate rather that the homogeneous sphere model is incapable of wholly explaining the phenomenon.

Measurements of the polarization of light scattered by non-spherical particles are described in the studies of Ivanoff [164] and Hodkinson [171]. Silicate, granite and calcite particles were examined in [164] with average diameters d from 0 to $50\,\mu m$ and in [171] suspensions in water of particles of quartz and flint whose relative refraction index was 1·16. The particles in [171] had an average diameter of $3·7\,\mu m$. In both studies, the data on $p(\gamma)$ were stable, slowly varying with respect to angle and size. According to the data of [171], $p(90°)$ is equal to 0·33 and according to [164] it decreases from approximately 0·24 for $\bar{d} = 3\,\mu m$ to 0·10 for particles with $\bar{d} > 15\,\mu m$. The data of [171] are close to the observations of A.A. Ivanoff. It follows from the above that in order to describe the polarization properties of a suspension, the homogeneous sphere model is not suitable.

We note in conclusion that the polarizing properties of ocean water are a simple property of the water masses, to be more exact of the hydrosol suspensions in them. In particular, A.A. Ivanoff notes [164] that in the Straits of Gibraltar, it is impossible to distinguish the inflowing waters of the Atlantic from the outflowing Mediterranean waters in terms of scattering, whereas they are significantly different in terms of polarization properties.

5.4.2. Light scattering matrices for ocean water

The experimental data for the scattering matrix of ocean water are usually given for cases in which the Stokes parameters of the beam are selected in the form given by Equation (4.39). We use this representation below. It differs from that for the scattering matrix for homogeneous spheres [Equation (4.41)].

The relationship between the two forms of notation for the parameters is given in Section 4.2 and makes it easy to go from one form of the scattering matrix to another.

The scattering matrix for ocean water is the sum of the molecular scattering matrices for pure ocean water and suspensions. The molecular scattering matrix has the form [75]:

$$\frac{b(90°)}{1 + \Delta}\begin{pmatrix} 2 - (1 - \Delta)\sin^2\gamma & -(1 - \Delta)\sin^2\gamma & 0 & 0 \\ -(1 - \Delta)\sin^2\gamma & (1 - \Delta)(2 - \sin^2\gamma) & 0 & 0 \\ 0 & 0 & 2(1 - \Delta)\cos\gamma & 0 \\ 0 & 0 & 0 & 2(1 - \Delta)\cos\gamma \end{pmatrix}.$$
$$(5.35)$$

The values of $b(90°)$ and Δ are given in Chapter 3.

The first measurements of the light scattering matrices for ocean water showed that they differ appreciable from the matrices for homogeneous spheres. According to the data of [133], the matrix obtained from samples of water from the northern Atlantic for scattering at an angle of $\gamma = 30°$ has the form (in some cases $A_3 = A_4$):

$$\begin{pmatrix} A_1 & B_1 & 0 & 0 \\ B_1 & A_1 & 0 & 0 \\ 0 & 0 & A_3 & 0 \\ 0 & 0 & 0 & A_4 \end{pmatrix}.$$
$$(5.36)$$

Scattering matrices were measured for the waters of the Black Sea and Baltic Sea, the Atlantic and Pacific Oceans [76].

In [33], the results are given of the statistical treatment of approximately 60 scattering matrices measured at different levels in the Atlantic and Pacific Oceans (from 0 to 2000 m) in the range of scattering angles 25–145° for a wavelength $\lambda_0 = 546$ nm [76]. The statistical treatment was carried out for the entire collection of data as well as separately for the 0, 10, 100–200 and 300–2000 m depth ranges. The average values for the different elements of the matrix as well as their dispersion were given also. We shall look at some of the results of [33]. We point out first that the matrix f_{ik} is often represented in terms of the normalized matrix $\tilde{f}_{ik} = f_{ik}/f_{11}$ where it is obvious that $|\tilde{f}_{ik}| \leqslant 1$. In Equation (4.39), the element f_{11} corresponds to the scattering function and thus the elements of the normalized matrix characterize the purely polarization aspects of the scattering. When the normalized matrix is used, the degree of polarization of the scattered light p is simply equal to $-\tilde{f}_{12}$. For the Pacific Ocean, normalized matrices of the following type were found to be characteristic:

$$\begin{pmatrix} 1 & \tilde{f}_{12} & \tilde{f}_{13} & \tilde{f}_{14} \\ \tilde{f}_{12} & \tilde{f}_{22} & \tilde{f}_{23} & \tilde{f}_{24} \\ 0 & \tilde{f}_{32} & \tilde{f}_{33} & \tilde{f}_{34} \\ \tilde{f}_{41} & 0 & \tilde{f}_{43} & \tilde{f}_{44} \end{pmatrix}. \tag{5.37}$$

Thus, all elements apart from $\tilde{f}_{31}$ and $\tilde{f}_{42}$ were found to be different from zero. Within the limits of accuracy of the measurements ($\pm 0{\cdot}1$) no dependence on depth was found in the elements of the normalized scattering matrix $\tilde{f}_{ik} = f_{ik}/f_{11}$. For the Atlantic Ocean, the average values of the elements $\tilde{f}_{13}, \tilde{f}_{14}, \tilde{f}_{23}, \tilde{f}_{24}, \tilde{f}_{32}, \tilde{f}_{34}$ and $\tilde{f}_{41}$ were found to be smaller (in the majority of cases they did not exceed $0{\cdot}2$) and the mean square deviations were significantly higher than for the waters of the Pacific Ocean (their values were $0{\cdot}1–0{\cdot}3$ for all components and over the entire range of angles). This means that the scattering of light by ocean water is essentially anisotropic and the nature of this anisotropy in the investigated range of depths does not change. However, the element f_{11} of the absolute matrix, ie the scattering function varied appreciably with depth, and its degree of elongation decreased. Figure 5.24 shows data for the elements $p(\gamma), \tilde{f}_{33}$ and $\tilde{f}_{43}$ illustrating some of the results of [33].

In [33], data are also given for measurements of the optical activity of seawater. Measurements on a standard saccharimeter with a base length of 400 mm showed that with an accuracy to $\pm 0{\cdot}03°$ there was no rotation of the plane of polarization.

As a whole, it should be pointed out that the available data on the scattering matrices differ from each other to a significant extent and these differences are related to the shape of the scattering matrix. Probably, the reasons for these differences lie in the different composition of the samples investigated. From the form of the scattering matrices, it is possible to form some ideas of the scattering medium that we are dealing with. Thus, for example, the fact that

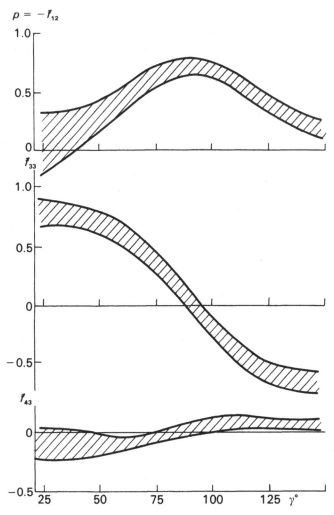

Fig. 5.24. Elements of the light scattering matrix for ocean water for four depths (0–2000 m) in the Atlantic and Pacific Oceans. The hatched area is the variation of each element.

the matrix components in the top right and bottom left corners are equal to zero indicates the isotropy of the medium and the absence of optical activity. The fact that the components $\tilde{f}_{11}$ and $\tilde{f}_{22}$, $\tilde{f}_{33}$ and $\tilde{f}_{44}$ are equal indicates that the particles are spherical. The fact that the components $\tilde{f}_{34}$ and $\tilde{f}_{43}$ are close to zero justifies the assumption that the particles are small compared to the wavelength (or have strong absorption). The matrices described in [33] provide evidence that marine suspensions contain appreciable amounts of particles which do not have mirror symmetry and which furthermore are oriented in space in some predominant manner (they cannot freely rotate about the horizontal axis).

The orientation of particles in seawater may take place under the action of gravity. In [134], two mechanisms are considered governing orientation involving gravity: (1) settling; (2) the centre of gravity not coinciding with the centre of pressure. Estimates made in [48] show that for seawater, matrices of the form Equation (5.36) would be expected for seawater rather than complex matrices of the form in Equation (5.37). Of course, it is impossible to exclude such cases but they seem untypical. As pointed out previously [110], there remains an important problem for future measurements, namely the correct determination of the 'instrument matrix' which has to be subtracted from the measured matrix or it is necessary to ensure that its elements are close to zero. It is also necessary to broaden significantly the range of theoretical calculations for particles with non-spherical shape and the experimental investigations on model media similar to that described in [171].

5.5. Optical characteristics of the waters of the world's oceans

5.5.1. Secchi depth field

The variation over space and time of optical characteristics of the water of the world's oceans have been studied by a large number of authors. There is a large literature in which general laws are studied relating to the ocean as a whole as well as the special characteristics of individual regions. We briefly look at some of the more important results (for more details see [5, 17, 25, 28, 29, 60a, 164]).

The distribution of optical properties of ocean water over space and time is determined by the distribution of the concentrations of dissolved organic matter and suspended substances. Both these characteristics are the result of complex interaction of a large number of factors. In the waters of the open ocean, they are determined by primary production, the cycle of organic matter, the dynamics of the waters and turbulence. Dynamics and turbulence control their propagation from the sites of formation. In coastal regions, river discharge, erosion of the shores and shelf play a significant role. In any region, aeolian transport is important. Thus, in the fifth voyage of the scientific research ship 'Dmitry Mendeleyev', particles of Sahara dust were observed in the waters of the Atlantic almost up to the shores of America itself [69]. A theoretical analysis of the problem presents great difficulties. The effect of turbulence and dynamics was studied by Virtki, Josef et al. (for references see [110]). A mathematical model of the ecosystems of the pelagic zone containing 10 elements (biogenic matter, detritus, phytoplankton and zooplankton, bacteria, etc) in the tropical regions of the ocean was studied in the work of M.E. Vinogradov et al. [15]. In this model, the relationship between the various elements of the community was investigated on the assumption that the system moves in space together with the water flow. Of course a full picture of the variation of optical characteristics may only be obtained by combining the analysis of dynamic and trophic factors. The difficulties here lie not only in the great complexity of the models but also in that both the equations governing the different processes and the values of the quantities determining the rates of these processes are poorly known. Therefore, calculations may only be carried out for extremely simplified idealized situations and their results give only a qualitative picture of the phenomena. The actual source of our knowledge of the fields of optical characteristics in the sea is the material of direct measurements during ship's voyages. The special 'optical' voyages are of particular importance: 'Albatross', fifth and tenth voyage of 'Dmitry Mendeleyev', 61st voyage of 'Vityazie', etc.

A detailed review of expeditionary optical research of the main oceans was carried out by V.I. Vojtov [17]. He points out that in all, there are about 5000 instrumental hydrooptical stations. From a map of the stations, it can be seen that the measurements are distributed extremely unevenly; in the southern hemisphere, especially south of 40°S, there are very few stations. There is significantly more Secchi depth (Zs) data. A map of the Zs field is given in Fig. 5.25. (in accordance with [110]). The similarity of this map to the map showing the distribution of suspended matter (Fig. 2.12) is striking. This is natural since the attenuation of visual radiation with which Zs is closely correlated is controlled by scattering by suspended matter. In the same way, as the distribution of suspended particles, the Zs field is subject to latitudinal and circumcontinental zonality. Turbid coastal waters can be identified which have a Secchi depth less than or equal to 10 m. They occupy an area of approximately 2·2 % of the ocean area of water. In the open ocean, c is associated with plankton and suspended matter. The distribution of the Secchi depth is determined there by the plankton content so that it is subject to latitudinal zonality. In the polar and temperate latitudes where the biomass of plankton is high, the Secchi depth is of the order of 10–20 m. The area occupied by the waters of this type is 31·7%. In tropical latitudes where there are few plankton, the Secchi depth is great and usually of the order of 30–40 m. These waters occupy 24·7% of the area. In the tropics, patches of especially transparent waters can be isolated with $Zs > 40$ m. They occupy an area of 5·1 %. The transitional gradation of 20–30 m is common for the subtropical zone and to some extent for temperate and equatorial latitudes. It is most widespread in the waters of the World's Ocean (36.3%). The average value of Zs for the World's Ocean as a whole is 27.5 m.

The distribution of Zs is closely related to the circulation of waters, with cyclic phenomena. In cyclones, a decrease in Secchi depth is observed and in anticyclones an increase. In particular, the highest values are found in the subtropical and tropical convergences. Examples of these are the Sargasso Sea in the Atlantic Ocean and the zone of the southern tropical convergence in the Pacific and Indian Oceans.

5.5.2. Attenuation coefficient field

An analysis of the distribution of the optical properties of the world ocean is usually carried out for the field of the attenuation coefficient c although it would be preferable to study the field of the coefficients b and a since these

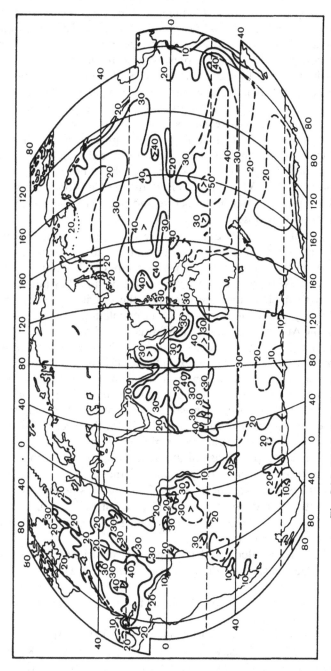

Fig. 5.25. Field of Secchi depth Z_S for the waters of the world's oceans.

quantities are directly related to the concentration of suspended and dissolved substances. This is simply due to the fact that the c data are more easily obtained. However, even for c, the data are inadequate, and in order to construct charts, methods of objective analysis of geophysical fields are used. These methods are based on the minimization of error and in the case of a sparse network make it possible to recommend the most probable values for the field.

The vertical distribution of the optical properties of the waters of the world's oceans was examined in [110]. In classifying the ocean mass based on the optical structure, the hydrological stratification of waters was naturally assumed [80] and bottom, deep and intermediate waters identified; reference data were given for each of these layers. As far as surface waters are concerned, the overall picture of their transparency distribution is shown by the chart for Zs (Fig. 5.25).

Charts and sections over extensive water areas have been produced in a number of studies [25, 28]. When the data from different authors are compared, it should be borne in mind that in temperate and polar latitudes, a significant seasonal variation in C is observed (Fig. 2.9).

From the studies of individual oceans, we note study [86] in which the waters of the Pacific Ocean are divided into regions according to the values of c (546). The authors divided the surface waters into four types and present a chart of the distribution of these types over the surface of the ocean. They note the close analogy between their chart and the charts of the distribution of plankton, suspensions, primary production and the water circulation system. A detailed investigation of the Atlantic Ocean for c (530) was carried out by G.G. Neuimin and N.A. Sorokina [64]. They used 98 stations in the northern Atlantic and 115 in the tropical Atlantic. In order to exclude the effect of seasonal variations, only winter observations were taken in the northern Atlantic. Using the methods of objective analysis, the authors constructed a field $c(\lambda, \phi)$ at the corners of a regular network with intervals of 2° of latitude and longitude for a number of levels. Their chart is shown in Fig. 5.26 for the level $z = 10$ m. This chart is similar to the Zs chart for the same region produced by Frederick which is given in [28]. The sharp increase in attenuation on the western shores of Africa is the result of the strong upwelling of the waters in this region. Furthermore, in the western part of the Gulf of Guinea, it is possible to see two turbidity areas with more detailed resolution, adjacent to the estuaries of the gigantic rivers of Africa, the Niger and the Congo. The streams of turbid water from them can be seen far to the west.

We present the extreme values for the attenuation coefficient for $\lambda_0 = 546$ nm. The maximum value measured in the oceans, $c = 2 \cdot 8$ m^{-1} was obtained on the Peruvian Shelf. The minimum values of $c = 0 \cdot 06$–$0 \cdot 07$ m^{-1} were measured in the Sargasso Sea and in the Pacific Ocean in the zones of the northern and southern tropical convergences.

The average values of c for the world's oceans are $0 \cdot 21$–$0 \cdot 30$ m^{-1}.

Of the regional investigations, we note the detailed study of G. Kullenberg et al. on the optics of the Sargasso Sea [155]. The authors determined the coefficients c, b and a as well as the scattering function in a small square 350

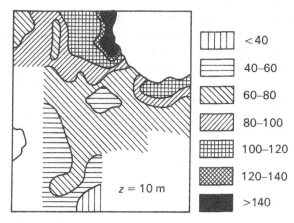

Fig. 5.26. Field of the attenuation coefficient in the waters of the Atlantic Ocean (in relative units).

miles to the south of the islands of Bermuda. Their data are of great interest in view of the uniqueness of the observed water. The coefficient a was determined from the characteristics of the light field using the equation of A.A. Gershun. Altogether, the Sargasso Sea is one of the most studied areas in optical oceanology. It was here that Kryummel observed $Zs = 66\cdot5$ m which was considered to be the record for a long time. Recently, A.P. Ivanoff *et al.* published the results of new measurements of $Zs, c, a, \beta(\gamma)$ and also characteristics of the light field at different depths in the sea [30]. As a whole, the data they obtained provide evidence that the optical properties of the waters of the Sargasso Sea are close to the typical properties of waters of transparent regions of the world's oceans. At the level of the seasonal thermocline (at a depth ~ 100 m), a layer with a step in c (increase of 5–20%) was noted in [30]. In the Sargasso Sea, a number of authors have noted special features of the optical structure related to eddies in the open ocean. Blurred layers of low transmittance were observed so that a patch of more transparent water occurs underneath the eddies. The position of the transparent patch is somewhat shifted relative to the axis of the vorticity.

Of the other regional investigations, we note the detailed analysis of the optical properties of the waters of the Black Sea carried out by G.G. Neuimin [63]. It is interesting that this sea is one of those regions of the world's oceans where the vertical profiles of the optical characteristics cannot be measured from samples. In such anoxic basins, there are extensive hydrogen sulphide zones. When a sample is lifted on board, it rapidly becomes cloudy due to the precipitation of colloidal sulphur. There are 14 such basins: the Black Sea, the Cariaco Trench in the Caribbean Sea, etc. Similar problems arise in the study of the optics of the Red Sea, where in the vicinity of the bottom, the water has a temperature of approximately 56°C and an anomalously high salinity. In terms of its optical properties, the waters of the Black Sea are close to those of ocean waters: they are markedly different from the other seas surrounding the Soviet Union. The special feature of the vertical structure of the sea's optical properties is an intense layer of low transmittance at a depth of 100–180 m; its thickness on average is 10–30 m. In [63], data are also given of the distribution of the optical characteristics over the surface as well as data on light fields in the sea.

Finally, it should be mentioned that more detailed data on the optical properties of actual ocean water are given in the monograph [60a].

Chapter 6

Inverse problems

6.1. General formulation

6.1.1. Ill-posedness, regularization and conditioning

The development of methods in which the properties of a medium are determined from its scattering characteristics has been a fundamental problem in physics for a long time. We come across this problem in quite different fields of physics. In the theory of elementary particles and in atomic physics, we are dealing with the determination of the field potential from the asymptote of the wave functions at infinity or in the classical formulation, from the given dependence of the effective cross-section upon the scattering angle.

In all cases, whether we are dealing with the scattering of photons or alpha particles, etc, we are confronted with 'direct' and 'inverse' problems. In the direct problem, the incident beam and the optical properties of the medium are known and the scattering characteristics are investigated. In the inverse problem, the characteristics of the beam are known before and after scattering so that the properties of the volume element of the medium in which the scattering took place may be deduced. Although the two problems are essentially different approaches to the same phenomenon, the methods for solving them are fundamentally different. Furthermore, although we always have a solution for the direct problem, this is far from being the case for the inverse problem. This is a common feature of the inverse problems.

The first question which arises here is whether the given problem is soluble or insoluble. The latter could mean that there is no data on the properties of the substance in the scattered beam and of course no amount of ingenuity will give us this information. In other words, the experiment has been set up incorrectly and the problem in theoretical study becomes the design of a proper experiment which will produce data about the medium of interest. In the language of analysis: a solution for the particular problem should exist.

The requirement that the experiment will produce data which lead to a solution to the problem is evident but is not specific enough. It is necessary that the experimental data lead to an unambiguous answer.

However, these two requirements, the existence of a solution and its uniqueness unfortunately do not exhaust the subject. In practice, when calculations are carried out, the question of the effect of errors in the initial data and in the calculation of the answer is also vitally important. Inverse problems are usually found to be ill-posed. This term introduced at the beginning of the century by the French mathematician J. Hadamard means that the calculation procedure for such problems is very sensitive to small errors in the initial data or calculation. A small, usually uncontrolled variation in the initial data or calculation error leads to a significant change in the answer and a strictly correct solution of the problem may lead to physically absurd results. In other words, ill-posed problems give rise to ambiguities.

In recent years, it has become clear that ill-posed problems have been encountered in analyses for a very long time and that there are a very large number of them. It was therefore important to develop a general method for solving such 'bad' problems. An analysis of the concrete procedures for solving ill-posed problems used in different areas of technology shows that all methods are based on the idea of incorporating extra information into the method of solution. During this process, the problem becomes well-posed. This means that in all cases where an algorithm determining an answer from the initial data has led to a large increase in errors, it is extended by additional considerations which make the problem stable. Previously, these considerations were invoked intuitively and each problem was handled differently on the basis of the physical meaning of the problem. Now, a general procedure has been worked out for solving such problems and has been rigorously formalized. This procedure is called regularization.

Strictly speaking, regularization, ie the incorporation of additional requirements, replaces the initial problem by another. The solutions of this new problem are called quasi solutions, and it is important that this substitution does not take us too far away from the initial problem, ie quasi solutions should be close to the solutions of the initial physical problem. In brief, it is necessary that regularization does not change the physical content of the problem but simply eliminates its calculation instability.

Regularization reduces to two operations: (1) the formulation of additional conditions which eliminate the physically absurd solution; (2) the construction of a formal algorithm for the problem which automatically includes the initial 'bad' problem together with the additional conditions. These additional conditions are similar to the superimposed ties in analytical mechanics. The method for handling them is very similar to the method of the Lagrange undetermined coefficients. There are a number of different methods for regularization. The best known method was suggested by A.N. Tikhonov. A systematic exposition is given in the monograph [83].

We illustrate how ill-posed problems are handled using some actual examples. We start with the simplest case. Suppose it is necessary to find $\bar{x}$, the root of the equation:

$$f(\bar{x}) = c \qquad (6.1)$$

It is obvious that $\bar{x} = f^*(c)$ where f^* is a function which is the inverse function of f. Using the well known theorem on the derivative of the inverse function, we easily find that:

$$d\bar{x}/\bar{x} = \phi(c) \, dc/c; \quad \phi(c) = c/[f^*(c)f'(c)] \qquad (6.2)$$

Equation (6.2) establishes the relationship between the 'input' errors (dc/c) and the 'output' errors $(d\bar{x}/\bar{x})$. The function $\phi(c)$ is the coefficient of error magnification. For those values of c for which $\phi(c)$ is large, it is impossible to use the suggested algorithm to find the solution: the problem is ill-posed. In particular, it cannot be done when $f'(\bar{x})$ is very small. This is understandable since in order to find the solution to Equation (6.1) it is necessary to determine the point of intersection of the straight line $y = c$ with the curve $y = f(x)$. However, where the derivative $f'(x)$ is very small, the curve $y = f(x)$ is almost parallel to the x-axis and naturally the coordinates of the point of intersection will be very sensitive to errors in c. In order to obtain acceptable values of $d\bar{x}/\bar{x}$, we either try to reduce dc/c as much as possible or, if this is impossible, to regularize the problem. This means that instead of $f(x)$, we introduce a new function $f_1(x)$ close to $f(x)$ but such that the new $\phi_1(c)$ is small in the range of c which is of interest to us. Sometimes, this is done quite crudely in the following manner. If it can be assumed, from physical considerations, that $f^*(c)$ is a 'good' function then its graph is simple to plot, calculating values only at 'smooth' points, bypassing any irregularities. This graph is used to calculate $\bar{x}$ for all values of c. The data on the quality of the function $f^*(c)$ also provide the additional considerations which make it possible to regularize the problem and the construction of the graph $f^*(c)$ constitutes in this case the regularization process.

In Equation (6.1), the well-posedness of the problem reduces to the question of the stability of the position of the point of intersection of two straight lines as the coefficients of the straight lines are varied. This is typical and the linearization of inverse problems often reduces to the solution of linear systems. Therefore, as a second example we will examine the system:

$$x - y = a; \quad x - (1 + 10^{-m})y = b \tag{6.3}$$

The determinant of this system $\Delta = 10^{-m} \neq 0$. According to Cramer's rule, there is a solution to the system given in Equation (6.3). It is unique and may be determined from Cramer's formulae if all the calculations prescribed by these formulae are carried out exactly. However, if $m \gg 1$, the quantity Δ is very small. In this case, the straight lines whose point of intersection we are seeking are almost parallel. For small changes in a and b, the position of the point of intersection will vary significantly. Since arithmetic operations always involve rounding and if they are obtained by experiment the numbers a and b are known with limited accuracy, it is impossible in practice to calculate the roots of Equation (6.3) when $m \gg 1$. This means that if the accuracy of the calculations or the accuracy of the numbers a and b are not increased, the system given in Equation (6.3) will be useless.

In the solution of a system of linear equations, in particular with a large number of unknowns, it is therefore very important to know in advance what sort of systems we are dealing with. Systems similar to the system given in Equation (6.3) are called ill-conditioned in algebra. In the general case, when we are dealing with the solution of a linear system:

$$\sum_{k=1}^{n} a_{ik}x_k = y_i, \quad i = 1, 2, 3 \ldots n \tag{6.4}$$

of n equations with respect to n unknowns, ill-conditioning of the system given in Equation (6.4) or the matrices of coefficients $\alpha = [\alpha_{ik}]$ means that among the hyperplanes constituting the system given in Equation (6.4), there are some, at least two, which are almost parallel. Usually, in inverse problems, systems of the type given in Equation (6.4) arise on linearization of an exact problem, and both the free terms y_i and the elements of the matrix α are known approximately. As a rule, the larger n is, the smaller is Δ. As a result of efforts to improve the accuracy of linearization by increasing the number of points n, the conditioning of the system given in Equation (6.4) is made worse. Every specific inversion problem has its own optimum n: any increase or decrease in n makes the situation worse. If the determinant of the matrix is small, then often variations in the elements at the limits of their accuracy will make Δ vanish. This means that the problem does not have a solution: the system given in Equation (6.4) is inconsistent.

Before attempting to solve the system, it is very important to know how to determine the degree of conditioning of its matrix. If the system is the result of linearization of a problem, it is often possible by varying the linearization points (which usually have been selected arbitrarily) to improve significantly the degree of conditioning of the matrix. In algebra, several different methods for estimating the degree of conditioning of a matrix have been suggested. For this, the so-called condition number is calculated: Todd, Turing, etc. For example, the Todd number is equal to $\max |\lambda_i|/\min |\lambda_i|$ where λ_i are the eigenvalues of the matrix $[\alpha_{ik}]$. The condition numbers characterize the magnification of errors in the vector y_i in the determination of x_k. They are equivalent to the quantity $\phi(c)$ in Equation (6.2). The various expressions for these numbers are related to the different methods of defining the closeness of the vectors x_k (in the sense of the root mean square, minimax, etc) and different methods for allowing for the errors in y_i and in the matrix $[\alpha_{ik}]$. The calculation of the conditioning numbers is extremely complicated and a simple estimate as described below is often sufficient. The simplest of these is based on the above considerations. The system is ill-conditioned if its determinant Δ is very small. This statement lacks one important qualification however. It is obvious that the solution of the system of equations does not change if each of the equations is multiplied by a constant number L. This means that the determinant of the system is multiplied by L^n, ie it may be made as big as desired. Therefore, the system of hyperplanes must first be normalized. For this, each of the n equations of the system given in Equation (6.4) must be multiplied by the quantity

$$\left(\sqrt{\sum_{k=1}^{n} \alpha_{ik}^2} \right)^{-1}.$$

The best conditioning will be obtained for a system of mutually perpendicular planes. The determinant of such a system, after the system has been normalized, will be equal to ± 1. Thus, the degree of conditioning of the system is determined by how close the modulus of the determinant of the normalized system is to 1.

For example, in Equation (6.3), it is easy to calculate that its normalized determinant is equal to $10^{-m}/4$. For large values of m, it is very small: the system will be ill-conditioned.

Physically, the ill-posedness of a problem is often related to extreme generality in its formulation; small changes in the initial parameters may lead to large changes in the answer. In order to exclude unacceptable solutions, it is necessary right from the start to somehow restrict the formulation of the problem. However, this contraction should not be exaggerated. We are faced with mutually conflicting requirements, and the achievement here of the correct proportion is a measure of the value of the method.

Thus, in formulating the inverse problem we must try to obtain: (1) a problem which has a solution; (2) a unique solution and (3) an algorithm as general as possible and as insensitive as possible to experimental and calculation errors.

6.1.2. Determination of the characteristics of a disperse medium from the scattering of light

After these general remarks, we turn to the inverse problem which is the subject of analysis in this chapter. We are concerned with the determination from the light scattering characteristics of the size distribution curve for particles suspended in a medium. This problem is of great interest in oceanography, meteorology, biology, technology, etc. The particles, with negligible total mass, have an enormous fragmented surface area and make a significant contribution to the properties of dispersive systems. For example, in the ocean it is small foreign particles which determine its optical properties and marine adsorption processes, etc.

The characteristics of light scattering depend on the properties of the medium. By measuring these characteristics for seawater and subtracting the molecular scattering which is known, it is possible to determine the scattering by suspended materials and to use the data obtained to determine the composition of these oceanic particles. We assume that the suspended particles are homogeneous spheres with radius a. We also assume that the volume element is irradiated by a parallel beam of light and that we are dealing with 'dilute' systems so that there is no interference or multiple scattering in the scattered beam. It is not difficult to show that in this case our problem reduces to the inversion of the linear integral equation of the first kind:

$$S(\gamma, \lambda) = N \int_{a_{min}}^{a_{max}} s(\gamma, \lambda, a) f(a) \, da \qquad (6.5)$$

where $S(\gamma, \lambda)$ is a measured characteristic of the light scattered by the ensemble of suspended particles (the total scattering coefficient or the scattering coefficient for a fixed angle) and $s(\gamma, \lambda, a)$ is the value of this characteristic for an individual particle of radius a; N is the numerical concentration; $f(a) = n(a)/N$ is the distribution function of the particles over size; a_{min} and a_{max} are the limiting radii of the suspended particles.

For example, the kernel $s(x, a)$ may be the scattering function of a monodispersive system of particles with radius a; in this case $S(x)$ is the polydispersive scattering function describing the scattering at an angle $\gamma = x$, etc. In all cases, the inverse theory problem consists of defining a method for calculating the unknown distribution function $f(a)$ from $s(x, a)$ and $S(x)$.

In such a complete formulation, there are great difficulties in solving the inverse problem and in fact the simplest procedure for solving Equation (6.5) is to replace it with an algebraic system of the type given in Equation (6.4). However, as a rule this approach does not achieve its goal, since the systems obtained in this way are ill-conditioned. For example, the strictly correct solution may contain negative roots, which are physically meaningless. In addition, the, higher the order of the system, the more sensitive is its solution to errors. Thus, the calculation is impossible in practice. The reason for the difficulties lies in the characteristics peculiar to the integral equations of the first kind: these equations are ill-posed.

It is often assumed that $a_{min} = 0$ and $a_{max} = \infty$, so that instead of Equation (6.5) we have the simpler equation:

$$S(x) = N \int_0^\infty s(x, a) f(a) \, da \qquad (6.6)$$

For some simple kernels, Equation (6.6) is not difficult to solve. Thus, if the kernel has the form exp (ixa) (the Fourier kernel), the value of $S(x)$ determined experimentally will be simply the Fourier transform of the function $f(a)$. In the more general case, if the kernel depends only on the product of the arguments:

$$s(x, a) = s(xa)$$

the formal solution of Equation (6.6) may be obtained with the help of the Mellin or Titchmarsh transformations [82].

The relationship between Equation (6.6) and the Fourier transformation makes it possible to understand the important feature of integral equations of the first kind in terms of their ill-posedness. It can be seen that the measurement errors always present in $S(x)$ limit the number of Fourier harmonics of the required function $f(a)$ which may be produced on solving Equation (6.6). The high harmonics, ie the fine details of the behaviour of $f(a)$, are lost in the noise of the function $S(x)$. This can be seen immediately if it is imagined that we add any term of the form $\Delta f(a) = Ce^{i\omega a}$ to the solution $f(a)$ of Equation (6.6). This changes the left-hand side of Equation (6.6) by the amount:

$$\Delta S = C \int_v^\infty s(x, a) \, e^{i\omega a} \, da$$

At high frequency ω, the value of ΔS, the Fourier component of the function $s(x, a)$ may be made very small so that, for example, ΔS falls within the measurement limits of S which are determined by its measurement accuracy. Thus, the values of $f(a) + \Delta f(a)$ will satisfy Equation (6.6) just as well and may be taken as its solutions, although with the appropriate choice of C the two expressions will differ from each other considerably. The measurement accuracy of S determines the critical frequency ω_0. The Fourier components with, frequency $\omega > \omega_0$ cannot be determined from Equation (6.6).

Due to the difficulties associated with the solution of Equations (6.5) and (6.6) in the complete formulation, restricted problems have gained wide acceptance in ocean optics. There are two types of restriction: (1) selected properties of the system are determined, for example the concentration of the particles N or the moments of the distribution of particles by size, etc; (2) additional conditions are imposed on the function $f(a)$ or on N. Sometimes, for example, it is assumed that the function $f(a)$ belongs to a family of gamma distributions or exponential distributions and the experimental data are used to determine the values of the parameters in these distributions. Such a restricted problem should be distinguished from the complete problem in which no conditions are imposed on N and $f(a)$ in advance and they are completely determined directly from the data of the optical experiment. We first consider the simplest case, the restricted problem.

6.2. The restricted problem

6.2.1. Determination of the numerical or mass concentration and other distribution moments

If it is assumed that the distribution function and the limiting dimensions of the suspended particles remain constant, then the measured scattering characteristic S may be used to determine the numerical concentration of the suspension N. In fact, transforming Equation (6.5) it is possible to write:

$$N = BS \tag{6.7}$$

where

$$B^{-1} = \int_{a_{min}}^{a_{max}} s(\gamma, \lambda, a)f(a)\,da.$$

If we are interested in the mass concentration rather than the numerical concentration, Equation (6.7) takes the form:

$$C = B'S, \quad B' = \rho\bar{v}B \tag{6.8}$$

where C is the mass concentration of the suspension, g/cm^3; ρ is the density of the material of the particles; and $\bar{v}$ is the average particle volume.

In [31], the transmittance of kaolin suspensions and clay silt in water was investigated. The linear relationship was studied between the concentration of the suspended matter C and the attenuation coefficient c determined for red light ($\lambda = 726$ nm):

$$c = \eta C \qquad (6.9)$$

The parameter η depends on the nature of the particles and their dimensions. When C was expressed in g/cm^3 and c in m^{-1}, then the quantity η varied from 0.6 to 0.19; its average value over 53 samples was 0.30.

For pure ocean waters (suspended particle concentration from 0.5 to 2 g/m^3) with a predominance of particles $a < 0.5$ μm, the following correlation relationship was obtained in [67] between the suspended particle concentration C and the coefficient c for $\lambda_0 = 546$ nm:

$$C = 1 \cdot 85c + 0.20 \qquad (6.10)$$

Many authors report a linear relationship between c and the suspended particle concentration [25].

When determining the suspended particle concentration C, the best results are from the relationship between the scattering coefficient for a given angle $b(\gamma)$ and the contribution of particles of different dimensions to C. The larger the particles, the smaller should be the angle γ chosen in the relationship:

$$y = m \cdot b(\gamma) + n \qquad (6.11)$$

In Equation (6.11), y may denote either a numerical concentration (N million/litre) or a mass concentration (C mg/m^3) of the suspended particle fraction. In [46], the optimum values of the regression coefficients m and n and the relative errors δ were determined for the fractions $a = 0.2$–0.5, 0.5–1.0 and > 1.0 μm using the scattering coefficients $b(\gamma)$ for the angles $\gamma = 45, 6$ and $1°$, respectively (for $\lambda_0 = 546$ nm). In calculating the mass concentration C, it was assumed that particles with $a = 0.2$–1.0 μm are mineral, $\rho = 2$g/cm^3, and that particles with $a > 1$ μm are organic, $\rho = 1$ g/cm^3. The results are given in Table 6.1. They were obtained from measurements in the Atlantic and Pacific Oceans and were verified by independent measurements in the Indian Ocean. In all cases, the errors in N and C did not lie outside the limits $3N\delta_N$ and $3C\delta_C$. Simple equations such as Equations (6.7) to (6.11) make it possible to determine rapidly the suspension field over large expanses of water using a towed transmittance meter or a nephelometer with fixed angles. The accuracy of such measurements is low but no less accurate than standard methods used by geologists and is significantly less cumbersome.

Sometimes, for simple estimates of the suspended particle concentration, Equation (6.5) is used in the monodispersive approximation. In this case, the integral in Equation (6.5) is replaced by the scattering characteristic of a monodispersive suspension of unit concentration with an 'effective' particle radius a_{eff}:

$$S(a_{\text{eff}}) = N \int_{a_{\min}}^{a_{\max}} s(a)f(a) \, da \qquad (6.12)$$

Table 6.1. Optimum values of the regression coefficients in Equation (6.11)

Particle fraction $a(\mu m)$	N			C			Number of samples	Voyage, region
	m	n	δ_N	m	n	δ_C		
0.2–0.5	3×10^4	−1.0	0.29	8.9×10^3	−3.0	0.16	29	Eighth voyage, Pacific Ocean
0.5–1.0	9.5	0.2	0.14	24	0.5	0.14	41	Eighth voyage, Pacific Ocean
>1.0	0.2	0.3	0.35	12	16	0.20	30	Fifth voyage, Atlantic Ocean and Pacific Ocean

If the concentration N is known, then a_{eff} is found by comparing the ratio S/N with the monodispersive value from the tables of [123]. We emphasize that the effective particle radius obtained in this manner depends on the property measured. The broader the distribution, the greater the differences between the various values for a_{eff}.

The monodispersive approximation may be used only for qualitative estimates.

If the monodispersive approximation is discarded, then it is possible to establish, from equations in Chapter 4, which distribution moments are determined in the measurements. With the help of some of the moments, it is possible to determine the general nature of the distribution. For complete recovery of the distribution function, it is necessary to know all the moments. The problem of restoring the function from its moments is called the problem of moments. It is examined in texts on applied analysis.

6.2.2. The fitting method

Another class of restrictive problems widespread in the study of the optics of turbid media is the selection of functions of a concrete distribution which describes the given suspension from a known set of functions. This selection is made by comparing the measured integral optical characteristic S, for example the scattering function, with a set of standard curves corresponding to the family studied. If the monodispersive characteristic is expressed by an explicit formula, the polydispersive characteristic can often be represented by some special function $S(\gamma_i)$ where γ_i is the scattering angle (or frequency) for which the measurements of S are carried out. Examples of such functions are given in Chapter 4 for different distribution families. In this case, the determination of the values of the parameters of the family v_k reduces to the solution of the system of equations:

$$S(\gamma_i, v_k) = C_i, \quad i = 1, 2 \dots n \qquad (6.13)$$

The number of equations n must not be less than k, the number of parameters in the required family. Despite its apparent simplicity, Equation (6.13) is not always convenient. Difficulties often arise similar to those discussed in connection with Equation (6.1) and Equation (6.13) turns out to be ill-conditioned.

In the general case, when it is necessary to use the Mie series for the monodispersive property $s(\gamma, \lambda, a)$, the values of S are represented in the form of tables, for example [123].

Suppose we have measured the scattering function of a sample of ocean water and by subtracting that for pure water have obtained the scattering function $\tilde{\beta}(\gamma)$ of the particle suspension. Different formulations of the problem are possible here. The simplest of them consists of determining which of the models corresponds to the measured data better than the others. We select some criterion for determining the differences between $\tilde{\beta}(\gamma)$ and the model values $\beta_k(\gamma)$. This is either the sum of the root-mean-square deviations or the maximum modulus of the deviation, etc. We shall look for example at the sum of the root-mean-square deviations $F^*(k) = \sum_{i=1}^{n} [\beta_k(\gamma_i) - \tilde{\beta}(\gamma_i)]^2$ over all n angles for which there are data. We must emphasize two points: (1) since the scattering functions $\beta(\gamma_i)$ for different scatering angles γ_i differ by several orders of magnitude, instead of the sum of the squares of the differences in $F^*(k)$ it is necessary to consider the relative quantities:

$$F(k) = \sum_{i=1}^{n} \left(\frac{\beta_k(\gamma_i) - \tilde{\beta}(\gamma_i)}{\tilde{\beta}(\gamma_i)} \right)^2 \tag{6.14}$$

Otherwise, the deviations in the region of small angles swamp those for medium and large angles. In Equation (6.14), all angles of comparison have become equivalent (a similar 'weighting' procedure for the contribution of different angles was used in [125] in the solution of a similar problem for an atmospheric aerosol); (2) in the determination of the optimum model, it is necessary to distinguish cases of sharp and flat minima of $F(k)$. We refer to Fig. 6.1. Both curves of $F(k)$ have a minimum at $k = k_0$ but only the first curve makes it possible to select a model with certainty. Taking measurement and calculation errors into account, the second curve gives no reason to prefer the k_0th model to the neighbouring model. We are faced again with an ill-conditioned problem and for a unique selection of k it is necessary to have additional information. If this is not available, the optical data only make it possible to recommend the superimposition (possible with weighting) of all those models for which the curve $F(k)$ does not lie outside certain limits $|\delta F|$ governed by the measurement accuracy of $\tilde{\beta}(\gamma)$. If we denote the error in $\tilde{\beta}(\gamma_i)$ by $\delta\tilde{\beta}(\gamma_i)$, then it is apparent that:

$$|\delta F| = 2 \sum_{i=1}^{n} \left| \left(\frac{\beta_k(\gamma_i)}{\tilde{\beta}(\gamma_i)} - 1 \right) \frac{\beta_k(\gamma_i)}{\tilde{\beta}(\gamma_i)} \frac{\delta\tilde{\beta}(\gamma_i)}{\tilde{\beta}(\gamma_i)} \right|$$

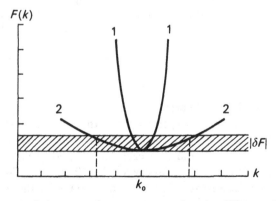

Fig. 6.1. Two characteristic types of curve representing the difference between actual and model distributions.

The area of undefined values of F is the shaded area in Fig. 6.1. For the case illustrated there, three models where $k = k_0 - 1$, k_0 and $k_0 + 1$ may be considered as optimum to the same extent.

As an example of using the fitting method we look at Equation (5.31). We take $b_w(\gamma)$ from [113] together with the data for 31 values of $b(\gamma)$ at $\lambda_0 = 546$ nm, in the region of small angles $\gamma = 0.33–10°$ and 10 values in the region of large angles $\gamma = 15 (15) 145°$. The measurements were carried out at 68 stations (32 in the open ocean and 36 at coastal stations) in the 0–25 m layer in the Pacific and Indian Oceans. Using the distribution charts for terrigenous and biogenic particles, each station was assigned to one of four classes designated b, bB, B and BB and t, tT, T and TT for biogenic and terrigenous suspensions, respectively (low, medium, high and very high concentrations). Since there were few stations, they were combined into a single Table 6.2. In this table, three cells were found to be empty. For the remaining 13 classes, average values $\bar{b}(\gamma)$ were calculated and the statistical discrimination of these classes was determined. It was found that only four groups were reliably discriminated, so the 13 classes were combined into four groups. In Table 6.2, these groups are indicated in parentheses. For each group, average values $\bar{b}_i(\gamma)$ were calculated. Furthermore, the root-mean-square deviations of the right- and left-hand sides of the equation:

$$b_i(\gamma) = \lambda^2 n_t b_{\text{ter}}(m_t, v, \rho_{\max}) + \lambda^2 n_b b_{\text{biog}}(m_b, \bar{\rho}, \rho_\sigma) \qquad (6.15)$$

were minimized for each group in a computer whose memory contained all of the polydispersive models considered in [123]. The factor $\lambda^2 n$ on the right-hand side of Equation (6.15) is associated with conversion of the dimensionless numbers α_r given in the tables in [123] to the coefficient b (Chapter 4).

Table 6.2. Distribution of the number of stations by suspension type

Gradation	t	tT	T	TT
b	0	3 (4)	6 (4)	0
bB	1 (4)	11 (4)	11 (3)	3 (2)
B	5 (2)	4 (2)	9 (2)	4 (2)
BB	0	8 (2)	2 (2)	1 (1)

This made it possible to determine the optimum values of the concentrations n_t and n_b, the refractive indices m_t and m_b and other parameters of the suspension for each $\bar{b}_i(\gamma)$. When considering the uncertainty with which the values of $\bar{b}_i(\gamma)$ are known to us, it was found that each corresponds approximately to 8–10 optimum models falling in the interval $|\delta F|$ (Fig. 6.1). The quality of the inversion is illustrated in Fig. 6.2, which shows data for the first group (where the scattering is greatest). Discrepancies are observed in the back scattering angles in the region of 105° and 45°. This means that at these angles, the fitting

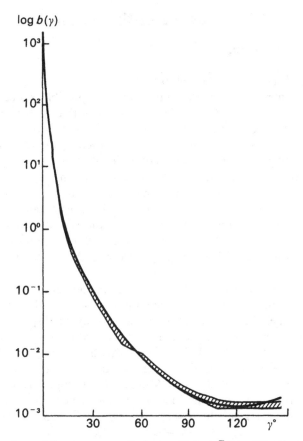

Fig. 6.2. Comparison of the average scattering coefficient, $\bar{b}_1(\gamma)$, of the first group (bold line) and the set of 10 'nearest' retrieved scattering coefficients (within the hatched area).

method is sensitive to the selection of the initial family of distributions. Models constructed in this manner may be useful for different problems. In particular, they make it possible to recommend an optimum value of $b(\gamma)$ for those angles γ for which measurements have not been made, for example at back scattering angles, etc.

Another example of using the fitting method is described in Section 6.7. There, it is used to determine the characteristics of terrigenous particles [from the values of $\beta(\gamma)$ in the interval 45–145°]. The minimax principle is used as the proximity criterion in Section 6.7. What effect has the proximity criterion used for the selection on the inversion results? For marine suspensions, this question is given special attention in [11]. It was found that the use of different criteria has practically no effect on the fitting result: the difference in the spectra of particles obtained in all the considered examples was less than 10%. In practice, this is the inversion accuracy which is determined by the measurement errors for the marine scattering functions (Section 6.6).

6.3 The method of fluctuations

6.3.1. Basic relationships

The method of dispersion was developed in the study [114] and is a simple concept. Measurement of the transmittance makes it possible to determine the optical thickness τ of the system. Observations show that if there are only a few particles in the beam, then the transmittance of the system undergoes appreciable fluctuations. These fluctuations are caused by random movements of the particles, the particles overlap each other in different ways. Usually, when measuring the transmittance we average these fluctuations supposing them to be noise. In fact, they contain valuable information on the properties of the dispersive system under study. The dispersion of the transmittance, as well as τ, depends directly on the number of particles in the studied volume so that simultaneous measurement of the transmittance and its dispersion gives us a method for determining both the average size and the concentration of the particles.

In order to construct a theoretical model of this phenomenon, we assume that the particles are situated completely randomly in space. We then consider the passage of a parallel beam of light through a layer of the dispersive medium, and, for the sake of simplicity, we assume that the medium consists of identical particles of spherical shape and that the attenuation of a parallel beam of intensity I is due only to the particles in the volume illuminated.

If the cross-sectional area of the beam S and the total projected cross-sectional area of the particle is y, then the intensity of the light passing through the dispersive medium, is given by $I = I_0 \cdot (S - y)/S = I_0 \cdot Y/S$ where I_0 is the intensity of the incident beam, $Y = (S - y)$. Calculating the average value and the dispersion of the random variable Y, we find the average value of the intensity of the light which has passed through the medium and its dispersion.

Thus, the statistical model of the phenomenon reduces to the problem of the random projection of circles on to a plane area A. Let this area be a unit square. We assume that the circles are projected independently and p_k is the probability that the centres of exactly k circles have fallen on to a square A. We are interested in the random variable $Y = (S - y)$, the area of the square not occupied by circles. The moments of this random variable are determined with the help of the theorem of G. Robins based on the general properties of conventional mathematical expectations established by A. N. Kolmogorov. It follows from this theorem that l the moment of the random variable Y will be:

$$E(Y^l) = \int\limits_{\substack{x_i \in A \\ 1 \leqslant i \leqslant l}} p(x_1 \ldots x_l) dx_1 \ldots dx_l$$

where $p(x_1 \ldots x_l)$ is the probability that not one of the points $x_i (1 \leqslant i \leqslant l)$ is covered by circles.

In order to calculate $p(x_1 \ldots x_l)$ it is sufficient to find the area of the common part of the circles and we limit ourselves to studying the average value and the dispersion of the random variable under consideration, ie the values $E(Y)$ and $E(Y^2)$.

We assume that the number of circles whose centres fall in square A are distributed according to the Poisson distribution: $p_k = e^{-\lambda}\lambda^k/k!$. Neglecting, for the sake of simplicity, phenomena at the boundary of the square [which is possible if the relative (dimensionless) area $s = s_0/S = \pi r^2$ of each of the incident circles $\ll 1$], we obtain:

$$E(Y) = e^{-\lambda s} \tag{6.16}$$

$$\sigma_y^2 = se^{-2\lambda s}\left[2\int_0^\pi a(\tau, \phi)\,d\phi + \frac{16r}{\pi}\int_0^\pi \cos\frac{\phi}{2}\cdot a(\tau, \phi)\,d\phi\right.$$
$$\left. - \frac{8r^2}{\pi}\int_0^\pi \cos^2\frac{\phi}{2}\cdot a(\tau, \phi)\,d\phi\right] \tag{6.17}$$

$$a(\tau, \phi) = \sin\phi\,\exp\left(\frac{\tau(\phi - \sin\phi)}{\pi} - 1\right) \tag{6.18}$$

where λ is the average number of circles falling on to the square ie $\tau = \lambda s$.

In our model, $s = s_0/S$ is the relative cross-section of the attenuation of light by the particles of the medium; $\lambda = \bar{n}lS$ is the average number of scattering centres in the illuminated volume of the medium and the optical thickness of the medium $\tau = \lambda s = l\bar{n}s_0$.

Assuming that $r \ll 1$ [so that the second and third terms in Equation (6.17) may be neglected], we obtain from Equations (6.16) to (6.18) (where $\lambda = \bar{N}$):

$$\bar{I} = I_0 e^{-\tau} \tag{6.19}$$

$$D = \overline{(\Delta I)^2} = I_0^2\,\phi(\tau)/\bar{N} \tag{6.20}$$

$$\phi(\tau) = 2\tau e^{-2\tau}\int_0^\pi a(\tau, \phi)\,d\phi \tag{6.21}$$

Equation (6.19) expresses the usual Bouguer-Lambert law.

Equation (6.20) can be used to find $\bar{N}$ (the average number of particles in the illuminated volume) from the measured dispersion D and the optical thickness τ [τ may be determined from Equation (6.19)] and consequently $\bar{n}$ and s_0.

We examine in more detail the special function $\phi(\tau)$ which relates the dispersion D with τ and $\bar{N}$ (see [114]). A graph of this function is shown in Fig. 6.3. At zero and at infinity, the values of $\phi(\tau)$ equal zero and have a maximum of 0.1770 at $\tau = 1.17$. The relationships $\phi(\tau) \to 0$ for $\tau \to 0$ or $\tau \to \infty$ can be physically explained by the fact that in the first case ($\tau \to 0$) there is no mutual shadowing and the detector records an unchange beam; in the second case ($\tau \to \infty$), the

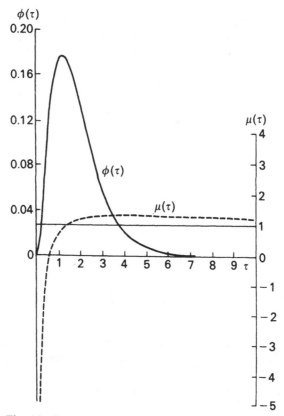

Fig. 6.3. Graphs of the special functions $\phi(\tau)$ and $\mu(\tau)$.

dispersion is small because there are many particles in the path of the ray and flux does not fluctuate significantly.

For small and large τ, we find the following asymptotic estimates of the function $\phi(\tau)$. When $\tau \to 0$, expanding the exponent under the integral sign in Equation (6.21) in a series and taking into account that:

$$\frac{1}{\pi^2} \int_0^\pi (\varphi - \sin \varphi)^2 \sin \varphi \, d\varphi = \frac{1}{2} - \frac{8}{3\pi^2} = 0.2298 \ldots$$

we obtain:

$$\varphi(\tau) = \tau^2 - 1.770\tau^3 + O(\tau^4) \tag{6.22}$$

When $\tau \to \infty$ the main contribution to the integral is in the region of the point

$\phi = \pi$. Denoting $\pi - \phi = \psi$ and assuming ψ small, we find that:

$$\phi(\tau) = 2\tau e^{-2\tau} \int_0^{\pi} \exp\left(-\tau \frac{\psi + \sin \psi}{\pi}\right) \sin \psi \, d\psi$$

$$= 2\tau e^{-2\tau} \int_0^{\infty} \exp\left(-\tau \frac{2\psi}{\pi}\right) \psi \, d\psi = \frac{\pi^2}{2} \frac{e^{-\tau}}{\tau} \qquad (6.23)$$

Taking into account that the integral in Equation (6.21) is a steadily increasing function of τ and that when $\tau \to 0$ it is equal to $\tau/2$, for any $\tau \geqslant 0$, we have $\phi(\tau) \geqslant \tau^2 e^{-2}$. On the other hand, when $0 \leqslant \phi \leqslant \pi$:

$$\exp\frac{\tau(\phi - \sin \phi)}{\pi} - 1 \leqslant e^{\tau}$$

Thus, for the function $\phi(\tau)$ we have the following estimates:

$$\tau^2 e^{-2\tau} \leqslant \phi(\tau) \leqslant 4\tau e^{-\tau}$$

The values of the function $\phi(\tau)$ are given in Table 6.3. The deviations of the asymptotic Equations (6.22) and (6.23) from the table are 2.5% when $\tau = 0 \cdot 1$ and 1.3% when $\tau = 10$.

We consider the variable coefficient γ of the intensity of the transmitted beam:

$$\gamma = \sqrt{D/I} = \sqrt{s}\,\theta(\tau), \quad \theta(\tau) = \sqrt{\phi(\tau)/\tau}.e^{\tau} \qquad (6.25)$$

Table 6.3. The special function $\varphi(\tau)$

τ	$\phi(\tau)$	τ	$\phi(\tau)$	τ	$\phi(\tau)$
0	0	1.6	1.585 −1	4.0	1.916 −2
0.1	8.380 −3	1.7	1.509 −1	4.2	1.542 −2
0.2	2.810 −2	1.8	1.428 −1	4.4	1.238 −2
0.3	5.301 −2	1.9	1.343 −1	4.6	9.922 −3
0.4	7.904 −2	2.0	1.257 −1	4.8	7.934 −3
0.5	1.037 −1	2.1	1.171 −1	5.0	6.345 −3
0.6	1.253 −1	2.2	1.087 −1	5.5	3.609 −3
0.7	1.432 −1	2.3	1.005 −1	6.0	2.046 −3
0.8	1.572 −1	2.4	9.259 −2	6.5	1.159 −3
0.9	1.672 −1	2.6	7.795 −2	7.0	6.568 −4
1:0	1.736 −1	2.8	6.499 −2	7.5	3.728 −4
1.1	1.767 −1	3.0	5.376 −2	8.0	2.119 −4
1.2	1.769 −1	3.2	4.415 −2	8.5	1.208 −4
1.3	1.749 −1	3.4	3.607 −2	9.0	6.899 −5
1.4	1.708 −1	3.6	2.932 −2	9.5	3.950 −5
1.5	1.652 −1	3.8	2.374 −2	10.0	2.266 −5

At small and large τ, we have respectively:

$$\gamma = \sqrt{s} \cdot \sqrt{\tau}; \quad \gamma = \sqrt{s}(\pi/\sqrt{2})(e^{\tau/2}/\tau) \tag{6.26}$$

Since the average number of circles λ falling on the square is constant, when $\tau \to 0$ there is no mutual shadowing and γ also tends to zero. When $\tau \to \infty$ the dispersion falls more slowly than the square of the intensity and $\gamma \to \infty$. Using Equations (6.24) for $\phi(\tau)$, we find that for any τ:

$$\sqrt{\tau} \leqslant \theta(\tau) < 2e^{\tau/2} \tag{6.27}$$

It can be seen from Equation (6.25) that for a given τ it is necessary to make $s = s_0/S$ as large as possible, ie to work with beams which are as narrow as possible.

We now present the final equations which make possible the determination of the average attenuation cross-section s_0 and the concentration $\bar{n}$ of the particles from the average intensity of the transmitted beam $\bar{I}$, the dispersion of the signal D and the cross-sectional area of the beam S:

$$s_0 = (D/I_0^2)[S\tau/\phi(\tau)]; \quad \bar{n} = \tau/(ls_0); \quad \tau = -\ln(\bar{I}/I_o) \tag{6.28}$$

6.3.2. Experimental verification and accuracy

The method was tested on model suspensions consisting of spherical and non-spherical particles. Experiments using polystyrene spheres (diameter 50–54 μm) suspended in water were described in [114]. The experimental apparatus consists of a parallel light beam source, a cell containing calibrated particles suspended in water and detector. A helium-neon laser was used as the light source. The instrument incorporated a DC amplifier and rectifier. The rectified signal was recorded giving the average amplitude of the fluctuation of the intensity of the transmitted light. This value is somewhat less than the dispersion but is considerably simpler to obtain experimentally.

In eight cases, the average diameter of the particles calculated from Equation (6.28) was 39 μm, ie 30% less than the actual size. This reduction was due to two reasons:

1. The substitution of the root-mean-square deviation $(\overline{(\Delta I)^2})^{1/2}$ for which Equation (6.28) was derived by the average linear deviation $\overline{|\Delta I|}$. This leads to a reduction of the results for the Gaussian distribution for example by 20%;
2. The presence in the recording system of a filter to eliminate oscillations in the illumination of the laser itself. This filter reduced the effect studied and led to a reduction in the measured fluctuations by approximately 5–10%.

A detailed assessment of the accuracy of the method and an optimum measurement system are given in [78]. Using Equation (6.28) it is not difficult to show that the relative error in the determination of ds_0/s_0 will be:

$$|ds_0/s_0| = |dD/D| + |[2 - \mu(\tau)]\, dI_0/I_0| + |\mu(\tau)\, d\bar{I}/\bar{I}| \qquad (6.29)$$

The function $\mu(\tau)$ in this equation can be simply expressed in terms of $\phi(\tau)$:

$$\mu(\tau) = 1/\tau - d\ln\phi(\tau)/d\tau \qquad (6.30)$$

The graph of $\mu(\tau)$ is shown in Fig. 6.3 and its table is given in [78]. It has an asymptotic representation:

$$\mu(\tau) = \begin{cases} -1/\tau + \beta + \gamma\tau + \dots & \text{when} \quad \tau \to 0,\ \beta = 1{\cdot}770\dots \\ 1 + 2/\tau + \dots & \text{when} \quad \tau \to \infty \end{cases} \qquad (6.31)$$

The function $\mu(\tau)$ has a root at $\tau = 0.52$ and a small maximum at $\tau = 4.4$ where it is approximately $1{\cdot}33$. The fact that $\mu(\tau)$ tends to infinity when $\tau \to 0$ is because at small τ the fluctuation method is ineffective.

The main sources of error are in determining the dispersion (the first term) and the error associated with the measurement of the intensity (second and third terms). The frequency spectrum of the fluctuations in intensity due to the motion of the particles occupies the region of low frequencies starting from 0Hz. This considerably hampers the determination of the dispersion when low-frequency components are present in the signal, arising from non-steady-state conditions in the source, noise in the receiver or other elements of the instrument circuit. It is also necessary to eliminate variations in the transmittance of the dispersive medium caused by processes occurring in it (particle settling, etc). Noises in the source have a specially strong effect when lasers are used. The dispersion of the intensity of the transmitted beam D' actually observed is the sum of the variable of interest D and the noise D_0.

The value of the variation of the dispersion D' is determined from the equation:

$$\sigma[D']/\bar{D}' = \sqrt{2/(n-1)}$$

where n is the number of readings obtained by processing the recordings of the fluctuations.

Calculating s_0, we substitute D' for D and the error in the determination of D will have two independent sources: the first is the error in the determination of D' and the second is the systematic error due to the contribution of the quantity D_0, which we have not allowed for:

$$\left|\frac{dD}{D}\right| = \frac{1}{D}\sqrt{\frac{2}{n-1}D'^2 + D_0^2} = \sqrt{\frac{2}{n-1}(1+\eta)^2 + \eta^2}, \quad \eta = D_0/D$$

At very small and large values of τ, when $\phi(\tau)/\tau \to 0$, then it follows from Equation (6.28) that the value of D is reduced and may become comparable with D_0. In these cases, the value of η and the error $|dD/D|$ increase. In the range of τ where $D \ll D_0$ and $\eta \ll 1$, the error is determined only by the number of readings n and is independent of τ. We also note that $D \sim s_0/S$ so that it is desirable, as we have already pointed out, to have S as small as possible. As far as the errors associated with the intensity measurements I_0 and $\bar{I}$ are concerned, then assuming that the relative errors of both variables are approximately equal and denoting $b[I]/I = A$, we derive the following expression for the total error in s_0:

$$\left|\frac{ds_0}{s_0}\right| = \sqrt{\frac{2}{n-1}(1+\eta)^2 + \eta^2 + A[2 - \mu(\tau) + |\mu(\tau)|]} \qquad (6.32)$$

As can be seen from Equation (6.32), it is also necessary to provide measurement conditions such that the parameters η and A are as small as possible. The value of A is determined by multiplicative noise and is a constant of the photometric apparatus. In contrast to it, η depends on the variable parameter D and when designing the apparatus it is necessary to make D as large as possible over the entire interval of τ.

The measurement system constructed on the basis of these ideas is described in [78] and illustrated in Fig. 6.4. A parallel beam from the light source S is modulated by the chopper 1, passes through a cell 2 containing the investigated medium and falls on to the light dividing mirror 3 which transmits the central part of the beam and reflects the remaining light to the photodetector 4. The diaphragm 5 forms a narrow beam from the transmitted light which is detected by a second photodetector 6. The angular apertures of both detectors are the same so that it may be assumed that the attenuation cross-section of any particle is the same for both channels. It is desirable to make the apertures as small as possible, to avoid having to make corrections for the light scattered by particles in the detector.

From the photodetectors the signals are compared and subtracted by the unit 7. There will be practically no fluctuations in the broad beam arriving at photodetector 4. On balancing and subtracting the two channels, the transmittance signal will be cancelled out but the fluctuations of interest will increase sharply. Unstable sources of light such as a laser may be used in the circuit. The difference signal is amplified by unit 8 and the synchronous detector 9, whose reference signal is provided by the photodiode 10. The latter is illuminated by the source light modulated by chopper 1. The fluctuation spectrum is recorded in unit 11.

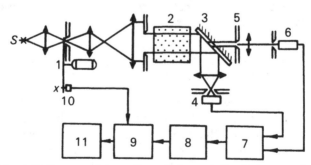

Fig. 6.4. Schematic diagram of the optics and the recording instrument for measurement using the fluctuation method.

The circuit described above was used to make measurements with calibrated monodispersive particles at different concentrations: lycopod spores [d_0 = (30 ± 1)μm] and polystyrene spheres [d_0 = (53 ± 1)μm]. The measurement results for 10 samples of each specimen are given in [78].

The quantity $\sqrt{D}/I_0$ characterizing the intensity of fluctuations was measured in these experiments from 0.45 to 5.5% (at high and low concentrations, respectively). The average diameters of the lycopod and polystyrene particles were 30.1 ± 1.8 and 52.4 ± 3.3 μm, respectively. A more complete representation of the accuracy of the method is given by the curve of relative errors $\Delta(\tau) = |\Delta d_0/d_0|$ for both models from the data of [78] shown in Fig. 6.5. In Fig. 6.5, the dependence $\Delta = \Delta(\tau)$ is also shown for particles with diameter $d_0 = 53\,\mu$m (curve 1) and 30 μm (curve 2) calculated from Equation (6.32). In these calculations, in accordance with the characteristics of the measurement system used in [78], it was assumed that $A = 10^{-2}$ and $D_0/I_0 = 10^{-5}$. 200 readings were obtained to produce the fluctuation spectrum and it can be seen from Fig. 6.5 that the experimental values of Δ agree closely with the values calculated from Equation (6.32). A general conclusion is that when the system is used in the range $0.5 \leqslant \tau \leqslant 2$ the error Δ does not exceed 6%. On changing to lower and particularly to higher values of τ, the error Δ increases.

The method of fluctuations may be also used for non-spherical particles. In this case, it gives the average cross-section of the particles in the beam. An example of such an application is described in [124] where suspensions of kaolin particles and cellulose fibre were investigated. The mass concentrations of kaolin and dry cellulose fibre were defined for producing the specimens. Under the microscope kaolin suspended in water looks like a collection of particles of all shapes with dimensions from 5 to 20 μm; a suspension of cellulose fibre forms clots and individual threads measuring several tens of micrometres. For polydispersive media with non-spherical particles it is possible to determine

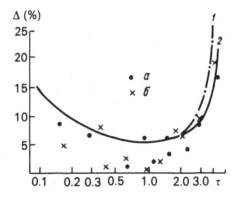

Fig. 6.5. Dependence of the relative error Δ on the optical thickness τ calculated from Equation (6.32) for lycopod particles (1) and for polystyrene particles (2). The experimental points shown are for lycopod particles (●) and for polystyrene particles (x).

from Equation (6.28) only the average size $\bar{d}_0$ and concentration $\bar{n}$ of some effective spheres for all possible cross-sections of the irregularly shaped particles. The mass concentration of the suspension in water can be calculated from the parameters obtained in this way and the known density of the suspended particles. The average error in determining the concentration of four samples was found to be approximately 20%. The results obtained make it possible to recommend the fluctuation method for working with actual suspensions.

In conclusion, we look at the application of the method of fluctuations to the study of the composition of marine suspended particles. Since the concentration of the fine suspension in the ocean is high and its relative variation is small, the method of fluctuations is convenient only for studying the large suspensions. In this case, the fine suspension may be assigned to an invariant background above the level of which the transmittance and its fluctuations are measured. We give an example of such an application in Section 6.7. Since the dispersion method is technically simple (probably simpler than measuring the small-angle scattering function), it can be used in constructing operational systems for observing the space-time dynamics of the variation of large fraction suspensions in the sea. It may also be used successfully for studying processes in the coastal zone, the dynamics of suspensions and their transport along the shore and the study of spray spectra, etc.

6.4. Method of small angles

6.4.1. Initial equation

We now return to the general Equation (6.5). For spheres, the kernel of Equation (6.5) is given by the Mie equations (Chapter 4). For three particular cases, when the kernel $s(\gamma, \lambda, a)$ may be represented as a simple analytical formula, it was possible to find exact solutions of this equation. They are called the method of small angles, the method of spectral transmittance and the method of the total scattering function [96, 174a, 132b].

We first look at these three methods. The solution here follows unambiguously from the experimental data. The study of the exact solutions is important to use for many reasons; they can be conveniently used to check different numerical regularization methods and to determine the necessary requirements for the measurement accuracy and the range and number of measurement points. Furthermore, these cases are important in themselves.

Regularization in exact methods arises as the result of the assumption that the necessary integral transformations may be performed on all the functions of the problem, ie that all the improper integrals involved do exist. These assumptions are analogous to those which are formulated in the theory of Fourier transforms. In contrast to the fitting method, these requirements are not at all onerous physically and therefore the exact solutions may almost always be used so long as the kernels used in them are suitable.

We expound the theory of the method of small angles following the studies of [94, 95] and describe its experimental application [95, 99]. In the procedure for calculating the composition of marine suspensions from light scattering data developed in [109], the method of small angles is combined with the fitting method: the former is used to determine the large fraction and the latter to determine the fine fraction.

The method of small angles is based on the investigation of the aureole seen in the direction of the source (Fig. 4.7). We turn to the equation for the intensity of light scattered by a large particle ($\rho = 2a/\lambda \gg 1$) at small scattering angles $\gamma(\gamma \ll 1)$ (Chapter 4):

$$I(\gamma) = I_0 a^2 J_1^2(\rho\gamma)/\gamma^2 \tag{6.33}$$

where I_0 is the intensity of the light incident on the particle.

The intensity of light scattered by a polydispersive system of particles will be:

$$\bar{I}(\gamma) = \frac{I_0}{\gamma^2} \int_0^\infty f(a)a^2 J_1^2\left(\frac{2\pi}{\lambda}\gamma a\right) da \tag{6.34}$$

The angular structure of the beam $\bar{I}(\gamma)$ corresponding to three types of distributions $f(a)$ is shown in Fig. 6.6. When the width of the distribution $f(a)$ is increased, the coronas gradually disappear: instead of dark and light rings, a gradual diminishing of the brightness of the scattered light is observed from the centre of the beam to its periphery.

Equation (6.34) is a particular case of the general Equation (6.5). Equation (6.34) may be inverted exactly. The inversion is based on the theory of generalized Fourier transforms. The general theory of these transforms is developed in the book by E. Titchmarsh [82]. The points of departure are the equations:

$$f(x) = \int_0^\infty k(xu)g(u) \, du \tag{6.35}$$

$$g(u) = \int_0^\infty h(uy)f(y) \, dy \tag{6.36}$$

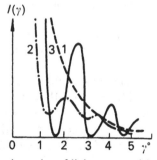

Fig. 6.6. Distribution of the intensity of light scattered by a polydispersive population of large particles ($\rho \gg 1$) at small angles ($\gamma \ll 1$) for populations with different width of distributions: 1, large; 2, medium; 3, zero.

It is shown in [82] that if in Equation (6.35):

$$k(x) = \frac{\sqrt{\pi}}{2} x \frac{d}{dx} [x J_v^2(x/2)]$$

then in Equation (6.36), the quantity $h(x)$ will be:

$$h(x) = -\sqrt{\pi} J_v(x/2) Y_v(x/2)$$

where $J_v(x)$ and $Y_v(x)$ are the Bessel and the Neumann functions of order v.

After simple transformations, we arrive at the equation:

$$f(\rho) = -\frac{2}{\rho^2} \int_0^\infty F(\rho\gamma)\varphi(\gamma)\, d\gamma \tag{6.37}$$

where the kernel:

$$F(x) = x J_1(x) Y_1(x) \tag{6.38}$$

and the function:

$$\phi(\gamma) = \pi \left(\frac{2\pi}{\lambda}\right)^3 \frac{d}{d\gamma}\left(\frac{\bar{I}(\gamma)}{\bar{I}_0}\gamma^3\right) \tag{6.39}$$

Hence it follows that having determined $\bar{I}(\gamma)$ by experiment and having calculated $\phi(\gamma)$ from Equation (6.39), we can use Equation (6.37) to find $f(\rho)$ for all ρ. The kernel of Equation (6.34) depends on the product of the variables. Integral equations with such kernels may be inverted using the Melinn transformation (see [82]). Of course, if this procedure is carried out with Equation (6.34) the same Equation (6.37) is obtained. Our shorter derivation follows [94]. Converting to the scattering coefficient $b(\gamma)$ we finally obtain:

$$f(\rho) = -\frac{4\pi^2}{a^2\lambda} \int_0^\infty F(\rho\gamma) \frac{d}{d\gamma} [b(\gamma)\gamma^3]\, d\gamma \tag{6.40}$$

The function $F(x)$ has the following asymptotic representations:

$$F(x) = \begin{cases} -x/\pi, & x \ll 1 \\ \cos(2x)/\pi, & x \gg 1 \end{cases} \tag{6.41}$$

Its graph is shown in Fig. 6.7. The four-figure table of $F(x)$ for $x = 0(0 \cdot 1)16$ is given in [94] and in more detail for $x = 0(0 \cdot 01)10$ in [23] and [4]. The first few roots x_k and the extreme value x_k^* and $F(x_k^*)$ are given in Table 6.4.

At large values of x, the function $F(x)$ coincides with a cosine curve. In this case, its root is $x_k = (k + 1/2)(\pi/2)$ and its extreme values are $x_k^* = k\pi/2$ (k is an integer) and $F(x_k^*) = 1/\pi$.

The inversion Equations (6.37) to (6.39) are based on Equation (6.33) for the monodispersive scattering function. It can be seen from the equations in

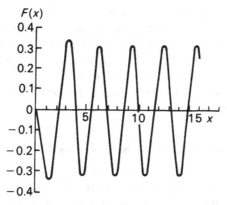

Fig. 6.7. Graph of the function $F(x)$.

Chapter 4 that they are suitable only for large phase shifts $\delta = 2\rho|m-1| \gg 1$. For a drop of water in air, this means that $\rho > 20$. However, for 'soft' biological particles, the limit with respect to ρ for Equation (6.33) is shifted to 130 ($m = 1 \cdot 05$) or even to 330 ($m = 1 \cdot 02$). However, we remember that in Chapter 4 it was shown that for large transparent spheres it is possible to extend considerably the area of application of Equation (6.33) if a correction factor $P(\rho, m)$ is introduced. When $P(\rho, m)$ is taken into account, instead of Equation (6.33) it is necessary to write:

$$I(\gamma) = P(\rho, m)I_0 a^2 J_1^2(\rho, \gamma)/\gamma^2 \tag{6.42}$$

It is important that the factor $P(\rho, m)$ does not depend on the angle γ. This means that the factor $P^{-1}(\rho, m)$ must be directly introduced into the new formula for $f(\rho)$. Thus, it now has the form:

$$f(\rho) = -\frac{1}{P(\rho, m)} \frac{4\pi^2}{a^2\lambda} \int_0^\infty F(\rho, \gamma) \frac{d}{d\gamma} [b(\gamma)\gamma^3] \, d\gamma \tag{6.43}$$

The correction factor for Equations (6.33) and (6.40) first appeared in [119] and

Table 6.4. Roots and extreme values of the function $F(x)$

k	x_k	x_k^*	$F(x_k^*)$
1	0.000	1.260	− 0.3730
2	2.197	3.015	+ 0.3304
3	3.832	4.636	− 0.3237
4	5.430	6.225	+ 0.3213
5	7.016	7.805	− 0.3202

[13]. It improves the results for any particles but it is especially useful in the case of soft biological particles in marine suspensions. Generally speaking, as we pointed out in Chapter 4, Equation (6.42) is less viable than the diffraction Equation (6.33) in the sense that its use demands a knowledge of the complex refractive index m of the particle. However, as has been shown by the calculations in [11] ignoring the correction factor may lead to two-fold and three-fold errors in the calculated spectra.

6.4.2. Experimental testing. Essential range of angles

In order to apply the procedure for determining $f(\rho)$ indicated by Equations (6.37) and (6.43), it is necessary to measure $I(\gamma)$ at small values of γ. In order to avoid the direct bright beam which makes it difficult to obtain small scattering angles γ, it was suggested in [95, 99, 100] that measurements be carried out in the focal plane of the receiving lens beyond the focus at which the direct beam is collected (Fig. 6.8). In this case, as was shown in [99, 100], within the limits of paraxial optics, the movement of particles along and perpendicular to the beam does not change the distribution of the intensity in the focal plane of the receiving lens. The light scattered at a given angle γ arrives at the focal plane at a given distance from the centre. $\bar{I}(\gamma)$ was recorded either on a photographic film with subsequent photometric evaluation or using a photomultiplier. The latter is moved in the focal plane along the radius from the centre to the periphery [23]. The dimensions of the focal point in the measurements of [100] corresponded to $\gamma_{\min} \approx 10'$. The measurements of $\bar{I}(\gamma)$ were carried out for angles $\gamma > \gamma_{\min}$ and covered a range to $\gamma \approx 5\text{–}6°$. The irradiance at the focal plane diminished very rapidly, by approximately an order of magnitude for every degree of the angle γ. For such large reductions (five to six orders of magnitude), for measurements in the region of small γ, it is worth using neutral density filters or to balance the measured signal at different values of γ by increasing the diameter of the aperture of the receiver as γ increases.

The scattering of light by the lenses and by the optically empty instrument produces a background reading $I_b(\gamma)$. This background value has to be subtracted and in Equation (6.39) $\bar{I}_{\text{meas}}(\gamma)$ must be substituted by $\bar{I}(\gamma) = \bar{I}_{\text{meas}}(\gamma) - I_b(\gamma)$ (Fig. 6.9).

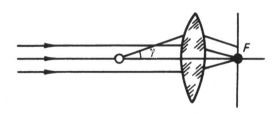

Fig. 6.8. Diagram illustrating the principle of measuring scattering at small angles.

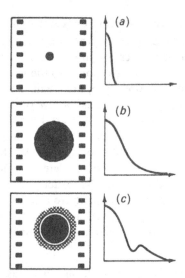

Fig. 6.9. Distribution of the light intensity in the focal plane of a receiving lens. (a) empty instrument (zero background); (b) particle population with wide size distribution; (c) particle population with narrow size distribution. On the left is the negative of the photographic film; on the right is the distribution $I(\gamma)$.

One of the main difficulties in applying the method of small angles is the impossibility of obtaining complete information from experiments, a strict requirement for the inversion. Thus, it can be seen from Equation (6.40) that in theory to calculate $f(\rho)$ we must have the function $\bar{I}(\gamma)$ for all angles γ from 0 to infinity. In fact, the scattering angles γ for which measurements of the scattering function are possible are limited from below by a certain γ_{min} and from above by a certain γ_{max}. This is a basic problem in all exact inversion methods: the effect on the precision of the inversion arising from the incomplete information which can be obtained from the experiment. The limits of the essential range of scattering angles γ in which it is necessary to measure $\bar{I}(\gamma)$ in order to determine $f(\rho)$ with a given accuracy naturally depends on the required particle size spectrum. These limits have been investigated [107] for particle populations with gamma size distributions. A general formula was derived for estimating γ_{max}. In particular, it was shown there that for all gamma systems with $\mu = 2\text{–}8$ with a modal radius of $5\,\mu m$, $\gamma_{max} \leqslant 6°$. The effect on the inversion of limiting the scattering angle to $\gamma_{max} \geqslant \gamma \geqslant \gamma_{min}$ was also studied. The physical nature of these limits is completely different. γ_{min} is related to the final dimension of the focal point at which the direct light beam is collected—it is impossible to measure the scattering function when $\gamma < \gamma_{min}$. Usually, the value of γ_{min} lies close to $10'$. γ_{max} is related to a number of factors which will now be considered.

First, the value of this angle must be such that the optical information contained in the range $0°$ to γ_{max} is sufficient for the retrieval of the size spectrum, and second it is important that the substitution of the exact scattering function by the approximate scattering function does not significantly alter the results obtained. In Chapter 4, we pointed out the range of angles for which Equation (6.33) is valid. An appreciable increase in error starts on approaching the first zero of the function $J_1(z)$ which corresponds to the value $z = \rho\gamma = 3\cdot83$. At low values of z, the errors in Equation (6.33) are small. The restriction imposed on the value of the product $\rho\gamma$ means that as the size of the particles increases, the range of angles in which Equation (6.33) is valid is reduced. Furthermore, it is necessary to take into account that as γ increases, the scattering intensity falls rapidly and at a certain γ_{max} (usually not above 6°), it becomes comparable to the noise of the detector so that any further determination of $\bar{I}(\gamma)$ is impossible. Therefore, even in the case of ideally exact measurements of the input information, the limits of its measurement are always restricted.

A gamma distribution with $\mu = 2$ was chosen for the calculation experiments in [103]. The distribution $f(\rho)$ was deduced for the nine cases shown in Table 6.5.

Table 6.5. Maximum and minimum scattering angles assumed in nine numerical experiments

No.	γ_{min}	γ_{max}	No.	γ_{min}	γ_{max}
1	0	∞	5	0.0002	0.101
2	0.004	0.101	6	0.0002	0.0802
3	0.006	0.101	7	0.0002	0.0702
4	0.008	0.101	8	0.0002	0.0602
			9	0.0002	0.0502

In [103], the function $\phi(\gamma)$ in the range $\gamma_{min} \leqslant \gamma \leqslant \gamma_{max}$ (Table 6.5, γ in radians) was calculated from the exact equations. Thus, the errors associated with the numerical differentiation in Equation (6.39) and with the restriction on the integration limits ρ_{min} and ρ_{max} in Equation (6.34) were eliminated.

The results of the calculations are given in Figs. 6.10 and 6.11. We first look at the role of γ_{max} (Fig. 6.10). Truncation at large angles means that all curves $f(\rho)$ tend to infinity when $\rho \to 0$. However, this does not cause any difficulties in restoring the spectra if there is a well-defined maximum of $f(\rho)$. It is necessary to ignore all values beyond the point of inflection and extrapolate the curve to zero. However, if γ_{max} is not chosen to be sufficiently large for the maximum of $f(\rho)$ to be defined, then it is also impossible to determine the left-hand side of the curve $f(\rho)$. This means that the experimental information is insufficient; the measurements were discontinued too soon. The descending slopes of the curves were determined correctly in all the examples illustrated in Figs 6.10 and 6.11.

We look at the effect of γ_{min} (Fig. 6.11). Both sides of the curves are appreciably deformed. If the contribution of the range $0.101-\infty$ is ignored, then at small ρ, the curve of $f(\rho)$ descends and intersects the ρ-axis. As γ_{min} increases, the entire $f(\rho)$ curve drops so that the right-hand side of the curve is also inaccurate. We

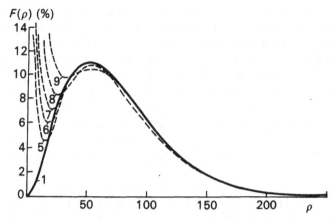

Fig. 6.10. Determination of $f(\rho)$ curves for limiting measurements at large angles. The figures on the curves correspond to the numbers in Table 6.5.

note that even in the worst of the cases given in Table 6.5, the modal value ρ_m is correctly restored. The restrictions at the small angle end may be practically eliminated if the asymptotic estimate of the function $\varphi(\gamma)$ is used when $\gamma \ll 1$. The function $\bar{I}(\gamma)$ has a maximum when $\gamma = 0$; consequently its Maclaurin series at small γ will take the form:

$$\bar{I}(\gamma) = \bar{I}(0) + c\gamma^2 + \ldots \tag{6.44}$$

Substituting in Equation (6.39), we find that:

$$\varphi(\gamma) = \alpha\gamma^2 + \beta\gamma^4 + \ldots; \quad \alpha = \frac{3\bar{I}(0)}{I_0}\,\pi\left(\frac{2\pi}{\lambda}\right)^3; \quad \beta = \frac{5c}{I_0}\,\pi\left(\frac{2\pi}{\lambda}\right)^3 \tag{6.45}$$

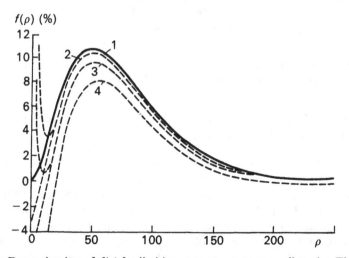

Fig. 6.11. Determination of $f(\rho)$ for limiting measurements at small angles. The figures on the curves correspond to the numbers in Table 6.5.

It follows from Equation (6.45) that when $\gamma \to 0$, it is possible to extrapolate the optical information from the equation $\phi(\gamma) = \alpha\gamma^2$. This makes it possible to avoid distortions associated with the limitations at small angles. As was pointed out above the limitation at large angles is not important when it is possible to determine the maximum of the function $f(\rho)$. Thus, it is necessary to emphasize that if the measurements cover a certain basic range of angles $\gamma_{min} \leqslant \gamma \leqslant \gamma_{max}$, then it is possible to restore reliably the spectrum. In [103], for example, this basic angle range will be 0·004–0·1. In this case, the errors in $f(\rho)$ do not exceed 5% over the interval $0{\cdot}4\rho_m < \rho < 5\rho_m$. Of course, the basic angle range depends on the acceptable size of the permissible error in the curve $f(\rho)$ over the given range of ρ.

For the inversion of marine scattering functions the method of small angles is used to determine the large particle size fraction of the marine suspension (Section 6.7). In this case, by virtue of Equation (6.45), it is assumed that $\gamma_{min} = 0$. As far as the selection of γ_{max} is concerned, since the form of the calculated spectrum is not known in advance, its selection must be guided by information based on the optical data. A large number of calculations on the inversion of model scattering functions using the method of small angles were carried out in [11] using the data given in the tables in [123] and these calculations showed that the elongation of the measured scattering functions at small scattering angles could be used as this optical information. This elongation can be characterized by the ratio $b(0.5°)/b(2°)$. The values of the optimum angle γ_{max} as a function of this ratio are given in Table 6.6.

Table 6.6. Values of γ_{max} for different values of scattering function elongation (in minutes)

$b(0.5)/b(2°)$	< 3.6	3.6–4.8	4.8–6.8	6.8–10.8	> 10.8
γ_{max}	100	90	80	70	60

It can be seen from the table that even for marine scattering functions with minimum elongation, the optimum value of the angle γ_{max} does not exceed 2°.

6.4.3. Use of actual and integral values of the scattering function. Accuracy of the method

In determining the errors caused by the limitation on the measurement interval of $\bar{I}(\gamma)$, we did not take into account the additional errors associated with the mathematical processing of optical information. In fact, the algorithm specified in Equations (6.37) to (6.39) has a drawback: it is necessary to differentiate the experimental function. This makes processing difficult and sometimes leads to a priori invalid results: the probability density is negative and the total probability exceeds unity, etc. The elimination of random errors of measurement, filtration and subsequent smoothing of the signal provide a satisfactory level of accuracy in some cases so that the calculated spectrum practically coincides with the true spectrum. However, this is sometimes inadequate. Although subsequent integration [in Equation (6.37)] improves the matter, it is desirable to avoid differentiation. With this aim, it was suggested in [104] that the actual and integral values of the scattering function are used to calculate the particle spectrum. We introduce the following notation:

$$T(\gamma) = \theta(\gamma) - \theta(\infty); \quad \theta(\gamma) = \gamma^3 \bar{I}(\gamma)/I_0; \quad S(\gamma) = \int_0^\gamma T(\tau)\,d\tau \qquad (6.46)$$

In terms of the functions $T(\gamma)$ and $S(\gamma)$, the distribution curve $f(\rho)$ is defined by the equations:

$$f(\rho) = -\frac{c}{\rho} \int_0^\infty h_T(\rho, \gamma) T(\gamma)\,d\gamma$$

$$f(\rho) = -c \int_0^\infty h_S(\rho, \gamma) S(\gamma)\,d\gamma \qquad (6.47)$$

where $c = (2\pi/\lambda)^3$ and the kernels $h_T(x)$ and $h_S(x)$ are given by:

$$h_T(x) = -2\pi Y_1(x)[2xJ_0(x) - J_1(x)] - 4 \qquad (6.48)$$

$$h_S(x) = 4\pi\{(1/x - x)J_1(x)Y_1(x) + J_0(x)[Y_0(x)x - Y_1(x)]\} - 4/x$$

In addition, as has been shown in [104], the asymptotic value $\theta(\infty)$ is simply:

$$\theta(\infty) = \lambda N \bar{a}/(2\pi^2) \qquad (6.49)$$

Thus, we find the average size of the particles of the required distribution from the value of $\theta(\infty)$. The functions $h_T(x)$ and $h_S(x)$ have the following asymptotic expressions:

$$h_T(x) = \begin{cases} 2, & x \ll 1 \\ 4\sin(2x), & x \gg 1 \end{cases}$$

$$h_S(x) = \begin{cases} 0, & x \ll 1 \\ -8\cos(2x), & x \gg 1 \end{cases} \qquad (6.50)$$

The graphs and tables of these functions are given in [106].

When the actual or integral values of the scattering functions are used, it is convenient to have sets of solutions which give these optical properties for standard types of distribution. For gamma distributions and the generalized gamma distributions it becomes necessary to study a certain class of special functions. These functions are investigated in [106] where formulae are given which express them in terms of eliptical integrals of the first kind $K(x)$ and the second kind $E(x)$; detailed tables of these functions are given, as well as asymptotic expressions at small and large scattering angles γ, etc. It is necessary to add that the application of Equation (6.47)—ie the use of the actual and integral values of the scattering function—give good results only for sufficiently broad distributions. This is the same condition as was specified in [104] in the derivation of these equations, ie:

$$\Delta\rho \gg \pi/(2\gamma_{max}) \tag{6.51}$$

where $\Delta\rho$ is the width of the required distribution and γ_{max} is the maximum scattering angle (in radians) used in the method of small angles.

If Equation (6.51) is not satisfied, then Equation (6.46) gives a lower accuracy than Equations (6.37) and (6.38). A numerical study of the two alternative methods by differentiating and using the actual values of the functions was carried out in [109]. For this purpose a model size spectrum similar to that found in marine suspensions was used. The problem was first solved directly and the calculated small angle scattering function was then inverted using the two procedures. Comparison showed that the spectrum obtained by differentiation showed better agreement with the model spectrum. Preference was given to this procedure in [109].

An important question concerns the accuracy of the method of small angles. This question has been studied experimentally by comparison with a known standard and with numerical experiments [96]. The experimental test showed that the method gives good accuracy. In [96], the first three distribution moments are compared for models and for natural suspensions and it is found that they may be calculated with an error not exceeding 10%.

We look at the accuracy of the method of small angles for analysing a marine suspension. This question was studied in [11], where with the help of the tables given in [123], a series of mathematical experiments was carried out. It was found that when Equation (6.43) is used with an optimum angle γ_{max} chosen from Table 6.6, the result agrees well with the initial spectrum. The accuracy of the recovery depends on the form of the initial spectrum. For an exponential distribution with $v = 3$, $\rho_{min} = 20$ and $\rho_{max} = 200$, the determined spectrum agrees with the initial spectrum with an accuracy of 25% over the range $20 \leqslant \rho \leqslant 130$; for a normal distribution, the large particle region of the spectrum is determined accurately (for $\bar{\rho} = 48$ and $\rho_\sigma = 24$, the accuracy is 20% over the region of ρ from 50 to 120).

Today, many authors are inclined to consider that the size distribution of the large particles in marine suspensions is, in many cases, close to an exponential distribution with $v \leqslant 4$. The calculations performed in [11] make it possible to conclude that for such distributions, the error of recovery does not exceed 25% for the range of values $20 \leqslant \rho \leqslant 120$. For $\lambda_0 = 0.546\,\mu m$ (in air) this range corresponds to values of r from 1.3 to 7.8 μm, ie it almost completely covers the range of sizes of large particles generally found in marine suspensions. In these calculations with model scattering functions, exact values were used. Furthermore, in order to estimate the effect of the measurement errors on the accuracy of the method, calculations were carried out on inversion of scattering functions distorted by a random error of 10% (which corresponds to the actual measurement accuracy). The results of these experiments showed that the spectra obtained differed by only 10–15% from the spectra deduced from the exact functions.

In conclusion, we make some remarks concerning the development of the method of small angles. We obtained the inversion Equation (6.37) in 1949 using the general equation given by E. Titchmarsh [82, p. 282]. In the same year, together with I.Z. Gordon at the Main Geophysical Observatory (MGO), we carried out measurements of the particle size spectra of a water fog by microphotography in parallel with the method of small angles. The method of small angles was expounded in our lecture at the review session of the MGO on 19 January 1955 and at the Fifth Cloud Conference on the 6 February 1956. A brief reference is made to it in [90, p. 274] and a detailed account is given in [94, p. 5, 153] and [95, p. 20, 21].

A similar method based on the inversion of the beam $\Phi(\gamma)$ scattered forwards in a cone with angle γ [see Equation (4.74)] was developed in a study by American authors [139]. The authors of [139] point out that the initial expression for the flux $\Phi(\gamma)$ was obtained in [145]. This equation, however, was previously given in the book [90, p. 149]. M. Kerker in his monograph [154, p. 395] points out that the two versions of the method of small angles are practically equivalent and were developed independently.

A remarkable feature of the method of small angles is that when it is used according to Equation (6.37) it is not necessary to know the refractive index of the material of the particles. As a consequence it has been applied to a large number of different problems in technology. A special book [4] has been devoted to it.

6.5. The method of spectral transmittance and the method of the total scattering function

6.5.1. Method of spectral transmittance

In this method, the van de Hulst Equation (4.85) is used for the scattering cross-section K. If a certain linear dimension is denoted by a_0, then introducing the dimensionless radius of the particles $r = a/a_0$, wave number $v = v^*a_0$, spectral transmittance of the polydispersive system $g(v\beta) = g^*(v^*)a_0$, $\beta = 2\pi(m - 1)$ and the distribution function $f(r) = f^*(a)a_0^4$ (the dimensional variables are marked with an asterisk), we obtain the following initial integral equation for the method of spectral transmittance:

$$g(v\beta) = \pi \int_0^\infty K(\beta vr) r^2 f(r)\, \mathrm{d}r \qquad (6.52)$$

Equation (6.52) is derived in [122] where the formal solution is given. The solution in an explicit form, suitable for application, is to be found, however, in the studies of [115]. These studies also gave a practical method for applying the method. A simple derivation of Equation (6.52) was later suggested in [108]. We explain it here and introduce the notation:

$$q(v\beta) = [g(v\beta) - g(\infty)]v\beta \tag{6.53}$$

It follows from Equation (6.52) that:

$$g(\infty) = \lim_{v \to \infty} g(v\beta) = 2\pi\overline{r^2} \tag{6.54}$$

Substituting Equation (6.53) in Equation (6.52), we find that:

$$q(x/2) = 2\pi \int_0^\infty p(xr)rf(r) \, dr, \quad x = 2\beta v \tag{6.55}$$

where:

$$p(z) = (1 - \cos z)/z - \sin z \tag{6.56}$$

It is shown in [108] that Equation (6.55) has the solution:

$$f(r) = \frac{1}{\pi^2 r} \int_0^\infty p(rx)q(x/2) \, dx \tag{6.57}$$

In order to calculate $f(r)$ from Equation (6.57), it is necessary to measure the spectral transmittance $g(v\beta)$. For polydispersive systems with not very narrow distributions, the curve $g(v\beta) = g(x/2)$ has only one maximum. For reliable inversion, it is necessary to obtain its variation in the region of the maximum and beyond it. We denote the width of this region by the range $\tau_{min} \leqslant \tau \leqslant \tau_{max}$. This is the basic bandwidth. We assume that in this region we have obtained m points for the function $g(x_j/2)$ ($j = 1, 2, 3 \dots m$). Beyond the region τ_{max}, we represent the function:

$$g(x/2) = C_0 + C_2/x^2, \quad x \geqslant \tau_{max} \tag{6.58}$$

The minimum of the experimental data is determined by the possibility of reliably determining the tail of the distribution—the constants C_0 and C_2.

The method of spectral transmittance was used to carry out a large number of calculations with gamma and exponential distributions which made it possible to define minimum requirements of the experimental data [115, 132a, 117]. An example of an inversion is given in Fig. 6.12. The calculations showed that a good particle size distribution may be deduced if a sufficient spectral bandwidth is covered. The requirements of the measurement accuracy for the transmittance $g(v)$ and the number of points m over the basic bandwidth were found to be moderate (accuracy $g \approx 5\%$, $m \approx 10$).

In the initial version of the method of spectral transmittance, the variation of $m(\lambda)$ was ignored. This was in contradiction to an essential feature of the

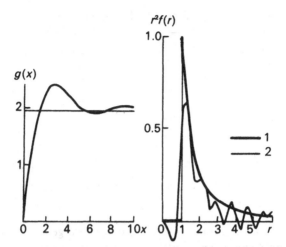

Fig. 6.12. Numerical experiments using the method of spectral transmittance for the Junge distribution; $a_1 = 0.1\,\mu m$; $a_2 = 7.5\,\mu m$. (a) curve of $g(x)$; (b) curve of $r^2f(r)$: 1, initial; 2, curve derived from 80 points using the rectangle formula. Using the Filon formula the same accuracy is obtained using only 5–6 points.

method, that the measurements of $g(v\beta)$ must be carried out over a wide spectral bandwidth. In [72], a procedure was developed which eliminated this drawback. Another important improvement in the method involves a special formula of integration, namely Filon's formula which takes into account the oscillating nature of the function under the integral [for calculating the integral in Equation (6.57) over the basic spectral interval]. Experimental verification of this method using calibrated models showed that it gives good results.

The main obstacle to using the method of spectral transmittance in ocean optics problems is that in the ocean, $g(v\beta)$ data may be obtained only over a relatively narrow spectral bandwidth where the water itself is transparent. In this respect, the studies [117] and [165a] are of great importance. They give an optimum procedure for extending the analysis of the measured data into the short-wave region. The new procedure in explicit form takes into account the analytical properties of the kernel of the equation and makes it possible to derive a solution to the problem for meaurements over a narrow spectral bandwidth. The estimates of the limits of the spectral bandwidth depend on the required distribution function. In [115] (for gamma distributions with $\mu = 2$), it was stated that:

$$\lambda_{\min} = 2\pi(m - 1)a_m; \quad \lambda_{\max} = 5\pi(m - 1)a_m \tag{6.59}$$

where a_m is the modal radius. For the Junge distribution, we have:

$$\lambda_{\min} = 4(m - 1)a_m; \quad \lambda_{\max} = 4(m - 1)a_{\max} \tag{6.60}$$

The spectral bandwidth for marine suspended particles is given in Table 6.7 from Equation (6.60).

Table 6.7. Spectral bandwidth for marine suspended particles

Type of suspension	Size (μm)			
	a_{min}	a_{max}	λ_{min}	λ_{max}
Fine, $m = 1.15$	0.1	1	0.06	0.6
	0.1	2	0.06	1.2
Large, $m = 1.02$	1	10	0.08	0.8
	2	15	0.16	1.2

We note that for both types of suspension, the range in which it is necessary to measure the spectral transmittance is practically the same, from the far ultraviolet to the near infrared. The method of spectral transmittance was experimentally verified on two-dimensional and three-dimensional models and it was found that the accuracy of the determinations was good. Due to the necessity of measuring the transmittance in the far ultraviolet, the method of spectral transmittance has not been applied yet in the study of marine suspensions. Nevertheless, we have given a brief description here since it may be used for fast measurements of individual suspension fractions in the range of radii 0.2–1 μm. When quartz cells are used, this can be done using standard apparatus.

6.5.2. Method of the total scattering function

For a polydispersive system of soft particles, we have:

$$\bar{I}(\gamma) = \int_0^\infty I(\gamma, a)f(a)\,da \tag{6.61}$$

where $I(\gamma, a)$ is the monodispersive function [see Equations (4.79) to (4.81)]. We introduce the notation (see [96, 116]):

$$r = a/a_0, \quad b = \sin(\gamma/2), \quad x/2 = 2\rho_0 b, \quad m(r) = r^2 f(r) \tag{6.62}$$

$$\psi(\gamma) = 2\pi^2 a^2 (1 + \cos^2 \gamma)/(1 - \cos \gamma)^2, \quad \bar{I}(\gamma) = I_0 \psi(\gamma)(1/a_0)g(x/2) \tag{6.63}$$

$$g(x/2) = \int_0^\infty K_1(rx/2)m(r)\,dr, \quad K_1(q) = \left(\frac{\sin q - q \cos q}{q}\right)^2 \tag{6.64}$$

The graphs of the functions $K_1(q)$ and $g(x)$ in Equation (6.64) are shown in Fig. 6.13. It can be seen from Equation (6.64) that $m(r)$ satisfies an equation of the type given by Equation (6.5). This equation may be inverted exactly as was

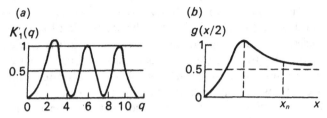

Fig. 6.13. The functions $g(x/2)$ for (a) a monodispersive system [where $g(x/2) = K_1(q)$] and (b) for a polydispersive system.

found in [116]. As in the method of spectral transmittance, it is possible to show (see [105]) that:

$$m(r) = \int_0^\infty h(rx)p(x/2)\,dx \qquad (6.65)$$

where:

$$p(x/2) = g(x/2) - g(\infty); \qquad (6.66)$$

$$h(z) = \frac{16}{\pi}\left[\left(\frac{2}{z^2} - 1\right)\left(\frac{\sin z}{z} - \cos z\right) - \frac{3}{4}\cos z\right] \qquad (6.67)$$

The properties of the function $h(z)$ and its tables are given in [105]. As in Section 6.5.1, we assume that the function $g(x/2)$ (the experimental scattering function) is defined at n points (for n scattering angles γ_j):

$$x_l = 4\rho_0 b_j; \quad b_j = \sin(\gamma_j/2) \qquad (6.68)$$

and that outside the limits of the interval $(0, x_n)$, it may be represented in the binomial form:

$$g(x/2) = C + D/x^2, \quad x > x_n \qquad (6.69)$$

A large number of calculations carried out using Equation (6.65) for different distributions lead to results similar to those which were obtained in the method of spectral transmittance. If an adequate range of values of x is covered, ie over the scattering angle γ, then the inversion recovers the particle spectrum adequately. It is necessary that x_n is twice as large as the value of x at the maximum of the function $g(x/2)$. In this case, it is sufficient to take n as approximately equal to 10 and to take the accuracy of photometric measurements equal to approximately 5%. As in the method of spectral transmittance, the basic range of angles over which experimental data of the function $g(x_j/2)$ must be obtained for the inversion to be successful depends on the distribution mode. Table 6.8 gives data for distributions with different modes a_m (for measurements with $\lambda = 0.55\ \mu m$). However, there is one curious characteristic. The scattering angle γ_{max} cannot be larger than 180°. It therefore follows that there is a certain minimum a_m below which inversion is impossible. For $\lambda = 0.55 \, \mu$m this limit is $a_m = 0.175 \, \mu$m. This means that for distributions with small particles, the scattering functions are so close to the Rayleigh functions that it is impossible to determine the composition of the disperse system from the form of the scattering function. If we wish to measure small particles we must shorten the wavelength (limiting $a_m \approx \lambda$). It can be seen from Table 6.8 that as the size of the particles is increased, information on the spectrum is limited to the region of small scattering angles. For large particles, it lies in the region of extremely small angles. Thus we arrive at the method of small angles which is effective for large particles. The method of the total scattering function is described in more detail in [96, 116].

Table 6.8. Range of scattering angles necessary to determine the particle size spectra for distributions with different modes

$a_m(\mu$m)	γ°_{min}	γ°_{max}
0.2	30	130
1.0	6	20
5.0	1	5

Returning to the investigation of marine suspensions, we note that in Section 6.7 we show that for scattering angles $\gamma > 30°$ the scattering function of marine suspensions may be described with the help of scattering at a 'soft' particle. This means that using this part of the scattering function it is possible to determine the spectrum of fine terrigenous suspensions from Equation (6.65). This agrees well with data given in Table 6.8.

If we use shorter λ for the measurements of the scattering function, for example $\lambda = 0.3 \, \mu$m, then the limit $a_m = 0.1 \, \mu$m and the limit to the application of the method are shifted by a factor of $0.55/0.30 = 1.8$ in the direction of finer particles. Nevertheless, since a considerable reduction in λ is difficult, it is possible that the finest part of marine suspensions should be considered as Rayleigh scatterers as is done in Section 6.7 and the method of the total scattering function should be used instead of the fitting method for the determination of the composition of marine suspensions.

In Chapter 5, we showed that in the approximation for optically soft particles it is possible to calculate the scattering of light by a system of particles of any shape. For such particles it is apparent that the method of the total scattering function also allows a solution to the inverse problem. We give this solution here, following [59].

The scattering function of a volume V containing particles of any shape is determined by Equations (5.25) and (5.26). We shall substitute for the variables in them: $x = b_1 - b_2$, $y = b_2$. We introduce the notation:

$$K_{\varphi_0}(x, y) = \overline{B(\mathbf{q}, x + y)B(\mathbf{q}, y)} \tag{6.70}$$

Let:

$$Q_{\varphi_0}(x) = \int_{-\infty}^{\infty} K_{\varphi_0}(x, y) \, dy \tag{6.71}$$

$$\tilde{Q}_{\varphi_0}(q) = \int_{-\infty}^{\infty} Q_{\varphi_0}(x) \, e^{-iqx} \, dx$$

Then:

$$I(\mathbf{q}) = I_0 a^2 k^4 [(1 + \cos^2 \gamma)/2] \tilde{Q}_{\varphi_0}(q) \tag{6.72}$$

Equation (6.72) makes sense when:

$$-2k \leqslant q \leqslant 2k \tag{6.73}$$

Equation (6.72) makes it possible to formulate the inverse problem, ie to estimate the quantity $\tilde{Q}_{\varphi_0}(q)$ from the measured values of the scattering functions of the light scattered by a volume V and to use it to estimate $Q_{\varphi_0}(x)$ which characterizes the distribution of particles of any shape by cross-sectional areas from Equation (6.71). Since, under experimental conditions, the range of values of λ is restricted, generally speaking there is an infinite set of values for $Q_{\varphi_0}(x)$ which satisfies Equation (6.71). Thus, the inverse problem involving the estimate of $Q_{\varphi_0}(x)$ from the scattering function has a large number of solutions. However, as the size of the particles is increased, the distribution $Q_{\varphi_0}(x)$ becomes wider and the relative fraction $\tilde{Q}_{\varphi_0}(q)$ lying outside Equation (6.73) decreases and at the limit a single solution to Equation (6.71) and (6.72) with respect to $Q_{\varphi_0}(x)$ is obtained. For particles large enough for the RHD approximation to apply it is possible to obtain a single solution for $Q_{\varphi_0}(x)$ of sufficient accuracy for practical purposes, provided that procedures can be used to regularize the solutions. It is known that for a randomly oriented population of particles, the limits of application of the RHD approximation are extended beyond those which are imposed upon a single particle. As was shown in [11], this limit is determined by the condition $2ka(m - 1) \leqslant 1$ with sufficient accuracy for practical problems, where a is the characteristic dimension of the particles. For randomly oriented particles (with no predominant orientation), the values of $Q_{\varphi_0}(x)$ and $I(q)$ depend only on the scattering angle γ and at constant γ do not depend on the direction of the incident radiation. As the degree of a predominant orientation increases, the range of variation of $Q_{\varphi_0}(x)|_{\gamma = \text{constant}}$ increases. This fact may be used for estimating the degree of predominant orientation of non-spherical particles of a given volume V. For very small particles the problem of determining $Q_{\varphi_0}(x)$ from the scattering function has a large number of solutions which is explained by the fact that Rayleigh scattering does not depend on the shape of the particles.

It can be determined easily from the above when it is possible to use the model of equivalent spheres and where it is possible to find the distribution density $f(a)$ over the range of radii a of spherical particles which has the same scattering function as our system of non-spherical particles. To this end, we equate Equation (6.72) to the value of the scattering function for spherical particles and from this we find $f(a)$. This equation is the same as that used to determine $f(a)$ by the method of the total scattering function. An analytical solution is given for this equation which for sufficiently large particles has a single solution by the RHD approximation. For small particles the inversion is impossible.

It follows from the above that for random orientation of the particles [where $\tilde{Q}_{\varphi_0}(q)$ depends only on the value of q and does not depend on the direction] a single solution to Equation (6.72) is obtained with respect to $f(a)$. If there is a predominant orientation, $f(a)$ that satisfies Equation (6.72), then the solution to Equation (6.72) will vary when the direction of the incident light with respect to V is changed because of the dependence $\tilde{Q}_{\varphi_0}(q)$ on the direction q_0.

6.6. Numerical inversion methods

6.6.1. Regularization in the family of smooth functions

The situations considered above when Equation (6.5) can be inverted in the general form are not always used. For any sphere the function $s(x, a)$ is given in tables. For non-spherical particles it is determined from the solution of the corresponding diffraction problem or by experiment where it is also obtained in the form of the matrix $s(x_i, a_j)$ and we are forced to change to a system of linear equations instead of the integral Equation (6.5). The solution of such systems is carried out numerically using different methods of regularization [83]. The physical meaning of regularization operators usually reduces to stabilization and smoothing of the solution by imposing additional requirements on the norm of the derivative of the function $f(a)$. Intuitively, it is assumed that if funtions $f(a)$ are obtained with rapid and, as a rule, irregular oscillations then this is a result of the incorrectness of the problem and such solutions should be discarded. There is a large literature on regularization methods (see [83] for a bibliography) and there is even a special reference work of computer programs. We cannot examine all the aspects of the large problem area here.

We shall just look at those studies which are directly related to ocean optics where the method of statistical regularization has been successfully used (for references see [83]). The essential feature of this is that a priori information on the required function is introduced by defining a certain distribution of probabilities. In this case, the solution of the problem is also obtained in the form of a set of functions with different probabilities and the answer is taken as the mathematical expectation calculated for this distribution. Depending on which a priori information is used, several versions of the method of statistical regularization have been developed: a solution in an ensemble defined by a finite sample or correlation matrix or in an ensemble of smooth functions, etc. In the first version, it is assumed that the solution belongs to a statistical ensemble of functions defined for example by a correlation matrix, mathematical expectation and other statistical characteristics of the ensemble. Of course, in this case it is necessary to be sure that the statistics used for regularization meet the requirements of the statistics to which the required function actually belongs. The application of 'alien' statistics leads to an increase in the errors in deduction and may considerably distort the form of the derived distribution. It will endeavour to correspond to imposed statistics. It is impossible to obtain the original distributions using such an approach. In the second version, use is made of the fact that the derived curve for the size distribution of particles is a continuous function. Depending on its smoothness this continuity determines a roughly good correlation of the values of the required function for close values of a. The regularizer is the probability density:

$$P_a(f) = C(\alpha)\exp[-\alpha(f, \Omega f/2]$$ (6.74)

where f is an n-dimensional vector defined by the components f_i, Ω is a symmetric non-negative matrix which is introduced so that the square shape $(f, \Omega f)$ is the norm of the derivative of the function f, for example, of second order. Thus, the quantity $(f, \Omega f)$ will be a finite-difference approximation to the integral:

$$\int |d^2 f(a)/da^2|^2 \, da$$ (6.75)

The smoothness parameter α, *a priori* unknown, is determined on the basis that there is a correlation between the values of $f(a)$ with close values of a. The degree of correlation is unknown in advance. It is this version of the method, seeking the solution in the statistical ensemble of smooth functions, which has been used for the inversion of the scattering functions of marine suspensions. The solution of the problem is represented by the equation:

$$\bar{f} = \int f_a P(\alpha|\bar{I})d\alpha$$ (6.76)

where $P(\alpha|\bar{I})$ is the *a posteriori* probability density for the parameter α for a known measurement result. In [20, 97], the integration in Equation (6.76) was not carried out and it was simply assumed that $f = f_a$ for a value of α for which $P(\alpha|\bar{I})$ has a maximum. In this procedure, the parameter α was determined from the measured function and a solution of the most probable smoothness was found.

A series of mathematical experiments was carried out to estimate the effect of different factors limiting the experimental information on the accuracy. In this case, the initial integral Equation (6.5) was substituted by a system of linear equations:

$$\bar{I}_j = \sum_{i=1}^{n} (I_{ji}f_i + \xi_i s_j), \quad j = 1 \ldots m$$ (6.77)

where f_i is the value of the distribution function defined at n points ($i = 1 \ldots n$); I_{ji} is the scattering matrix (monodispersive function); s_j^2 is the dispersion of the random error of the function; ξ_i is the normally distributed random variable with mathematical expectation zero and dispersion 1. Thus, the variables $\bar{I}_j$ imitate the result of the experiment. They were used to determine f_i. The experiments were carried out with distributions similar to those which are found in ocean optics: exponential and gamma distributions.

We look briefly at some of the results of the calculations. The effect of random measurement errors can be seen from the curves in Fig. 6.14. In all cases, the determined function reproduces the initial function with an accuracy as high as that of the calculation error with the exception of the section of ρ from 0.5 to 9 where it is represented by an absolutely straight line. In order to understand this, we note that for small values of ρ the elements of the matrix $I(\gamma, \rho)$ in the range of ρ from 0.5 to 6 at all scattering angles are relatively small and this means that there is little information about the initial points of the function $f(\rho)$. The determination of this part of the $f(\rho)$ curve depends on additional information in the form of the regularizing functional $\Omega(f)$ which represents the norm of the second derivative of $f(\rho)$. The absence of information about the initial points in the function $\bar{I}(\gamma)$ means that at these points the derived function is only determined by the regularization process, ie by minimizing the functional $\Omega(f)$. This functional assumes a minimum (zero) value for a linear function $f(\rho)$. Therefore, in the region where there is little or no information, the regularized solution is a linear extrapolation from the region containing the information. In order to obtain information on the function $f(\rho)$ at small values of ρ, it is necessary to increase the accuracy of $\bar{I}(\gamma)$.

As can be seen from Fig. 6.14, the retrieved curve agrees better with the actual curve for an error of 1 and 3% than for an error of 10% although in the case 3 shown in Fig. 6.14 the function $f(\rho)$, at small values of ρ will also be linear [for an accuracy of 1%, it is better 'corrected' by the results of 'measurement' of

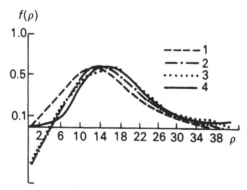

Fig. 6.14. Determination of the function $f(\rho)$ for the root-mean-square error S equal to 1, 10%; 2, 3%, 3, 1% $\bar{I}$. Distributions: 4, initial; 1, 2, and 3, calculated (for $\rho < 9$ curves 2 and 3 are indistinguishable).

$\overline{I}(\gamma)$ than for an accuracy of 10%: the 'correction' takes place at smaller values of ρ].

Crudely speaking, all this means that small particles are eclipsed by large particles; they are simply not seen behind them. Particle sizes which make the most contribution to scattering are determined with the least error. In all cases, for a fraction to be revealed, it is necessary that its contribution is larger than the measurement error $\overline{I}(\gamma)$. For an error in the scattering function equal to 10%, only the large particle fraction is deduced; details of the distribution with a fine structure are not revealed. The theoretical error of the calculation lies between 30 and 12%.

On increasing the measurement accuracy for the scattering function, the accuracy of the retrieval also increases. If $\overline{I}(\gamma)$ is known with an accuracy of 1%, then it is possible to determine fine particles of the distribution and also the fine structure with an error of $7 \pm 2\%$ (the large particle fraction of the distributions is determined more accurately). It is natural that by increasing the number of measurement points for the scattering function the retrieval error is decreased. For example, it is reduced by a factor of 2 on increasing the number of points from 6 to 20. However, when the number of points continues to rise (up to 31), the quality of the calculation does not substantially improve. The retrieval depends on the shape of the distribution functions, ie on the relationships between the number of large and fine particles in the scattering volume. As the number of large particles decreases so that the relative contribution of fine particles increases, the fine particle end of the distribution function is derived more accurately. In the determination of exponential functions, the distribution obtained is somewhat smooth compared to the actual distribution and is only close to it in the smooth middle section.

Studies on the effect of errors in the value of the complex refractive index m of the particles on the accuracy of calculation are of great importance. Usually, m is known only approximately for the inversion. In order to determine the effect of inaccuracy in m, calculations were carried out in which the direct problem was calculated for one value of m and for the inversion, other values of m were used. The effect of inaccuracy in n and k is different. The imaginary part of the refractive index has a larger effect. Thus, if the imaginary part is small ($k \approx 6 \times 10^{-3} - 10^{-4}$, the direct problem is solved with the real part of m) the determination is satisfactorily made. If it is large ($\approx 0.05-0.01$), the determination leads to distribution functions which are considerably different from those of the actual distributions. Large amplitude oscillations arise and there are large errors (up to 300%). Variation in n even by $\delta n = 0.18$ leads only to a shift in the maximum. We may add that polydispersive scattering functions with different values of m systematically differ from each other. These differences may be significant (up to 120%) when the refractive indices are appreciably different in their imaginary parts. This also leads to systematic errors in the determination. Where it makes no sense to increase the measurement accuracy: if the kernel is incorrectly selected, then the errors do not decrease when the measurement errors for the scattering function are reduced.

The effect of the value of m on the calculation makes it possible to formulate the problem of deducting the actual value of m for marine suspended particles. If the inversion is carried out successively with several kernels, then it can be seen that when the refractive index is close to the actual value, the form of the recovered function differs considerably from that which is obtained in the opposite case. When an incorrect m is used (and k is most significant here) the solution oscillates strongly and there is a sharp increase in the retrieval error. If the kernel is correctly selected, there is a sharp reduction in the theoretical error and a consequent stabilization of the solution. The value of the refractive index of the particles, may be estimated in this way.

The method of statistical regularization may also be used when the suspension consists of a mixture of different sorts of particles [98]. The mathematical problem of determination and resolution of the various components in the multi-component system may be formulated in the same way as in the single-component system. It can be solved in the usual manner if the required solution is taken as a vector consisting of **R** the vectors of the components of the distribution and an operator matrix $\hat{K}$ is taken which consists of the matrix elements of the operators $\hat{K}^{(R)}$ corresponding to individual components of the mixture. In this case, the sequence of the components is unimportant. When the two-component version of the method is used the requirements of the accuracy of the determination of the scattering functions are considerably increased. The determination error in the two-component version is higher than in the one-component version by a factor of approximately 1.5–2. Of the two components, the one which has the more informative kernel (eg fewer absorbing particles) is more accurately determined. An increase in the number of scattering function points from 16 to 33 appreciably improves the determination of the less informative component of the mixture.

For direct inversion of the scattering functions of marine suspensions, the measured data from the fifth and tenth voyages of the scientific research ship 'Dmitry Mendeleyev' were used. One-component and two-component versions of the method were used. The determination of the fine fraction was carried out with a kernel corresponding to the refractive index 1·15 and for the large fraction a kernel was used for $m = 1.02$ and $m = 1.05$. Optimum conditions for the inversion were established in the calculations: the ranges of angles and the equation kernels. The optimum procedure for determination was also established. It was assumed that in order to determine the composition of the fine suspension a range of angles 40–145° should be used; the refractive index was assumed to be 1.15; the large fraction was determined from the scattering function in the range of angles 30′–35° and refractive indices equal to 1.02 and 1.05 were chosen. These conditions for inversion are optimum when the one-component version of the inversion method is used. The theoretical error of determination for the composition of the terrigenous components of the suspension varied from 90 to 18% and the error of determination of the biological component from 100 to 11%. With such an error, the different regions of the distribution function are determined, if the scattering function is known, with an error of 10%.

The two-component version of the method was also used, ie it was assumed that there was a mixture of two sorts of particles (with refractive indices 1.02 and 1.15) and the total scattering function was used for the inversion (ie no assumption was made about the separation of these sorts of particle nor about the demarcation of their zones of influence in the scattering function). However, in this version of the method, the solution was smoother and the theoretical error of the method became 1.5–2 times larger than in the one-component version. Apparently, the 10% measurement accuracy of the inverted scattering functions was insufficient for the application of the two-component version. With the higher measurement accuracy of $\bar{I}(\gamma)$, the application of the two-component version will be more effective. The inversion of actual scattering functions made it possible to estimate the number of particles of fine terrigenous marine suspensions ($a \leqslant 1.25\,\mu m$) from 1.2×10^7 to 5.4×10^7 particles/litre and the number of particles of the large organic marine suspension ($a > 1\cdot 25\,\mu m$) from 0.1×10^6 to 1.8×10^6 particles/litre.

6.6.2. Consideration of additional optical information

In the previous analysis, the smoothness of the solution was assumed. The smoothness parameter α was determined from the measured data with the requirement that the solution be as smooth as possible. As calculations have shown due to its excessive generality, this regularization method is sometimes inadequate. In a number of cases the *a posteriori* probability $P(\alpha|\bar{I})$ has several maxima. This may lead to an incorrect selection of α and $f(\rho)$. Mathematical experiments confirm that for the function $f(\rho)$ with a large number of inflections, the selected α is too high, the solution becomes smooth and part of the information about the particle size spectrum is lost. Therefore another regularization method was developed in [20] based on the consideration of additional optical information. First, a method was used which explicitly takes into account the non-negativity of the distribution function. This assumption in itself is a strong regularizer. Second, the possibility was investigated of additional definition of the solution of Equation (6.5) by a new equation analogous to Equation (6.5):

$$\bar{I}_2(\gamma) = \int_0^\infty I_2(\gamma,\,\rho)f(\rho)\,\mathrm{d}\rho \qquad (6.78)$$

but for a scattering function $\bar{I}_2(\gamma)$ measured at another significantly different wavelength of light. The new kernel $I_2(\gamma,\rho)$ should be taken for the corresponding refractive index and $f(\rho)$ is the same in both equations. The solutions of each of these systems taken individually and found without additional conditions form two ensembles of non-regularized solutions. Each of these ensembles contains the true solution but the average function of the ensemble, which is considered to be the solution of the problem, may be very far from the true solution.

For a combined solution of the system in Equation (6.78), for two λ, we obtain the ensemble of solutions which is the intersection of the ensembles for λ_1 and for λ_2:

$$P(f|\bar{I}_1, \bar{I}_2) = P(f|\bar{I}_1)P(f|\bar{I}_2) \qquad (6.79)$$

Obviously, ensemble (6.79) will be narrower than the ensembles corresponding to the individual wavelengths. Mutually incompatible functions were eliminated but the true function was retained. It is obvious that this narrowing of the ensemble (regularization) will depend on the extent to which $I_1(\gamma, \rho)$ and $I_2(\gamma, \rho)$ as well as $\bar{I}_1$ and $\bar{I}_2$ are different. The results of calculating the distribution function by using the new regularization procedure are given in Fig. 6.15 where the determination of $f(a)$ is shown without smoothing ($\alpha = 0$) but with the simultaneous utilization of different pairs of scattering functions. We can see that the utilization of a pair of functions significantly improves the determination. The improvement is especially significant when a pair of functions $\bar{I}_1$ and $\bar{I}_2$ corresponding to significantly different refractive indices is selected for the inversion. Thus, the function $f(\rho)$ determined from the scattering function pair of $m = 1\cdot29 - 0\cdot05i$ and $m = 1\cdot30 - 0\cdot1i$ (a significant difference in the imaginary parts of m) gives a distribution close to the true distribution, whereas determination from only one value of m considerably differs from the actual distribution. On inversion of a pair of scattering functions with similar values of m (1.33 and $1\cdot31 - 10^{-4}i$), we obtain a less satisfactory solution but nevertheless it gives a better idea of the $f(a)$ distribution than the result of inverting a single $\bar{I}(\gamma)$. In these experiments, the error in the scattering functions was 10%. As far as the theoretical determination error is concerned, when a pair of scattering functions is used it remains significantly large although it is considerably reduced compared to the inversion of one $\bar{I}(\gamma)$. We remember that the theoretical

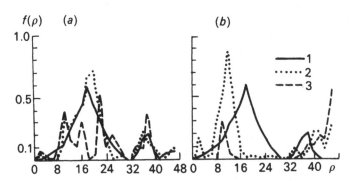

Fig. 6.15. Determination of the function $f(\rho)$ when $S = 10\%$ and $\alpha = 0.1$, initial function; 2, function deduced from a pair of functions (a) $m_1 = 1.29 - 0.05i$ and $m_2 = 1.30 - 0.1i$ and (b) $m_1 = 1.33$ and $m_2 = 1.31 - 10^{-4}i$; 3, from one function (a) $m_1 = 1.29 - 0.05i$ and (b) $m_1 = 1\cdot33$.

determination error in the statistical regularization method is the root-mean-square deviation from the mean of the functions in the ensemble, ie a value which directly indicates the width of the ensemble. A reduction in the determination error when a pair of scattering functions is used indicates the narrowing of the ensemble, ie regularization. In the case of close refractive indices (kernels), the ensembles of non-regularized solutions are close and the combined inversion of the functions $\bar{I}_1(\gamma)$ and $\bar{I}_2(\gamma)$ have a low efficiency; there is not sufficient stabilization of the solution and the error remains very high—up to 500%. However, this is almost three times smaller than in the case with one $\bar{I}(\gamma)$. In the case in which m is considerably different, the ensembles of non-regularized solutions differ significantly. On combined inversion of such functions there is a significant narrowing of the ensemble. The error is reduced by an order of magnitude but still remains high ($\approx 100\%$). Despite this, the solution obtained gives a good description of the initial function since the weighting of the functions close to the solution is increased significantly ie regularization occurs due to the exclusion of functions which are not common to the two non-regularized ensembles. It follows from these experiments that even with a 10% error in the initial data, simultaneous inversion of two scattering functions with significantly different values of m makes it possible to regularize the solution sufficiently and to obtain a correct representation of $f(a)$ without resorting to smoothing. The involvement of a second scattering function also improves the recovery significantly when a search algorithm is used for the most probable solution in the ensemble of smoothing functions. This result is obvious and is explained by the fact that new informative terms are added to Equation (6.77). This is well illustrated by the curves in Fig. 6.16.

In conclusion, we emphasize that the new procedure for solving the integral equation of the first kind by the statistical regularization method without resorting to *a priori* information about the smoothness of the solution will only

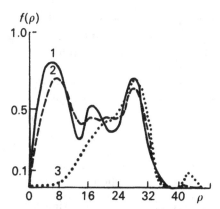

Fig. 6.16. Determination of a complex distribution with simultaneous utilization of both algorithms: optimum selection of α and two scattering functions. Curve 1, initial function; curve 2, determination from $m_1 = 1\cdot33$ and $m_2 = 1\cdot29 - 0\cdot05i$. For the sake of illustration, curve 3 determined for one scattering function ($m = 1.33$) is also shown; $S = 1\%$.

be effective if the functions $\bar{I}_1(\gamma)$ and $\bar{I}_2(\gamma)$ selected by us differ from each other to such an extent that the intersection of the two probability distributions is narrower than each distribution taken separately. The degree of regularization depends on how successfully the pair (or triplet, etc) of values of m is selected. The advantage of such a regularization procedure is obvious: the solution does not depend upon smoothing which may not be appropriate to it; it is only necessary that the required function fits the experimental data.

In order to apply the suggested regularization procedure in practice, it is necessary to determine the criterion for successful selection of pairs of wavelength λ_1 and λ_2 and correspondingly the pairs of scattering functions. It is obvious that two (or more) functions differing at least in terms of measurement accuracy must be taken as the inverted optical information. Thus, scattering functions which have values close to that of the (dimensionless electrodynamic) absorption coefficient k complement each other poorly and it is ineffective to combine them. It makes better sense to choose scattering functions relating to different absorption coefficients. Apparently, it is sufficient for these differences in k to be as little as 0.05.

6.7. Determination of the composition of marine suspension

6.7.1 Principle of the method

The method of determining the quantity and distribution by size of particles suspended in seawater, combining the fitting method and the method of small angles, was proposed in [109]. It is based on the following four assumptions:

1. There are two types of particle in marine suspensions: mineral particles with $m = 1.15$ and organic particles whose refractive index is not subject to any hypotheses.
2. Each of these types of particle belongs to different fractions in terms of size; the mineral particles form the fine fraction (radius $< 1\,\mu$m) and the organic particles form the large fraction (radius $> 1\,\mu$m).
3. The large particles do not make an appreciable contribution to the scattering at angles larger than 15°. The angular range 15–145° is therefore used to determine the fine fraction by the selection method.
4. In order to determine the size distribution of the large particles by the method of small angles, a range of angles from 20′ to 10° is used. The scattering coefficient of the large particles is found from the difference: $b_L(\gamma) = b(\gamma) - b_F(\gamma)$ where $b(\gamma)$ is the measured scattering function and $b_F(\gamma)$ is that calculated for the fine fraction.

Preliminary analyses of several hundred scattering functions measured in the northern part of the Indian Ocean, various regions of the Pacific Ocean, Sargasso Sea and Black Sea have indicated that the exponential distributions in the range of radii $r_1 \leqslant r \leqslant r_2$ (Chapter 4) are good models for the size distributions of terrigenous particles. The coefficient v varies from 4 for relatively turbid waters (Black Sea, waters of the Pacific Ocean, South Equatorial Current, the eastern part of which is subject to the action of the Peruvian upwelling) to 6 for more transparent waters (eg deep waters of the Sargasso Sea). A more detailed analysis showed that as a rule, good agreement is obtained for medium scattering angles. For large angles the theoretical curves usually lie below the experimental curves. The low values in the region of large scattering angles in [109] were explained by the fact that very small particles of $r < r_1$ give rise to Rayleigh scattering but are neglected in theoretical models. As a result, for angles $> 15°$, the equation:

$$b_{susp}(\gamma) = b(\gamma) - b_{water}(\gamma) = [n_F z(m, v, r_1, r_2) + \Delta]c(1 + \cos^2 \gamma) \quad (6.80)$$

was used to find the scattering coefficient of the suspension in [109].

In Equation (6.80), n_F is the concentration of the fine particles; $c = (32\pi^5/\lambda^4)|(m^2 - 1)/(m^2 + 2)|$ is a constant depending on the wavelength of the light and the refractive index of the particle m; $z(m, v, r_1, r_2)c(1 + \cos^2 \gamma)$ is the scattering coefficient for the fine suspension of unit concentration which has an exponential size distribution with parameters v, r_1 and r_2; $\Delta = (\overline{nr^6})_R$ is the contribution of the Rayleigh scattering particles. For these particles, the optical data enables only the product of the concentration and the average radius to the power 6 to be found. Since the upper limit of the exponential distribution has only a weak effect on the form of the scattering function, it is assumed to be constant $(r_2 = 1, 2 \mu m)$ and the problem reduces to selecting 4 parameters: n_F, v, r_1 and Δ from the experimental data. For this purpose polydispersive scattering functions were calculated from the equations in Chapter 4 for $r_2 = 1, 2 \mu m$, $v = 4, 5, 6$, $r_1 = 0.01(0.01)0.20$ for six λ for a total of 756 models. The proximity criterion was taken as the minimum of the maximum deviation of the experimental and theoretical scattering function in the range 15–145°. In order to assess the quality of the selection in [109], a comparison of the experimental and determined scattering functions was made. Data from six different regions of the world's oceans were analysed and are given in Table 6.9 in the order of increasing b. Three open ocean regions (cases 1, 2, 3), the Black Sea and two closed basins (cases 5, 6) were considered. In all examples, it was possible to select theoretical scattering functions close to those found by experiment. The accuracy of the approximation practically coincided with the measured accuracy. For the sake of illustration, two cases are reproduced in Fig. 6.17: the transparent deep waters of the Sargasso Sea and the turbid surface waters of the Black Sea. In four other examples, the picture is approximately the same. The agreement may seem very good but it should be borne in mind that the scale along the y-axis is logarithmic and that the procedure involves five fitting parameters. The suspension characteristics corresponding to all the examined cases are given in Table 6.9. It can be seen from these data that the positioning order of the examples is not completely successful. It is likely that the high values of b in Kieta Bay and in Tarawi Atoll Lagoon are connected with biological particles. A more correct order would be 2, 1, 5, 6, 3, 4 as is confirmed by the particle size data for the large fraction.

Table 6.9. Properties of suspensions in different regions of the ocean

No.	Region	Depth, zm	$b \times 10^3$ (m^{-1})	v	$r_1\mu m$	$r_2\mu m$	$n_f \times 10^{-12}m^{-3}$	$\Delta \times 10^{30}\,m^3$
							Selected parameters	
1	Pacific Ocean to the north-west of the island of Rarotonga	10	66	5	0.14	1	0.029	0.89
2	Sargasso Sea	500	75	6	0.13	2	0.86	2.1
3	Pacific Ocean, South Equatorial Current	5	160	4	0.03	1	31	0.27
4	Black Sea, Karadag	10	210	4	0.05	1	12	1.6
5	Pacific Ocean, Solomon Islands, Kieta Bay	5	540	0.18 5	5	2	1.0	3.6
6	Pacific Ocean, Hilbert Island, Tarawi Atoll Lagoon	5	1000	5	0.10	2	17	4.5

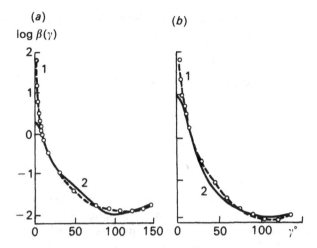

Fig. 6.17. Comparison of the experimental (1) and the theoretical (2) scattering functions. (a) Sargasso Sea, 500 m depth $b = 0.075\,\text{m}^{-1}$; $v = 6$; $r_1 = 0.13\,\mu\text{m}$; $r_2 = 2\,\mu\text{m}$; $n_f = 8.6 \times 10^{11}\,\text{m}^{-3}$; $\Delta = 2.1 \times 10^{-30}\,\text{m}^3$; (b) Black Sea, Karadag region, 10 m depth; $b = 0.21\,\text{m}^{-1}$; $v = 4$; $r_1 = 0.05\,\mu\text{m}$; $r_2 = 1\,\mu\text{m}$; $n_F = 12 \times 10^{12}\,\text{m}^{-3}$; $\Delta = 1.6 \times 10^{-30}\,\text{m}^3$

The particle size spectrum for the large fraction is determined by the method of small angles (Section 6.4). An example is illustrated in Fig. 6.18 which shows a calculated particle spectrum for the large fraction of a marine sample.

In order to determine the absolute numerical concentration of particles, the size spectrum for the large fraction $f_L(r)$ was normalized with the total scattering coefficient for the large fraction b_L. It was calculated as the difference between the total scattering coefficient for seawater b and the sum of the scattering

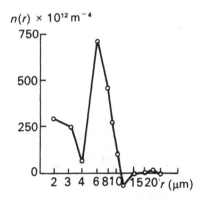

Fig. 6.18. Example of a particle size spectrum for the large fraction determined by the method of small angles (south-western part of the Pacific Ocean).

coefficients for the fine fraction (b_f), the Rayleigh scatters (b_R) and pure water (b_w):

$$b_L = b - (b_F + b_R + b_w) \tag{6.81}$$

For a polydispersive suspension of large particles:

$$b_L = 2\pi n_L \int_{r_1}^{r_{max}} f_i(r) r^2 \, dr \tag{6.82}$$

from which it is easy to find the concentration of large particles n_L.

The distributions of particles of the large fraction obtained in this way joined up well with the distributions of the fine fraction in terms of both relative variation and absolute values. There was enormous interest in comparing the optical data on the suspension with data which were obtained at the same time by geologists using filtration and separation. Examples of the comparison are shown in Fig. 6.19 and Table 6.10. In both cases, the same sample of water was used for optical and for geological measurements. Table 6.10 refers to a depth of 50 m in the Pacific Ocean to the north-west of the island of Rarotonga in spring 1971. The method used by the geologists based on the use of standard microscopes did not permit estimation of the numerical concentration of very fine particles; the actual lower limit of particle size that can be determined depends on the resolving power of the microscope and the individual characteristics of the observer. It is, as a rule, unknown. Usually, geologists assign all particles with a radius $r < 1{\cdot}25\,\mu m$ to the group of minimum size.

As can be seen from Fig. 6.19 and Table 6.10, the optical data are close to the geological data over the range of particle size for which the geological methods are roughly reliable. Both types of data correctly reflect the main feature of the size distribution of suspended particles: the significant predominance of fine particles. As can be seen from the optical data, there is an enormous quantity of fine particles: the number of particles with dimensions

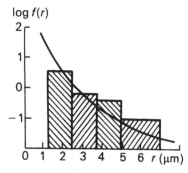

Fig. 6.19. Comparison of the particle size spectrum of a marine suspension obtained by inversion of the scattering function (continuous curve) with a microscopic histogram (hatched area); Black Sea in the Karadag region, summer 1969.

Table 6.10. Comparison of optical and geological data

Radius of the particles (μm)	Concentration, thousand particles per litre	
	Optical data	Geological data
<0.5	2.8×10^5	—
0.5–1.25	1.7×10^3	92
1.25–2.5	4.1	11
2.5–5.0	10	5.6
5.0–12.5	5.2	2.8
>12.5	0.1	1.1

greater than 1 μm is less than 1 % of the total quantity of particles. The numerical concentration of suspended particles may be estimated from optical data to the order of 10^8–10^{10} particles per litre of seawater, in contrast to 10^4–10^6 particles per litre from the geological data. The difference in the mass concentrations will not of course be so great.

We carry out an estimate calculation from the data given in Table 6.10. We assume the density of the particles of the fine fraction to be $2.65\,g/cm^3$ and of the large fraction to be $1.0\,g/cm^3$. We find that the concentration of the suspension is $C = 0.036\,mg/litre$ from the optical data and $C = 0.014\,mg/litre$ from the geological data. Thus, the sieves catch approximately 40% of the suspended particles in seawater. Since this difference is essentially related to the small particles and the division into fractions will be different for different substances, errors in the standard geological method may be the reason also for inaccuracies in the estimate of the distribution of the chemical composition of suspensions in the world's oceans.

6.7.2. Improved version

The method described above has been improved in several ways (see [11]). The possibility of replacing the fitting of an ensemble of smoothing functions by a numerical analysis was examined. The results of numerical inversion of actual marine scattering functions using the procedure of Section 6.6 revealed the following:

1. When kernels with $m = 1.15$ are used for the entire range of angles 20′–145° the retrieved function strongly oscillates and has a large error. The best result is obtained for the range 30–145°, though, in a number of cases, the result has significant error (80–90%). A stable solution with a minimum error (30–40%) was obtained when a range of angles 40–145° was used.

2. When the scattering functions are inverted with a kernel for a refractive index of 1·02, a large error is incurred at scattering angles greater than 10°.
3. The total number of particles approximately agrees with the values obtained by the procedure of [109]. It significantly exceeds the number obtained by microscopic counting.

The main result of comparing the numerical inversion of marine scattering functions by the methods given in Section 6.6 and [109] is that the effect of the large fraction is not restricted to the region of angles up to 15° but also applies to angles 30–40°. Furthermore, certain systematic differences were revealed between the particle size spectra obtained by the two methods. In this connection, an improved inversion procedure was developed in [11] based on the method in [109], which we now consider in detail.

We start with the determination of the composition of the fine fraction. It has already been pointed out that the large particles may make some contribution to the scattering for angles of 15–40° which will affect the inversion result. This assumption was verified in two ways. First, for different marine scattering functions the contribution of the large fraction (whose composition was determined by the method of small angles) to the scattering at angles larger than 15° was calculated directly. It was found that for an angle of 15° this contribution may be $\approx 15\%$ but above 45° it is less than 3% in all cases. Second, for these scattering functions the results of inversion were compared using the ranges 15–145° and 45–145°. It was found that in the majority of cases, the spectra of the fine fraction did not differ significantly but appreciable differences were observed for individual samples. It is therefore recommended in [11] that an angular range of 45–145° be used for determining the composition of the fine fraction.

An important question relates to the use of the RHD approximation for the kernel of the equation. In Chapter 4, we showed that for monodispersed particles with $m = 1.15$ it may be reliably used only up to $\rho \approx 2$. However, in Chapter 4 we pointed out that for polydispersive systems with exponential distribution $(v = 3 - 6)$ for $\rho_{max} = 20$ the error in the RHD approximation does not exceed 15% in the angular range 45–145°. Therefore, this approximation can probably be used although we note that the maximum ρ_2 used in the models in [109] was 36. A similar situation also occurs in the method of small angles and in the methods given in Section 6.5. From experience in using these methods it has been shown that if the polydispersive scattering function is correct then the inversion also gives valid results. Of course, this question requires further careful investigation.

Finally, in [11], the stability of the fitting procedure was investigated with respect to measurement errors. For this purpose the scattering functions calculated for certain model distributions in accordance with the actual measurement accuracy were distorted by a random error of 10%. For these distorted curves, models were selected using the normal procedure, and it was found that the distribution functions obtained differed from the initial distribution functions by 10–15%.

We now turn to the determination of the composition of the large fraction. In this case, a considerable improvement was made by replacing Equation (6.40) by Equation (6.43). As a result, the final equation used for determining the particle size spectrum for marine suspensions, using the method of small angles, has the following form:

$$f(\rho) = - \frac{4\pi^2}{\lambda a^2 P(\rho, m)} \int_0^{\gamma_{max}} F(\rho, \gamma) \frac{d}{d\gamma} [b(\gamma)\gamma^3] \, d\gamma \qquad (6.83)$$

We note that without additional normalization Equation (6.83) gives values for the numerical concentration of the particles directly in a single range of sizes.

We saw in Section 6.4 that for a marine suspension the basic range of angles is $\gamma_{max} \leqslant 2°$. As a result of this, we can use the scattering function of seawater directly for the inversion, since the contribution of the fine terrigenous suspension to the scattering at angles 0–2° does not exceed 1–3%. When Equation (6.83) is used it is necessary to know m. In [109], it is recommended that $m = 1.02$ is used. In order to investigate how the differences in the actual value of m from the assumed value affect the inversion, mathematical experiments were carried out in [11] in which the scattering functions calculated for different refractive indices were inverted using a kernel for $m = 1.02$. These calculations showed that in cases where the differences between the actual value of m and the assumed value do not exceed 0·03, they do not have a significant effect upon the result. If these differences exceed 0·1, the derived spectrum considerably differs from the initial spectrum. This result agrees with that described in Section 6.6.

In conclusion, we turn to Table 6.11. This gives the results from [11] of comparing the spectra of particles obtained by three methods: (1) by the improved version of [11]; (2) by the numerical inversion of marine scattering functions by the method of statistical regularization; (3) by counting particles under the microscope. The measurements were carried out in the summer of 1979 in the Indian Ocean. The exact position of the stations is shown on the chart in [9]. As can be seen from Table 6.11, the number of particles detected by optical methods appreciably exceeds the number of particles counted under the microscope.

A comparison of the two optical methods shows that there is a range of dimensions where the data of both methods agree very well and at the same time there are areas of appreciable difference. It is interesting to investigate these discrepancies. They are associated with both the deviations of 'real' marine suspensions from the homogeneous sphere model which are difficult to analyse and with errors in the methods which are used. We have already discussed the errors in the fitting and small angle methods and we considered the errors in the statistical regularization method in Section 6.6. We remember that in the calculation by this method of such sharply varying functions as the exponential distribution, distributions are obtained which are smooth in comparison with the initial distributions and are close to it only in the middle smooth region. Considering all errors inherent in each method, it is possible to conclude that

Table 6.11. Number of particles (million/litre) in the range of sizes Δr (μm) at a depth of 10 m

Δr (μm)	Subequatorial divergence (station 770)			Bay of Bengal in the vicinity of the Ganges Delta (station 783)			Andaman Sea (station 788)		
	1	2	3	1	2	3	1	2	3
0.2–0.3	64	2.0		370	9.0		62	1.9	
0.3–0.4	11	1.8		63	8.3		15	1.7	
0.4–0.5	3.0	1.6		17	7.5		5.4	1.5	
0.5–0.6	1.0	1.4		6.1	6.5		2.3	1.4	
0.6–0.7	0.45	1.3		2.6	6.0		1.2	1.3	
0.7–0.8	0.22	1.2		1.3	5.2		0.68	1.1	
0.8–0.9	0.11	1.0		0.68	4.6		0.41	0.98	
0.9–1.0	0.67	0.90		0.39	3.7		0.26	0.84	
1.0–1.25	0.77	0.65	0.28	4.4	2.8	0.70	0.30	0.64	0.41
1.25–2.5	1.1	0.18	0.026	6.3	0.90	0.10	0.45	0.11	0.037
2.5–5.0	0.29	1.1	0.12	1.8	0.75	0.094	0.12	0.15	0.012
5.0–10.0	0.026	0.020	0.078	0.30	0.20	0.050	0.028	0.020	0.012
0.2–0.5	78	5.4		450	25		82	5.1	
0.5–1.0	1.8	5.8		11	26		4.9	5.6	
1.0–10.0	2.2	2.2	0.44	13	4.7	0.94	0.90	0.92	0.47

1, according to the procedure in [11]; 2, statistical regularization method; 3, microscopic counting; the first figures in column 3 correspond to the total number of particles whose radii are less than 1.25 μm.

with a measured accuracy for the scattering functions of 10%, two three-fold differences in results which are to be observed in individual size ranges should not cause any surprise. The reasons for low particle counts on filters are discussed in Chapter 2. We note here that the concentration of suspensions obtained by the settling method and the method of small angles agree satisfactorily (compare columns 2 and 4 in Table 2.3).

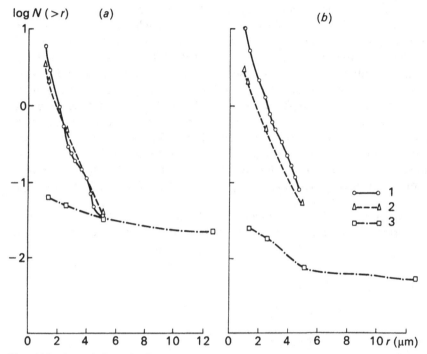

Fig. 6.20. Cumulative distribution curves for particles by size at depths of (a) 10 m and (b) 20 m. 1, Coulter counter; 2, light scattering method; 3, counting under the microscope.

Figure 6.20 shows a comparison of the cumulative distribution curves for particles by size with the data obtained by inverting scattering functions, counting on filters and by using the Coulter counter as described in Chapter 2. Measurements with the latter are apparently the best of the methods available to geology. Unfortunately, no previous data of simultaneous measurements of scattering functions and particle counting by Coulter counter were to be found. Such a comparison was therefore made in [110] from samples taken from an equatorial area of the Pacific Ocean in which seasonal variations were known to be small. In one case this was station 361–5th voyage of the 'Dmitry Mendeleyev', 0°07′N, 88°26′W, and in the other case it was station VPT-41 of the 'Yaqina', 0°07′N, 90°09′W. Results are compared in Fig. 6.20, where it can be seen that the optical data agree well with the Coulter counter data, whereas the microscopic analysis gives considerably lower values.

6.7.3. Combined application of the statistical regularization and fluctuation methods

An interesting study is described in [3] in which the authors have combined the methods described in Sections 6.3 and 6.6. The optical characteristics of marine suspensions were investigated in the coastal region of the Black Sea (Fig. 6.21). The measurements were made using a special apparatus which recorded the scattering function, the fluctuations in the intensity of scattering at fixed angles, the transmittance, the scattering angle and the depth. The radiation source was a laser ($\lambda = 632.8$ nm). The apparatus is described in detail in [3]. The investigations were carried out at depths of 1 and 3 m and at the bottom; the scattering function, transparency and fluctuations in the intensity of the direct beam were measured (scattering functions are shown in Fig. 6.21) and the plotted results differ appreciably from the smooth type of curve illustrated in Fig. 6.17. They display a number of extreme values; furthermore they have a significantly more extended 'bow'. We have already discussed the extrema of $b(\gamma)$ curves in Chapter 5 where they were noticed because the apparatus of [3] makes it possible to measure $b(\gamma)$ in the form of a continuous function of the angle γ. The large elongation at small angles indicates the high concentration of large particles.

The scattering functions were inverted by the method of statistical regularization described in Section 6.6. For the inversion of the terrigenous suspension

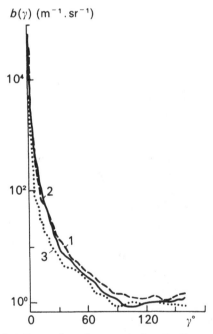

Fig. 6.21. Scattering functions of marine suspensions in the coastal region of the Black Sea at different depths. 1, 1 m; 2, 3 m 3, bottom.

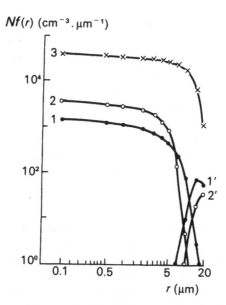

Fig. 6.22. Particle size spectra calculated from the scattering functions of Fig. 6.21 by statistical regularization, method. Curves 1–3 were obtained by inversion of the corresponding scattering functions in Fig. 6.21. Curves 1' and 2' are for large organic particles.

fraction, a refractive index of 1·15 was used and for the organic fraction a refractive index of 1·02 was used. In the calculation of the particle size distribution of the scattering function in the angular range $37° \leqslant \gamma \leqslant 170°$ was used for the mineral fraction and $0.2 \leqslant \gamma \leqslant 8°$ for the organics. Figure 6.22 shows the particle size spectra for coastal suspensions calculated from the scattering functions. It can be seen from Fig. 6.22 that coastal suspensions are considerably different from ocean suspensions. For terrigenous particles with radius $<5\ \mu m$ the distribution may be described by an exponential law with a coefficient v of only 0.15 and for the steeply descending part of the curves $(r > 5\ \mu m)$ a value of $v = 5.5$. It was not possible to establish completely the spectra of large (organic) particles, since the curves did not fit into the calculated size range. The authors of [3] assume that these curves follow a logarithmic normal distribution. The bottom sample (curve 3) is distinguished by a high concentration of particles in all size ranges. The error in the calculation of the size spectra of large particles with a diameter of more than $20\ \mu m$ is 6% and increases to 30–50% with a reduction in diameter to $1–0.5\ \mu m$.

The results of using the method of fluctuations for the large function are given in Table 6.12. The diameter of the light beam was 2 mm and the length of the instrument 98 cm.

It can be seen from Table 6·12 that the large fraction consists of particles with effective radii $7–14\ \mu m$; the numerical concentration varies in the range 0·75–2·6 million/litre. The mass concentrations of the large particle size

Table 6.12. Characteristics of large suspended particles

Depth (m)	Attenuation coefficient C	$\sqrt{D/I_0}$ (%)	Effective radius (μm)	Concentration Numerical (million/litre)	Mass (mg/litre)
1	0.43	0.47	8.5	1.4	2
2	0.57	0.72	12	0.75	4
3	0.61	0.4	7.2	2.6	2.5
4	0.72	0.7	14	0.95	5.6

suspensions given in the last column of the table were calculated by the method of fluctuations using the equations given in Section 6.3. Comparison of mean concentrations for large particles ($\geqslant 10\,\mu$m — obtained by integration of the curves in Fig. 6.22) with those indicated in the table, gives values of the same order. Thus, the fluctuation method gives additional optical information and results which agree with those obtained independently from inversion of the scattering functions. The fluctuation method has also been used for measurements in surface waters of the tropical Atlantic (in the thermocline layer), and according to the data on the transmittance and its fluctuations, the authors found that the average radius of the suspended particles lies in the range 4.5–7 μm and the average numerical concentration in the range 0.2–1.5 million/litre. Both the size of the particles and their concentration were smaller than in the coastal waters of the Black Sea [2a].

References

1. Angot André; *Compléments de mathematiques*, 3rd Edition, Paris, 1957.
2. Aponasenko A. D., Frank N. A., Sidko F. Ya. Spectrophotometer for hydrooptical investigations DSFG-2. In *Optical Methods of Studying Oceans and Inland Water Bodies*, USSR Academy of Sciences Publishers, Novosibirsk, 1979, pp. 194–297.
2a. Afonin E. I., Basharin V. A.; Apparatus and procedures for measuring the scattering functions and fluctuations of scattered light. In *Optics of the Sea*, Nauka, Moscow, 1983, pp. 212–16.
3. Afonin E. I., Solov'ev M. V., Basharin V. A.; Determination of the characteristics of marine suspensions from optical measurements. In *Integrated Hydrophysical and Hydrochemical Studies of the Black Sea*, Sevastopol, 1980, pp. 61–7.
4. Bayvel L. P., Lagunov A. S.; Measurement and verification of the dispersion of particles by the method of light scattering at small angles. *M. Energiya*, 1977.
5. Berezkin V. A., Gershun A. A., Yanishevsky Yu. D.; *Transmittance and the Colour of the Sea*, Military Naval Academy Publishers, Lenningrad, 1940.
6. Bogdanov Yu. A., Licitsyn A. P.; Suspensions and colloids. In *Oceanology: Chemistry of the Ocean, Vol. 1*, Nauka, Moscow, 1979, pp. 325–36.
7. Bogdanov Yu. A., Kopelevich O. V.; Granulometric investigations of the finely dispersed matter in ocean water. In *Forms of the Elements and Radionuclides in Sea water*, Nauka, Moscow, 1974, pp. 119–23.
8. Born M., Wolf E.; *Principles of Optics*, 2nd Edition, Pergamon Press, Oxford, 1964.
9. Brekhovskikh L. M.; Main projects of the tenth voyage of the scientific research ship 'Dmitry Mendeleyev'. In *Hydrophysical and Optical Research Works in the Indian Ocean*, Nauka, Moscow, 1975, pp. 2–6.
10. Burenkov V. I., Vasil'kov A. P., Kel'balikhanov B. F.; Spectra of the scattering of light by seawater. In *Optical Methods of Studying Oceans and Inland Water Bodies*, Estonian Academy of Sciences Publishers, Tallinn, 1980, pp. 54–7.
11. Burenkov V. I. *et al.*; Comparison of the different methods for determining the composition of marine suspensions. In ref. 9, pp. 74–82.
12. Burenkov V. I. *et al.*; Brightness spectra for emergent radiation and their variations with the height of observation. In ref. 2, pp. 41–57.
13. Burenkov V. I., Kopelevich O. V., Shifrin K. S.; Scattering of light by large particles with a refractive index close to unity. In *Physics of the Atmosphere and the Ocean*, Nauka, Moscow, pp. 828–35.
14. Burkov V. A., Koshlyakov M. N., Stepanov V. N.; General information on the world ocean. In *Oceanology: Physics of the Ocean, Vol. 1*, Nauka, Moscow, 1978, pp. 11–84.
15. Vinogradov M. E. *et al.*; Mathematical model of the ecosystem of the pelagic zone of the tropical regions of the ocean. *Oceanology*, 1973, **13** (5), pp. 852–66.
16. Vojtov V. I.; On the optical properties of Antartic intermediate waters. In ref. 9, pp. 49–51.

17. Vojtov, V. I.; Optical expedition studies in the Atlantic, Indian and Pacific Oceans. In *Optics of the Ocean and the Atmosphere*, Elm Publishers, Azerbaijanian Academy of Sciences, Baku, 1983, pp. 163–93.

18. Vojtov V. I., Kopelevich O. V., Shifrin K. S. Problems and basic results of investigating the optical properties of the waters of the Indian Ocean. In ref. 9, pp. 32–41.

19. Vuks M. F.; *Scattering of Light in Gases, Liquids and Solutions*, The State University of Leningrad Publishers, Leningrad, 1977.

20. Gashko V.A., Shifrin K. S.; Determination of the distribution of particles by size from the spectral scattering functions. *Physics of the Atmosphere and the Ocean*, 1976, **12** (10), pp. 1045–52.

21. Gershun A. A.; *The Light Field Theory*, ONTI, Leningrad, 1936.

22. Gillem A., Stern E.; *Electronic spectra of organic compounds absorption*.

23. Golikov V. I.; Apparatus for measuring the size spectrum of spherical particles and mist drops. In *Proceedings of the Main Geophysical Observatory*, Gidrometeoizdat, Leningrad, 1961, No. 109, pp. 76–89.

24. Gurevich I. Ya., Shifrin K. S.; Reflection of the visible and infrared radiation of oil films on the sea. In ref. **10**, pp. 166–83.

25. Jerlov N. G.; *Optical Oceanography*, Elsevier, Amsterdam, 1968. *Marine Optics*, Elsevier, Amsterdam, 1976.

26. Efimenko I. D. *et al.*; Illuminance of natural light in the waters of the Indian Ocean. In ref. 9, pp. 104–8.

27. Zolotarev V. M., Demin A. V.; Optical constants of water over a wide spectral range. *Optics and Spectroscopy*, 1977, **43**(2), pp. 271–9.

28. Ivanoff A. A.; *Introduction à l'oceanographie*, Paris, Vol I, 1972, *Vol. II*, 1975

29. Ivanov A. P.; *Physical Principles of Hydrooptics*, Nauka i tekhnika, Minsk, 1975.

30. Ivanov A. P. *et al.* Optical properties of the waters of the Sargasso Sea. *Physics of the Atmosphere and the Ocean*, 1980, **16** (3), pp. 313–20.

31. Ivanov K. I. On the dependence between the constant of suspended matter and the attenuation coefficient. In *Proceedings of the State Institution of Oceanography*, Moscow, No. 15, 1950, pp. 39–48.

32. *Isotopic analysis of water*. 2nd Edition, The USSR Academy of Sciences Publishers, Moscow, 1957.

33. Kadyshevich E. A., Lyubovtseva Yu. S., Rozenberg G. V.; Light scattering matrices for the waters of the Pacific and Atlantic Oceans. *Physics of the Atmosphere and the Ocean*, 1976, **12** (2), pp. 186–195.

34. Kamenkovich V. M., Monin A. S.; Basic positions of the thermohydromechanics of the ocean. In *Physics of the Ocean, Vol. 1*, Nauka, Moscow, 1978, pp. 85–112.

35. Karabashev G. S., Solov'ev A. N., Zangalis K. P.; Photoluminescence of the waters of the Atlantic and Pacific Oceans. In *Hydrophysical and Optical Research Works in the Atlantic and Pacific Oceans*, Nauka, Moscow, 1974, pp. 143–53.

36. Karabashev G. S., Yakubovich V. V.; Space–time variation of the coefficient of scattering of light at an angle of 45° in the active layer of the Indian Ocean. In ref. 9, pp. 82–93.

37. Koblents-Mishke O. I., Konovalov B. V.; Spectral absorption of radiant energy by marine suspensions. In ref. 35, pp. 280–92.

38. Kozlyaninov M. V.; On the calculation of the visibility of a white standard disc. *Oceanology*, 1980, **20** (2), pp. 329–34.

39. Kozlyaninov M. V.; Basic principles of optical measurements in the sea and some hydrophotometric calculations. In ref. 17, pp. 96–162.

40. Konovalov B. V. Some characteristics of the spectral absorption by marine suspensions. In ref. 2, pp. 58–65.

41. Kopelevich O. V.; Optical properties of pure water. *Optics and Spectroscopy*, 1976, **41** (4), pp. 666–8.

42. Kopelevich O. V., Burenkov V. I.; On a nephelometric method for determining the

scattering coefficient. *Physics of the Atmosphere and the Ocean*, 1971, **7** (12), pp. 1280–9.

43. Kopelevich O. V., Burenkov V. I.; Statistical characteristics of the scattering functions. In *Optics of the Ocean and the Atmosphere*, Nauka, Leningrad, 1972, pp. 126–36.

44. Kopelevich O. V., Burenkov V. I.; On the relationship between the spectral values of the absorption coefficients of light for a seawater, phytoplankton pigments and yellow substance. *Oceanology*, 1977, **17** (3), pp. 427–33.

45. Kopelevich O. V., Gushchin O. A.; On statistical and physical models for the light scattering properties of seawater. *Physics of the Atmosphere and the Ocean*, 1978, **14** (9), pp. 967–73.

46. Kopelevich O. V., Mashtakov Yu. A., Burenkov V. I.; Investigation of the vertical stratification of the scattering properties of water. In ref. 9, pp. 54–60.

47. Kopelevich O. V. *et al.*; Apparatus and methods for investigating the optical properties and absorption of light in seawater. In ref. 35, pp. 97–112.

48. Kopelevich O. V., Shifrin K. S.; Modern ideas on the optical properties of seawater. In ref. 17, pp. 4–55.

49. Landau L. D., Lifshits E. M.; *Electrodynamics of Continuous Media*, Pergamon Press, 1957.

50. Lanczos C.; *Applied Analysis*. Prentice Hall, 1956.

51. Levin I. M.; On the distribution functions of cloud drops by size. *Geophysics*, 1958, No. 10, pp. 1211–18.

52. Levin I. M. On the theory of the white disc. *Physics of the Atmosphere and the Ocean*, 1980, **16** (9), pp. 926–32.

53. Lisitsyn A. P.; Sources and regularities of the deposition of minerals in the oceans. In *Geology of the Ocean*, Nauka, Moscow, 1979, pp. 164–80.

54. McCartney E.; *Optics of the Atmosphere*, John Wiley and Sons, New York, 1977.

55. Man'kovsky V. I.; Fine structure of the light scattering functions in sea and ocean waters. In *Marine Hydrophysical Research*, Proceedings of the Marine Hydrophysical Institute of the Ukranian Academy of Sciences, Sevastopol, 1972, **58** (2), pp. 126–40.

56. Man'kovsky V. I. Extreme values of the light scattering function of seawater. In ref. 55, 1973, **62** (3), pp. 100–8.

57. Man'kovsky V. I.; Experimental and theoretical data on the point of intersection of the light scattering function of marine suspensions. *Physics of the Atmosphere and the Ocean*, 1975, **11** (12), pp. 1284–93.

58. Man'kovsky V. I.; Empirical formula for estimating the attenuation coefficient of light in seawater from the depth of visibility of a white disc. *Oceanology*, 1978, **18** (4), pp. 750–3.

59. Mikulinsky I. A., Shifrin K. S.; Inverse problem for the ensemble of 'soft' non-spherical particles. In ref. 10, pp. 74–7.

60. Monin A. S.; Preface to ref. 74.

60a. Monin A. S. (Ed); *Optics of the Ocean*, Nauka, Moscow, 1983. *Physical Optics of the Ocean, Vol. 1, Applied Optics of the Ocean, Vol. 2.*

61. Monin A. S., Kamenkovich V. M., Kort V. G.; *Variation in the World's Oceans* Gidrometeoizdat, Leningrad, 1974.

62. Monin A. S., Shifrin K. S.; Optics and hydrodynamics of the ocean. *Priroda*, 1972, **12**, pp. 66–71.

63. Neuimin G. G.; Optical characteristics of the waters of the Black Sea. In *Integral Oceanographical Investigations of the Black Sea*, Naukova Dumka, Kiev, 1980, pp. 199–215.

64. Neuimin G. G., Sorokina N. A., Ermolenko A. I.; Objective analysis of the transmittance field of the waters of the tropical Atlantic. In ref. 55, 1974, **65** (2), pp. 64–72.

65. Monin A. S. (Ed); *Oceanology: Biology of the Ocean, Vol. 1*, Nauka, Moscow, 1977.

66. Monin A. S. (Ed); *Oceanology: Chemistry of the Ocean, Vol. 1*, Nauka, Moscow, 1979.

67. Ochakovsky Yu. E.; On the dependence of the attenuation coefficient of light on the suspensions contained in the sea. In Proceedings of the Institute of Oceanology, Nauka, Moscow, 1965, **77**, pp. 35–40.

68. Ochakovsky Yu. E.; On the comparison of measured and calculated functions. In ref. 67, 1965, **77**, pp. 125–30.

69. Pavlov V. M.; Transmittance of seawater. In ref. 35, pp. 127–39.

70. Pelevin V. N., Rutkovskaya V. A.; On the optical classification of ocean waters from the spectral attenuation of solar radiation. *Oceanology*, 1977, **1** (17), Edition 1, pp. 40–5.

71. Perel'man A. Ya., Shifrin K. S.; Utilization of light scattering to determine the structure of dispersed systems with exponential distribution. *Optics and Spectroscopy*, **26** (6), pp. 1013–18, **27** (1), pp. 137–43.

71b. Application of the transmittance method in the case of a gamma distribution. *Optics and Spectroscopy*, 1969, **27** (1), pp. 137–43.

72. Perel'man A. Ya., Shifrin K. S.; Determination of particle spectra from the data on the attenuation coefficient taking into account the dispersion of light. *Physics of the Atmosphere and the Ocean*, 1979, **14** (1), pp. 66–73.

73. Polyakova E. A., Shifrin K. S.; Microstructures and the transmittance of rain. In ref. 23, 1953, No. 42, pp. 84–96.

74. Popov N. I., Fedorov K. N., Orlov V. N.; Seawater. Nauka, Moscow, 1979.

75. Rosenberg G. V. The Stokes vector parameter. *Advances in Physical Sciences*, 1955, **56** (1), pp. 77–110.

76. Rosenberg G. V. *et al.*; Investigation of the light scattering matrices of seawater. In ref. 35, pp. 304–25.

77. Romankevich E. A. *Geochemistry of the Organic Matter in the Ocean*, Nauka, Moscow, 1977.

78. Sakharov A. N., Shifrin K. S.; Determination of the average size and concentration of suspended particles from the fluctuations in intensity of transmitted light. *Optics and Spectroscopy*, 1975, **39** (2), pp. 367–72.

79. Sid'ko F. Ya. *et al.*; Investigation of the optical properties of populations of single-celled algae. In *Continuous Control of Culturing Microorganisms*, Nauka, Moscow, 1967, pp. 38–69.

80. Stepanov V. N. *The World's Oceans. Dynamics and Properties of Water*, Znanie, Moscow, 1974.

81. Tables on the scattering of light. *Gidrometeoizdrat*, Leningrad; *Vol. 1*, 1966, *Vol. 2*, 1958, *Vol. 3*, 1968, *Vol. 4*, 1971.

82 Titchmarsh E. C.; *An Introduction to the Theory of Fourier Integral*, Oxford University Press, 1937.

83. Tikhonov A. N., Arsenin V. Ya.; *Methods for Solving Incorrect Problems*, Nauka, Moscow, 1974.

84. Fabelinsky I. A.; *Molecular Light Scattering*, Fizmatgiz, Moscow, 1965.

85. Fedorova E. O.; Study of the functions of the scattering of light by large transparent particles of spherical and arbitrary shape, Thesis, Goi, Leningrad, 1952; In ref. 31, 1957, **25** (151), Oborongiz, Moscow.

86. Khalemsky E. N., Voytov V. I.; Zoning of the waters of the Pacific Ocean in terms of the transmittance. In ref. 43, pp. 181–6.

87. Van de Hulst H.; *Light Scattering by Small Particles*, John Wiley and Sons, New York, 1957.

88. Shemshura V. E., Fedirko V. I.; On the relationships between certain hydrooptical parameters. *Oceanology*, 1981, **21** (1), pp. 51–4.

89. Shurkliff W.; *Polarized Light*, Harvard University Press, Cambridge, MA, 1962.

90. Shifrin K. S.; *Scattering of Light in a Turbid Medium*, NASA Technical Translation F-477, Washington, DC, 1968.

91. Shifrin K. S.; Scattering of light by two-layer particles. *Geophysics*, 1952, **2**, pp. 15–20.

92. Shifrin K. S.; On the theory of the radiation properties of clouds. *Compt. Rend. USSR Acad. Sci.*, **94** (4), 1954, pp. 673–6; in ref. 23, 1955, **46** (108), pp.5–33.

92a. Shifrin K. S.; On the calculation of the radiation properties of clouds. In ref. 23, 1955, **46**(108), pp. 5–33.

93. Shifrin K. S.; On the theory of optical methods for the investigation of colloidal systems. In Proceedings of the All-Union Extra-Mural Forest Institute, Leningrad, 1955, No. 1, pp. 33–42.

94. Shifrin K. S.; Calculation of a certain class of definite integrals containing the square of the Bessel function of the first order. In ref. 93, 1956, No. 2, pp. 153–62.

95. Shifrin K. S.; Optical investigations of cloud particles. In *Investigation of Clouds, Precipitates, and Storm Electricity*, Gidrometeoizdat, Leningrad, 1957, pp. 19–25.

96. Shifrin K. S.; Study of the properties of matter from single scattering. In *Theoretical and Applied Problems in the Scattering of Light*, Nauka i tekhnika, Minsk, 1971, pp. 228–44.

97. Shifrin K. S. et al.; Recovery of the distribution of particles over size. *Physics of the Atmosphere and the Ocean*, 1972, **12**, pp. 1268–78.

98. Shifrin K. S. et al.; Determination of the microstructure of the scattering medium. *Physics of the Atmosphere and the Ocean*, 1974, **4**, pp. 416–20.

99. Shifrin K. S., Golikov V. I.; Determination of the spectra of drops by the method of small angles. In *Investigation of Clouds, Precipitates and Storm Electricity*, Nauka, Moscow, 1961, pp. 266–77.

100. Shifrin K. S., Golikov V. I.; Investigation of the apparatus for measuring particles by the method of small angles. In ref. 23, 1965, **170**, pp. 127–39.

101. Shifrin K. S. et al.; Investigation of an aerosol above the sea. *Physics of the Atmosphere and the Ocean*, 1980, **16** (13), pp. 254–60.

102. Shifrin K. S., Zel'manovich I. L.; Matrix for calculating the coefficient of back scattering. In ref. 23, 1965, **170**, pp. 61–70.

103. Shifrin K. S., Kolmakov I. B.; Effect of limiting the interval of measuring the function. *Physics of the Atmosphere and the Ocean*, 1966, **2** (8), pp. 851–8.

104. Shifrin K. S., Kolmakov I. B.; Calculation of the particle size spectrum. *Physics of the Atmosphere and the Ocean*, 1967, **3** (12), pp. 1271–9.

105. Shifrin K. S., Kolmakov I. B.; Calculation of the size of particles from the volume scattering function. *Physics of the Atmosphere and the Ocean*, 1971, **7** (5), pp. 554–6.

106. Shifrin K. S., Kolmakov I. B. Resolvent tables. Investigation of special functions. *Trudy Vycokogornogo Geofiz. Instituta, Gidrometeoizdat*, 1968, **8**, pp. 126–51.

107. Shifrin K. S.; Significant region for scattering angles. *Physics of the Atmosphere and the Ocean*, 1966, **2** (9), pp. 928–32.

108. Shifrin K. S., Kolmakov I. B., Chernyshev V. I.; On the calculation of the size of particles of a dispersive system for the data on its transmittance. *Physics of the Atmosphere and the Ocean*, 1969, **5** (10), pp. 1085–9.

109. Shifrin K. S. et al.; Utilization of the light scattering functions for investigating marine suspensions. In ref. 43, pp. 25–44.

110. Shifrin K. S. et al.; Optics of the ocean. In *Physics of the Ocean, Vol. 1*, Nauka, Moscow, 1978, pp. 340–96.

111. Shifrin K.S., Mikulinsky I. A.; Scattering of light by a system of particles in the Rayleigh–Hans approximation. *Optics and Spectroscopy*, 1982, **52** (2), pp. 359–66. Scattering of light by an ensemble of 'soft' non-spherical particles. In *Optics of the Sea*, Nauka, Moscow, 1983, pp. 17–32.

112. Shifrin K. S., Moisseev M. M.; Calculation of molecular scattering of light by

seawater. In ref. 10, pp. 102–5.

113. Shifrin K. S., Moisseev M. M.; Calculation of the coefficient of molecular scattering of light by seawater; empirical formula for the relative refractive index of seawater. In ref. 17, pp. 248–64.

114. Shifrin K. S., Moroz B. Z., Sakharov A. N.; Determination of the characteristics of a dispersive medium from data on its transmittance. *Compt. Rend. USSR Acad. Sci.*, 1971, **199** (3), pp. 581–98.

115. Shifrin K. S., Perel'man A. Ya.; Determination of the particle spectrum of a dispersive system from data on its transmittance. *Compt. Rend. USSR. Acad. Sci.*, 1963, **151**, pp. 326–9; *Optics and Spectroscopy*, 1963, **15** (4–6), 1964, **16** (1), 1965, **20** (1); Calculation of the particle spectrum of a dispersive system from data on its transmittance. In ref. 23, 1965, **170**, pp. 3–36.

116. Shifrin K. S., Perel'man A. Ya.; Inversion of functions for 'soft' particles. *Compt. Rend. USSR Acad. Sci.*, 1964, **158** (3), pp. 158–60.

117. Shifrin K.S., Perel'man A. Ya., Volgin V. N.; Calculation of the distribution density of particle radii. *Optics and Spectroscopy*, 1981, **51** (6), pp. 963–71.

117a. Shifrin K. S., Perel'man A. Ya., Kokorin A. M.; Scattering of light by a coated sphere. In *Optics of the Sea and the Atmosphere*, State Institute of Optics, Leningrad, 1984, pp. 98–9.

118. Shifrin K. S., Perel'man A. Ya., Punina V. A.; Structure of the light field at small angles. In ref. 23, 1966, **183**, pp. 3–26.

119. Shifrin K. S., Punina V. A.; On the light scattering functions in the region of small angles. *Physics of the Atmosphere and the Ocean*, 1968, **4** (7), pp. 784–91.

120. Shifrin K. S., Raskin V. F.; On the theory of the Rocard function. In ref. 23, 1960, **100**, pp. 3–14.

121. Shifrin K. S., Raskin V. F. The function for the generalized Young distribution. In ref. 23, 1967, **203**, pp. 155–8.

122. Shifrin K. S., Raskin V. F.; Spectral transmittance and the inverse problem of scattering theory. *Optics and Spectroscopy*, 1961, **11** (2), pp. 268–71.

123. Shifrin K. S., Salganik I. N.; Scattering of light by seawater models. *Tables on Light Scattering, Vol. 5*, Gidrometeorizdat, Leningrad, 1973.

124. Shifrin K. S. *et al.*; Determination of the weight concentration of suspensions in colloidal systems with particles of any shape using the fluctuation method. *Compt. Rend. USSR. Acad. Sci.*, 1974, **215** (5), pp. 1085–6.

125. Shifrin K. S., Chaya nova E. A.; Functions for Young and Young-type distributions. In ref. 23, 1965, **170**, pp. 93–104.

125a. Shifrin K. S., Shifrin Ya. S., Mikulinsky I. A.; Scattering of light by an ensemble of large particles of any shape. 1984, *Letters to the Journal of Technical Physics*, **2**, pp. 68–72. *Compt. Rend. USSR. Acad. Sci.*, 1984, **277** (3), pp, 582–5.

126. Shifrin Ya. S.; *Statistical Antenna Theory*, Golem, Boulder CO, 1971.

127. Shuleikin V. V.; *Physics of the Sea*, Nauka, Moscow, 1968.

128. Einstein A.; *Collection of Scientific Works, Vol. 3*, Nauka, Moscow, 1966, pp. 216–37.

129. Yakubovich V. V.; *Experimental Investigations on the Space-time Structure of the Light Scattering Coefficient Field in the Ocean*, Thesis, IOAN, Moscow, 1977.

130. Aden A. L., Kerker M.; Scattering of electromagnetic waves from two concentric spheres. *Journal of Applied Physics*, 1951, **22**, (10) pp. 1242–6.

131. Acquista Ch.; Light scattering by Irregularly Shaped particles, Edited by Schuerman D. Plenum Press, London, 1980.

132. Armstrong F., Boalch G. T.; UV absorption of seawater. *Nature*, 1961, **192**, (4805), pp. 858–9.

132a. Bakhtiyarov V. G. *et al.*; The inversion of accurate data on the extinction coefficient by the transparency method. *Pure and Applied Geophysics*, 1964, **64** (2), pp. 204–11.

132b. Bayvel L. P., Jones A. R.; *Electromagnetic Scattering and its Applications*, Applied Science Publishers, London, 1981.

133. Beardsley G. F., Jr.; Mueller scattering matrix of seawater. *Journal of the Optical Society of America, 1968*, **58** (1), pp. 52–7.
134. Byalko A. V.; Polarization relations and fluctuations of scattering on suspensions. Preprint L. D. Landau. *Institute of Theoretical Physics*, 1977.
135. Cabannes J.; *Diffusion moleculaire de la lumiére*, Paris, 1929.
136. *Chemical Oceanography, Vol. 1*, edited is Riley, J. P., Skirrow. G., Academic Press, London and New York, 1965.
137. Chen C. T., Miller F. J.; The specific volume of seawater. *Deep-Sea Research*, 1976, **23** (7), pp. 595–612.
138. Chester K., Stoner J.; Concentration of suspended matter in surface seawater. *Nature*. 1972, **240** (5383), pp. 552–3.
139. Chin J. H., Sliepcevich C. M., Tribus M.; Determination of particle size distribution in polydispersed systems. *Journal of Physical Chemistry*, 1955, **59** (9), pp. 845–8.
140. Clarke, G. L., James H. R.; Laboratory analysis of the selective absorption of light by seawater. *Journal of the Optical Society of America*, 1939, **29** (1), pp. 43–5.
141. Coulter W. H.; High speed automatic blood cell counter and cell size analyzer. *Proceedings of the National Electronic Conference*, 1957, **12**, pp. 1034–42.
142. Dorsey N. E.; *Properties of Ordinary Water Substance*, Hafner, New York, 1968.
143. Dreisch Th.; Absorp. koel. einiger Flussig. und ihrer Dämpfe in Ultraroten. *Zeitschrift fuer Physik*, 1924, **30**, (200), S. 342.
144. Friedman D.; Infrared characteristics of ocean water. *Applied Optics*, 1969, **8** (10), pp. 2073–8.
145. Gumprecht R. O., Sliepcevich C. M.; Scattering of light by large spherical particles. *Journal of Physical Chemistry*, 1953, **57** (1), pp. 90–5.
146. Gutter A.; Die Miesche Theorie der Beugung durch dielectrische Kygeln mit absorbierendem Kern und ihre Bedeutung Probleme der interstellaren Materie und des atmosphärischen Aerosols. *Annalen der Physik*, 1952, **11** (2), S. 65–98.
147. Harris F. S., Jr.; Light diffraction patterns. *Applied Optics*, 1964, **3** (8), pp. 909–13.
148. Hass M., Davisson J. W.; Absorption coefficient of pure water at 488 and 541.5 nm by adiabatic laser calorimetry. *Journal of Optical Society of America*, **67** (5), pp. 622–4.
149. Hojerslev N. K.; On the origin of yellow substance in marine environment. *Rapport Inst. Fysisk Oceanogr.*, Copenhagen, 1980, **42**, pp. 57–81.
150. Irvine W. M., Pollack J. B.; Infrared optical properties of water and ice spheres. *Icarus*, 1968, **8**, pp 324–36.
151. James H. R., Birge E. A.; A laboratory study of light by lake waters. *Transactions of the Wisconsin Academy of Sciences, Arts and Letters*, 1938, **31**, pp. 1–54.
152. Kalle K.; Über die gelösten organischen Komponenten in Meer-Wasser. *Kieler Meeresforsch*, 1962, **18** (3), S. 128–31.
153. Kalle K.; The problem of the Gelbstoff in the sea. *Oceanography and Marine Biology, Annual Reviews*, 1966, **4**, pp. 91–104.
154. Kerker M.; *The Scattering of Light and Other Electromagnetic Radiation*, Academic Press, New York, 1969.
155. Kullenberg G., Lundgren B., Malmberg Sv. A.; Inherent optical properties of the Sargasso Sea. *Rept Inst. Phys. Ocean.*, Copenhagen, 1970, **11**.
156. Latimer P., Rabinovitch E.; Selective scattering of light by pigments *in vivo. Archives of Biochemistry and Biophysics*, 1959, **84**, pp. 428–441.
157. Lenoble J.; L'absorption du rayonment ultraviolet par les ions presents dans la Mer. *Revue d'Optique*, 1956, **35** (10), pp. 526–31.
158. Lochet R.; Contribution ál'etude de la diffusion moleculaire de la lumiére (effect Rayleigh). *Annales de Physique*, 1953, **8**, pp. 14–60.
159. Morel A.; Interpretation des variations de la forme de l'indicatrice de diffusion de la lumiere par les eaux de mer. *Annales de Geophysique*, 1965, **21** (2), pp. 281–4.
160. Morel A.; Indicatrices de diffusion calculees par la theorie de Mie. *Rap. Lab. Oceanogr. Phys.*, Univ. de Paris, 1973, **10**.

161. Morel A.; Diffusion de la lumiere par les eaux de mer. Result, experimentaux et approach theorique. *AGARD Lecture Series*, 1973, **61**, 3.1-1-71.
162. O'Connor C. L., Schlupf J. P.; Brillouin scattering in water. *Journal of Chemical Physics*, 1967, **47** (1), pp. 31-8.
163. Ogura N., Hanya T.; UV absorption of the seawater in relation to organic and inorganic matter. *International Journal of Oceanology and Limnology*, 1967, **1** (2), pp. 91-102.
164. *Optical Aspects of Oceanography*, Edited by Jerlov N. G., Steeman-Nielsen, E. Academic Press, London and New York, 1974.
165. Pearson K., *Tables of the Incomplete Beta-functions*. London University College, 1948.
165a. Perelman A. Ya., Shifrin K. S.; Improvements to the spectral transparency method for determining particle size. *Applied Optics*, **19** (11), pp. 1787-93.
166. Perelman A. Ya.; Application of Mie series of soft particles. *Pure and Applied Geophysics*, 1978, **116**, pp. 1077-88.
167. Perelman A. Ya.; On the scattering from soft particles. *Journal of Colloid and Interface Science*, 1978, **63** (3), pp. 593-6.
168. Pethica B. A., Smart C.; Light scattering of electrolyte solutions. *Transactions of the Faraday Society*, 1966, **62**, pp. 1890-9.
169. Pinkley L., Williams D.; Optical properties of seawater in the infrared. *Journal of Optical Society of America*, 1976, **66** (6), pp. 554-8.
170. Preisendorfer R.; *Hydrologic Optics. Vol. 1-6*, NOAA, Honolulu, Hawaii, 1976.
171. *Proceedings of the Interdisciplinary Conference on Electromagnetic Scattering*. edited by Kerker M., Pergamon Press, New York and Oxford, 1963.
172. Querry M. R., Cary Ph. G., Waring R. C.; Split-pulse laser method for measuring attenuation coefficients of transparent liquids. *Applied Optics*, 1978, **17** (22), pp. 3587-92.
173. Rocard I.; Sur les proprietes optiques de l'atmosphere: diffusion, absorption. *Rev d'Optique*, 1930, **3**, pp. 97-111.
174. Rousset A., Lochet R.; Sur la diffusion anisotrope des solutions aqueouses. *Comptes Rendus de l'Academie des Sciences*, 1955, **240**, pp. 70-3.
174a. Shifrin K. S., Perelman A. Ya.; Inversion of light scattering data for the determination of spherical particle spectrum. In *Electromagnetic Scattering, Vol. II*, edited by Rowell R. L., Stein R. S., 1967, Gordon and Breach, New York, pp. 131-67.
175. Shifrin K. S., Perelman A. Ya.; Calculation of particle distribution by the data on spectral transparency. *Pure and Applied Geophysics*, 1964, **58** (2), pp. 208-20.
176. Shifrin K. S., Perelman A. Ya.; Determination of particle spectrum of atmospheric aerosol by light scattering. *Tellus*, 1966, **18** (2), pp. 566-72.
177. Sullivan S. A.; Experimental study of the absorption in distilled water, artificial water and heavy water in the visible region of the spectrum. *Journal of the Optical Society of America*, 1963, **53**, pp. 962-7.
178. Tam A. C., Patel C. K.; Optical absorptions of light and heavy water. *Applied Optics*, 1979, **18** (19), pp. 3348-58.
178a. Vašiček Antonin.; *Tables of Determination of Optical Constants from the Intensities of Reflected Light*, 1964, Česk. Akad. VED, p. 113.
179. Yentsch Ch. S.; The influence of phytoplankton pigments on the colour of sea water. *Deep-Sea Research*, 1960, **7** (1) pp. 1-9.
180. Yentsch Ch. S. Measurements of visible light absorption by particulate matter in the ocean. *Limnology and Oceanography*, 1962, **7** (2), pp. 207-17.
181. Zaneveld J. R., Pak H.; Method for the determination of the index of refraction of particles suspended in the ocean. *Journal of the Optical Society of America*, 1973, **63** (3), pp. 321-4.

Index

Printed in the United States
By Bookmasters